Liposome Technology

Volume I

Preparation of Liposomes

Editor

Gregory Gregoriadis, Ph.D.

Clinical Research Centre

Middlesex, England

CRC Press

Taylor & Francis Group

Boca Raton London New York

CRC Press is an imprint of the

Taylor & Francis Group, an **informa** business

First published 1984 by CRC Press
Taylor & Francis Group
6000 Broken Sound Parkway NW, Suite 300
Boca Raton, FL 33487-2742

Reissued 2018 by CRC Press

Library of Congress Cataloging in Publication Data
Main entry under title:

Liposome technology.

Bibliography: v., p.
Includes indexes.
Contents: v. 1. Preparation of liposomes -- v. 2. Incorporation of drugs, proteins, and genetic material -- v. 3. Targeted drug delivery and biological interaction.
1. Drugs--Vehicles. 2. Liposomes. I. Gregoriadis, Gregory.
RS201.V43L55 1983 615'. 191 83-7088
ISBN 0-8493-5316-5 (v. 1)
ISBN 0-8493-5317-3 (v. 2)
ISBN 0-8493-5318-1 (v. 3)

A Library of Congress record exists under LC control number: 83007088

ISBN 13: 978-1-315-89500-0 (hbk)
ISBN 13: 978-1-351-07410-0 (ebk)

Visit the Taylor & Francis Web site at http://www.taylorandfrancis.com and the
CRC Press Web site at http://www.crcpress.com

To My Parents

Christos and Athena

PREFACE

Targeting of drugs through carrier systems that would, ideally, ignore the normal part of the body and home to diseased areas in need of treatment, has been for the most part of this century only a little more credible than the philosophers' stone. Recently, however, parallel but separate developments have began to transform drug targeting from a striking intellectual concept to a powerful exact science. First, a wide array of biological molecules are now recognized to possess ligands with specific affinity for respective receptors on the cell's surface. Second, the technical breakthrough in the production of monoclonal antibodies has made available in large quantities immunoglobulins that are highly specific for individual antigens. Such molecules form the basis of the modern approach to receptor (including cell surface antigens in a wider sense) mediated drug delivery. Further, they are being instrumental to a third development namely, the emergence of sophisticated artificial carriers exemplified by liposomes.

Liposomes are unusually versatile, notably in size, surface charge, lipid composition, the plethora of pharmacologically active molecules they can accommodate in both the aqueous and lipid phase and permeability to entrapped molecules. This not only has led to a number of ingenious techniques for the preparation of liposomes with practical advantages, it has also helped to adjust the system to a multitude of uses in cell biology, pharmacology, immunology, genetic engineering, and therapeutic and preventive medicine. A significant advance has also been made by our ability to control the fate of liposomes and the effect of their contents in vivo. This was achieved through knowledge of factors in the body that influence liposomal behavior. Modification of the structure of liposomes in a variety of ways enables us to harness such behavior to suit particular needs. The drug-carrier potential of liposomes has now acquired a new important dimension through the use of targeted molecules. Receptor-recognizing antibodies, glycoproteins, and glycolipids anchored onto the surface of liposomes are able to mediate uptake of the liposomal moiety and its drug contents by target cells. There are clear indications that targeting of liposomes can occur in vivo, at least for cells to which the carrier has immediate access.

Although the role of liposomes in drug targeting has been discussed extensively in several reviews and books, there has been no comprehensive coverage of related methodology. This book constitutes the first attempt to put together all aspects of liposome technology as applied to medical sciences. Contributors were encouraged to place emphasis on methodology as experienced in their own laboratory and in certain cases, the same methodologies are described by more than one laboratory. This was a deliberate policy: several aspects of liposome technology are still not an exact science and personal experience coming from more than one source was thought likely to help everyone, especially those who have just entered the field. A typical chapter includes (1) an introductory section directly relevant to the author's subject with concise coverage of relevant literature; (2) a detailed methodology section presenting experiences from the author's laboratory and a few examples of actual application of the method presented; (3) a critical discussion to enable the reader to appreciate the advantages and disadvantages of the method and compare it with those developed by other workers. The fifty chapters contributed have been distributed logically into three volumes. Volume I deals directly with methods for the preparation of liposomes and auxiliary techniques. Volume II describes procedures for the entrapment of a number of drugs, including genetic material, into selected types of liposomes. These two volumes contain virtually all methods available for efficient drug entrapment in the presence or, preferably, absence of organic solvents. Finally, Volume III is devoted to the growing variety of techniques yielding targeted liposomes and to approaches of studying liposomal behavior in the biological milieu both in vitro and in vivo.

It has been a pleasure for me to edit this book and to come into close contact with the thoughts and experiences of so many liposomologists. It is hoped that in view of the great and ever increasing number of workers in the field and the significance that liposomes have attained in a wide spectrum of disciplines, this multi-authored book in liposome technology will serve a useful purpose. I take this opportunity to thank all authors for their enthusiastic response to my invitation to contribute. I am also grateful to Mrs. M. Moriarty for editorial assistance and to CRC Press, Inc. personnel for their valuable help and truly professional cooperation.

Gregory Gregoriadis
June 1983

THE EDITOR

Gregory Gregoriadis, Ph.D., is a senior member of staff at the Medical Research Council's Clinical Research Centre, Harrow, Middlesex, England. He received his first degree in Chemistry from the University of Athens and his M.Sc. and Ph.D. in Biochemistry from McGill University. He has carried out research in as diverse fields as the metabolism of trace metals and the interaction of plasma glycoproteins with hepatic receptors. Since 1970 he has worked, published, and lectured extensively on the targeting of drugs via liposomes. Dr. Gregoriadis' interest in drug targeting is reflected in his founding in 1978 the Gordon Research Conference Series on "Drug Carriers in Biology and Medicine" of which he was the first Chairman and in 1981 the NATO Advanced Studies Institute Series "Targeting of Drugs" of which he is the Director. He has also been the editor of *Drug Carriers in Biology and Medicine* and the senior editor of *Liposomes in Biological Systems* and *Targeting of Drugs*.

CONTRIBUTORS

T. M. Allen, Ph.D.
Associate Professor
Department of Pharmocology
University of Alberta
Edmonton, Alberta, Canada

Carl R. Alving, M.D.
Department of Membrane Biochemistry
Walter Reed Army Institute of Research
Washington, D.C.

B. K. Bachhawat, Ph.D.
Department of Enzyme Engineering
Indian Institute of Chemical Biology
Calcutta, India

John D. Baldeschwieler, Ph.D.
Professor of Chemistry
California Institute of Technology
Pasadena, California

Jacques Barbet, Ph.D.
Section on Analytic Immunology
Centre d'Immunologie INSERM-CNRS
Marseille, France

Gillian M. Barratt, Ph.D.
Research Fellow
Department of Biochemistry
Charing Cross Hospital Medical School
London, England

Edward A. Bayer, Ph.D.
Department of Biophysics
Weizmann Institute of Science
Rehovot, Israel

Pierre Baudhuin, Ph.D.
Professor
Laboratoire de Chimie Physiologique
University of Louvain
Brussels, Belgium

Robert Blumenthal, Ph.D.
Senior Investigator
Laboratory of Mathematical Biology
National Cancer Institute
National Institutes of Health
Bethesda, Maryland

Vicente J. Caride, M.D.
Associate Clinical Professor
School of Medicine
Yale University; and
Director, Department of Nuclear Medicine
Hospital of St. Raphael
New Haven, Connecticut

Dennis Chapman, Ph.D.
Professor
Department of Biophysical Chemistry
Royal Free Hospital School of Medicine
University of London
London, England

Pierre Chatelain, Ph.D.
Laboratoire de Recherche
Continental Pharma
Mont-Saint-Guilbert, Belgium

Theodore C. Cree, Ph.D.
Research Associate
Department of Human Oncology
University of Wisconsin
Madison, Wisconsin

Amelia Cudd, Ph.D.
Centre de Biophysique Moléculaire
Centre National de la Recherche Scientifique
Orleans, Cedex, France

Jan Damen, Ph.D.
Division of Cell Biology
Netherlands Cancer Institute
Amsterdam, The Netherlands

Pijush K. Das, Ph.D.
Visiting Fellow
Developmental and Metabolic Neurology Branch
National Institutes of Health
Bethesda, Maryland

David W. Deamer, Ph.D.
Professor
Department of Zoology
University of California
Davis, California

Fabienne Defrise-Quertain, Ph.D.
Laboratoire de Chimie Physique des Macromolécules aux Interfaces
Université Libre de Bruxelles
Brussels, Belgium

Michel Delmelle, Ph.D.
Department of Atomic and Molecular Physics
University of Liege
Liege, Belgium

Paul Dragsten, Ph.D.
Proctor and Gamble, Inc.
Cincinnati, Ohio

Hiroo Endoh, Ph.D.
Research Member
Kureha Chemical Company
Tokyo, Japan

Eva Eriksson, Ph.D.
Research Scientist
Department of Biochemistry
Chemical Center
University of Lund; and Ferrosan
Malmö, Sweden

Hakan Eriksson, Ph.D.
Research Scientist
Department of Pure and Applied Biochemistry
Chemical Center
University of Lund
Lund, Sweden

Michael W. Fountain, Ph.D.
Senior Research Scientist
The Liposome Company
Princeton, New Jersey

Jürgen Freise, M.D.
Senior Research Scientist
Department of Internal Medicine and Gastroenterology
Medizinische Hochschule Hannover
Hannover, West Germany

Sven Frøkjaer, Ph.D.
Head, Physical-Chemistry Laboratory
Novo Research Institute
Bagsvaerd, Denmark

Nigel Gains, Ph.D.
Research Scientist
Department of Biochemistry
Swiss Federal Institute of Technology
Zurich, Switzerland

Prahlad C. Ghosh, Ph.D.
Research Associate
Department of Microbiology
Uniformed Services University of the Health Sciences
Bethesda, Maryland

Gregory Gregoriadis, Ph.D.
Head, Laboratory for Drug Targeting
Clinical Research Centre
Harrow, Middlesex, England

Pierre Guiot, Ph.D.
Research Worker
International Institute of Cellular and Molecular Pathology
Brussels, Belgium

Luke S. S. Guo, Ph.D.
Group Leader
Research and Development
Tago, Inc.
Burlingame, California

Robert L. Hamilton, Ph.D.
Senior Staff
Cardiovascular Research Institute
Associate Professor of Anatomy
University of California Medical Center
San Francisco, California

Yoshiyuki Hashimoto, Ph.D.
Professor and Chairman
Department of Hygienic Chemistry
Pharmaceutical Institute
Tohoku University
Sendai, Japan

Helmut Hauser, Ph.D.
Department of Biochemistry
Swiss Federal Institute of Technology
Zurich, Switzerland

Amalia S. Havaranis, Ph.D.
Technical Specialist
AMF-CUNO Microfiltration Division
Meriden, Connecticut

Pierre Henkart, Ph.D.
Immunology Branch
National Cancer Institute
National Institutes of Health
Bethesda, Maryland

Stuart M. Heywood, Ph.D.
Professor and Head
Department of Genetics and Cell Biology
University of Connecticut
Storrs, Connecticut

Ejnar L. Hjorth, B. Pharm., M.Sc.
Analytical Chemist
Department of Analysis
Bagsvaerd, Denmark

Anthony Huang, Ph.D.
Biology Division
Oak Ridge National Laboratory
Oak Ridge, Tennessee

Leaf Huang, Ph.D.
Associate Professor of Biochemistry
University of Tennessee
Knoxville, Tennessee

Karl J. Hwang, Ph.D.
Associate Professor
School of Pharmacy
University of Southern California
Los Angeles, California

Vilma K. Jansons, Ph.D.
Associate Professor of Microbiology
UMDNJ-New Jersey Medical School
Newark, New Jersey

D. S. Johnston, Ph.D.
Biochemistry and Chemistry Department
Royal Free Hospital Medical School
University of London
London, England

Stephen J. Kennel, Ph.D.
Biology Division
Oak Ridge National Laboratory
Oak Ridge, Tennessee

Christopher J. Kirby, Ph.D.
Senior Research Scientist
Division of Clinical Sciences
The Clinical Research Centre
Harrow, Middlesex, England

Richard Kirsh, Ph.D.
Department of Immunology
Smith Kline and French Laboratories
Philadelphia, Pennsylvania

Richard D. Klausner, Ph.D.
Laboratory of Biochemistry and Metabolism
National Institute for Arthritis, Metabolic, and Digestive Diseases
National Institutes of Health
Bethesda, Maryland

Thomas Koestler, Ph.D.
Department of Tumor Biology
Smith Kline and French Laboratories
Philadelphia, Pennsylvania

A. W. T. Konings, Ph.D.
Department of Radiopathology
University of Groningen
Groningen, The Netherlands

R. Lazo, B.S.
Department of Experimental Pathology
Roswell Park Memorial Institute
Buffalo, New York

Peter I. Lelkes, Ph.D.
Scientist
Laboratory of Cell Biology and Genetics
National Institute of Arthritis, Diabetes, Digestive and Kidney Diseases
National Institutes of Health
Bethesda, Maryland

Lee D. Leserman, M.D., Ph.D.
Section on Biological Applications of Liposomes
Centre d'Immunologie INSERM-CNRS
Marseille, France

Paul F. Lurquin, Ph.D.
Associate Professor of Genetics
Program in Genetics and Cell Biology
Washington State University
Pullman, Washington

Patrick Machy, Ph.D.
Section on Biological Applications of Liposomes
Centre d'Immunologie INSERM-CNRS
Marseille, France

Richard L. Magin, Ph.D.
Assistant Professor of Electrical Engineering and Bioengineering
University of Illinois
Urbana, Illinois

Bo Mattiasson, Ph.D.
Associate Professor
Department of Pure and Applied Biochemistry
Chemical Center
University of Lund
Lund, Sweden

Inger Mattsby-Baltzer, Ph.D.
National Research Council Research Associate
Department of Membrane Biochemistry
Walter Reed Army Institute of Research
Washington, D.C.

E. Mayhew, Ph.D.
Cancer Research Scientist V
Roswell Park Memorial Institute; and
Associate Research Professor
State University of New York
Buffalo, New York

Toshiyuki Nagata, Ph.D.
Faculty of Science
Department of Biology
Nagoya University
Nagoya, Japan

Claude Nicolau, Ph.D.
Professor and Director
Centre de Biophysique Moléculaire
Centre National de la Recherche Scientifique
Orleans, Cedex, France

John T. O'Loughlin, Ph.D.
Postdoctoral Researcher
Marine Biological Laboratory
Woods Hole, Massachusetts

Mitree M. Ponpipom, Ph.D.
Senior Research Fellow
Membrane and Arthritis Research Department
Merck Sharp and Dohme Research Laboratories
Rahway, New Jersey

George Poste, D.V.M., Ph.D.
Vice President
Director of Research and Development
Smith Kline and French Laboratories; and
Research Professor
Department of Pathology and Laboratory Medicine
University of Pennsylvania
Philadelphia, Pennsylvania

Evelyn Ralston, Ph.D.
Laboratoire de Chimie Biologique
Universite Libre de Bruxelles
Gensie, Belgium

L. S. Rao, Ph.D.
Pharmaceutical Development
The Wellcome Foundation Ltd.
Dartford, Kent, England

J. M. Ruysschaert, Ph.D.
Associate Professor
Laboratoire de Chimie Physique des Macromolécules aux Interfaces
Universite Libre de Bruxelles
Brussels, Belgium

Brenda E. Ryman, Ph.D.
Professor
Department of Biochemistry
Charing Cross Hospital Medical School
London, England

Gerrit L. Scherphof, Ph.D.
Professor of Physiological Chemistry
State University
Groningen, The Netherlands

Judith Senior, B.Sc.
Research Biochemist
Clinical Research Centre
Harrow, Middlesex, England

T. Y. Shen, Ph.D., D.Sc.
Vice President, Membrane and Arthritis Research
Merck Sharp and Dohme Research Laboratories
Rahway, New Jersey

Shigeki Shichijo, Ph.D.
Department of Immunology
Kurume University
Kurume, Fukuoka, Japan

Harm Snippe, Ph.D.
Associate Professor of Immunology
Department of Immunology
Laboratory of Microbiology
State University of Utrecht
Utrecht, The Netherlands

Issac Soliemani
Department of Genetics and Cell Biology
University of Connecticut
Storrs, Connecticut

Henry D. Sostman, M.D.
Assistant Professor
Department of Diagnostic Imaging
Yale University School of Medicine
New Haven, Connecticut

George Strauss, Ph.D.
Professor of Chemistry
Rutgers, The State University of New Jersey
New Brunswick, New Jersey

Minoru Sugawara
Student
Department of Hygienic Chemistry
Tohoku University
Sendai, Japan

Glenn M. Swartz
Department of Membrane Biochemistry
Walter Reed Army Institute of Research
Washington, D.C.

Inga-Mai Tegmo-Larsson, Ph.D.
Visiting Assistant Professor
Department of Chemistry
Hobart and William Smith Colleges
Geneva, New York

Vladimir P. Torchilin,Ph.D.
Head, Laboratory of Enzyme Engineering
Institute of Experimental Cardiology
National Cardiology Research Centre
Moscow, U.S.S.R.

Nilden S. Tüzel*
Postgraduate Research Student
Department of Biochemistry
Charing Cross Hospital Medical School
London, England

W. J. Vail, Ph.D.
Professor and Chairman
Department of Biology
Frostburg State College
Frostburg, Maryland

Catherine Vakirtzi-Lemonias, Ph.D.
Senior Research Staff
Biology Division
Nuclear Research Center "Demokritos"
Greek Atomic Energy Commission
Attikis, Greece

Arend Jan van Houte, Ph.D.
Research Fellow
Department of Immunology
Laboratory of Microbiology
State University of Utrecht
Utrecht, The Netherlands

Hans Georg Weder, Ph.D.
Professor
Swiss Federal Institute of Technology
Zurich, Switzerland

* Present address: Department of Biochemistry, Royal Free Hospital Medical School, London, England.

John N. Weinstein, Ph.D.
Senior Investigator
Laboratory of Mathematical Biology
National Cancer Institute
National Institutes of Health
Bethesda, Maryland

Meir Wilchek, Ph.D.
Professor
Department of Biophysics
Weizmann Institute of Science
Rehovot, Israel; and National Institutes of Health
Bethesda, Maryland

Jan Wilschut, Ph.D.
Assistant Professor
Laboratory of Physiological Chemistry
University of Groningen
Groningen, The Netherlands

Ole Wørts, B. Pharm., M.Sc.
Research Manager
Department of Pharmaceutics
Novo Research Institute
Bagsvaerd, Denmark

Po-Shun Wu, Ph.D.
Postdoctoral Researcher
Division of Chemistry and Chemical Engineering
California Institute of Technology
Pasadena, California

Milton B. Yatvin, Ph.D.
Professor of Human Oncology and Radiology
University of Wisconsin
Madison, Wisconsin

Othmar Zumbuehl, Ph.D.
Assistant Professor
Swiss Federal Institute of Technology
Zurich, Switzerland

TABLE OF CONTENTS

Volume I

Volume II

Volume III

Chapter 1

PHYSICOCHEMICAL PROPERTIES OF PHOSPHOLIPIDS AND LIPID-WATER SYSTEMS

Dennis Chapman

TABLE OF CONTENTS

I. INTRODUCTION

The study of myelin formation and lamellar structures of lipid-water systems has a long history. Even ideas for exploiting these structures were proposed as long ago as 1945:

"L'on pourrait peut-être avoir dans les formations bulleuses de savons dévelopées dans la masse du solvant un modèle de membranes lipidiques biologiques abordable à l'expérimentation. Leur étude au micromanipulateur pourrait peut-être apporter quelques données expérimentales sur leurs propriétés physiques et sur leur rôle dans la diffusion des divers ions ou molécules."

— Bernard Teitel.[1]

In the early 1960s, great emphasis was placed upon the study of the properties of lipid-water systems as model biomembrane structures. Dervichian, Bangham, Luzzati, and Chapman each studied these systems from different points of view, aiming to reveal insight into biomembrane properties.

The study of the submicroscopic and dynamic properties of these model biomembranes such as fluidity,[2] phase transitions, and change to other mesomorphic structures was for some time carried out distinctly and separately from the study of more macroscopic properties such as their ion trapping and release characteristics.[3] The former has been useful to provide insight into the dynamic structure of biomembranes, such as the fluidity of the lipid matrix, their diffusion characteristics,[4] etc., while the latter has led directly to a potential tool for drug delivery systems.[5]

Considerable interest and excitement was created by this possibility of in vivo application of liposomes in medicine and research. Sessa and Weismann[6] and Gregoriadis et al.[5,7] considered the liposome as a vehicle for replacement therapy in genetic deficiencies of lysosomal enzymes. The mushrooming interest since then in liposomes as carriers can be viewed as part of a long-term pharmacological search for ways to microencapsulate therapeutic agents.

II. THERMOTROPIC PHASE TRANSITIONS

Phospholipid molecules form the bilayer matrix of natural biomembranes and various classes occur. These include the lecithins (phosphatidylcholines — PC) and the phosphatidylethanolamines (PE) and phosphatidylserines (PS). Sphingolipids also occur. The structures of two classes of phospholipid are shown in Figure 1.

A. Solid-State Structure

An X-ray crystal structure has been reported of dimyristoyl lecithin.[8] The molecular conformation of this lipid is shown in Figure 2, where the molecular conformation of two molecules is shown. The mean planes containing the lipid chains are essentially parallel to one another. The intermolecular packing produces the classic lipid bilayer structure. Two water molecules are located between the phosphate oxygens of molecules A and B. The hydrocarbon chain packing shows that chain planes are alternately parallel and perpendicular to each other.

The capillary melting points of a number of pure phospholipids have been determined and are shown to be quite high, e.g., the values of the capillary melting points for the diacyl PEs are about 200°C, while for the PCs these capillary melting points are about 230°C. It is important to note that these values are independent of both the chain length and the degree of unsaturation of the fatty acid residues associated with the phospholipid. These capillary melting points can be compared with those of other compounds having long chains. The capillary melting points of fatty acids containing the same length of hydrocarbon chain are much lower, e.g., stearic acid melts at 69.7°C. On the other hand, the capillary melting

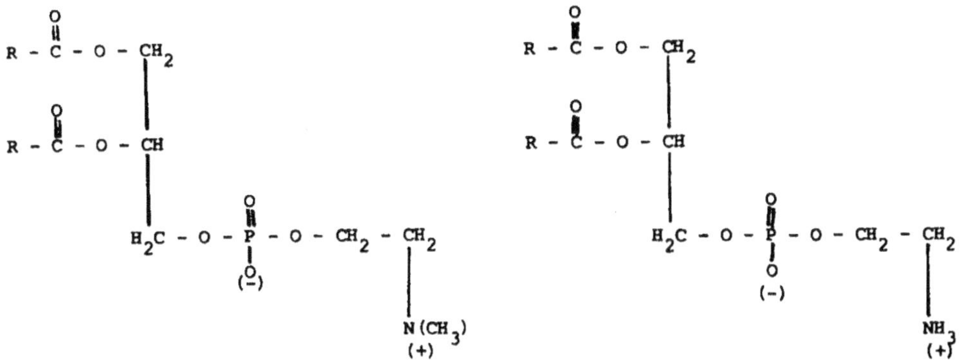

Phosphatidyl choline (PC) **Phosphatidyl ethanolamine**

FIGURE 1. Molecular structures of some natural phospholipids.

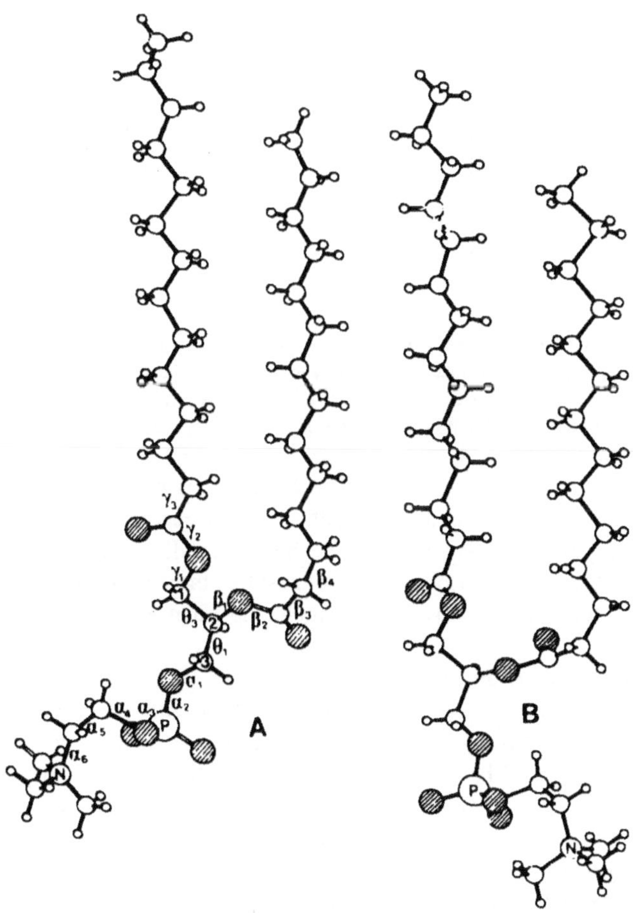

FIGURE 2. Molecules A and B of dimyristoyl lecithin (DMPC) in a view along the unit cell a-axis. (From Pearson, R. H. and Pascher, I., *Nature (London)*, 281, 499, 1979. With permission.)

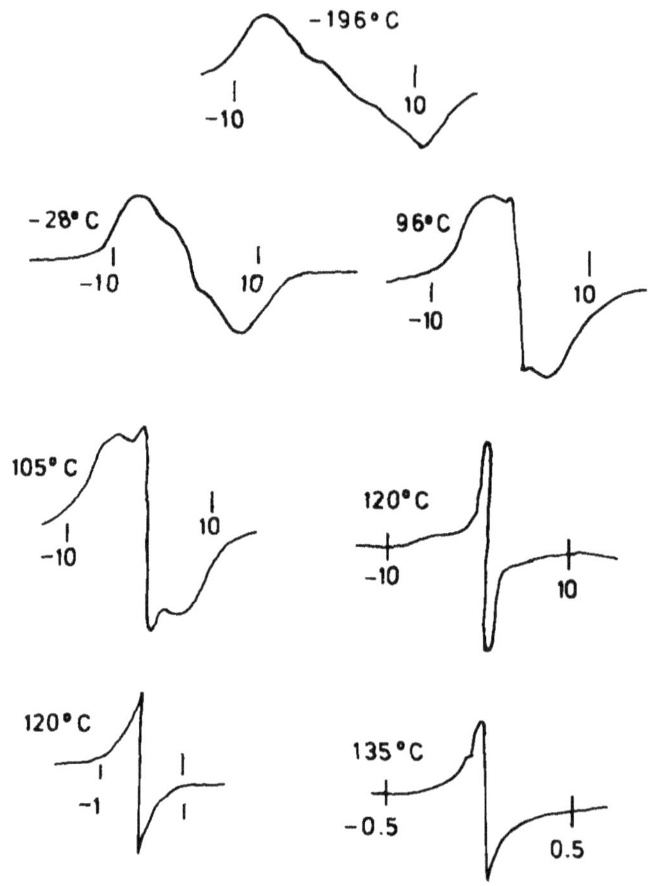

FIGURE 3. PMR derivative absorption spectra of 1,2-dimyristoyl-DL-phosphatidylethanolamine at various temperatures. The abscissae are marked in gauss. (From Chapman, D. and Salsbury, N. J., *Trans. Farad. Soc.*, 62, 2607, 1966. With permission.)

point of sodium stearate is ≈ 300°C. The high values of the capillary melting points of phospholipids are, therefore, consistent with the occurrence of ionic linkages associated with the polar groups of the phospholipid. The higher capillary melting points of the sodium soaps, compared with the phospholipids, suggest that there is possibly greater ionic character associated with the polar groups of these molecules than with the diacyl PCs and PEs.

B. The Mesomorphic State

1. Liquid Crystals Formed by the Action of Heat (Thermotropic Mesomorphism of Lipids)

In addition to the capillary melting point, other phase changes occur with phospholipids at lower temperatures. When a pure phospholipid, dimyristoyl phosphatidylethanolamine (DMPE) containing two fully saturated chains, is heated from room temperature up to the capillary melting point, a number of thermotropic changes occur (i.e., phase changes caused by the effect of heat).

Differential thermal analysis (DTA) shows that a marked endothermic transition (absorption of heat) occurs[9] at about 120°C. An additional heat change occurs at ≈ 135°C, and only a small heat change is involved near the capillary melting point of the lipid. This behavior is similar to that which occurs with compounds which form liquid crystals, e.g., *o*-azoxyanisole or cholesteryl propionate, which give nematic and cholesteric liquid crystalline phases, respectively.

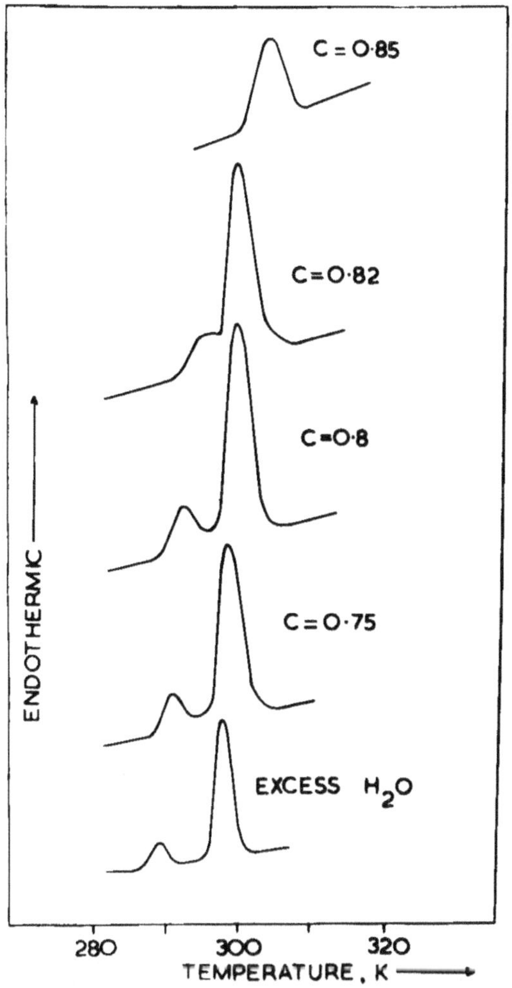

FIGURE 4. DSC curves for the dimyristoyl (dimyristoyl lecithin) DMPC-H_2O system as a function of lipid concentration c = wt lipid/(wt lipid + wt water).

The X-ray long spacings show a dramatic reduction to some two thirds of their original value at the first transition temperature with a further small reduction at the second transition temperature. The X-ray short spacings change at the first transition point from sharp diffraction lines to a diffuse spacing at ≈ 4.6 Å. Nuclear (proton) magnetic resonance (PMR) studies show a gradual reduction in line width from about 15 G at liquid nitrogen temperature until, at the first transition temperature, there is a sudden reduction in the line width (to ≈ 0.09 G). PMR spectra of DMPE are shown in Figure 3. These show that molecular motion increases gradually as the temperature increases until, at the transition temperature, a considerable increase in the molecular motion takes place,[10] as revealed by an appreciable reduction in line width.

The main conclusions are the following. (1) Even with the fully saturated phospholipid at room temperature, some molecular motion occurs in the solid. This is evident from the PMR spectra and from the IR spectra taken at liquid nitrogen and room temperatures. (2) When the phospholipid is heated to a higher temperature, it reaches a transition point at ≈ 120°C, a marked endothermic change occurs, and the hydrocarbon chains in the lipid "melt"

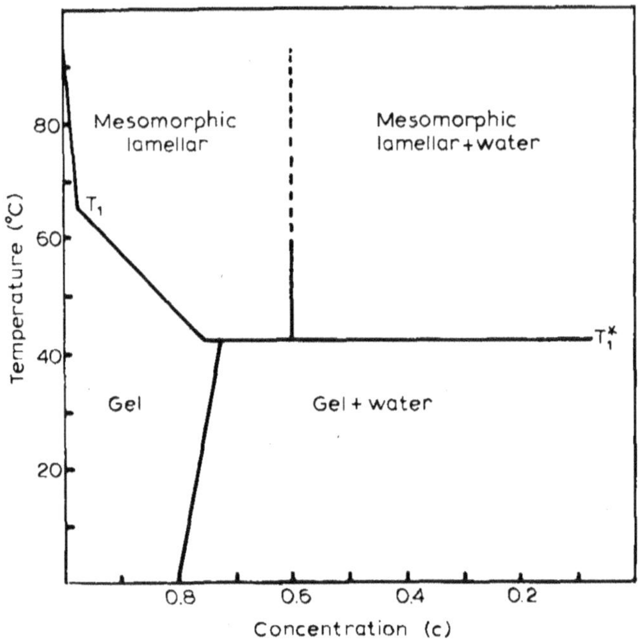

FIGURE 5. Simple phase diagram of 1,2-dipalmitoyl lecithin-water system. (From Chapman, D., Williams, R. M., and Ladbrooke, B. D., *Chem. Phys. Lipids*, 1, 445, 1967. With permission.)

and exhibit a very high degree of molecular motion. This is evident both in the appearance of the IR spectrum and also in the narrow PMR line width at temperature of 120°C and higher. On the one hand the broad diffuse appearance of the IR spectrum is consistent with the chains flexing and twisting sand with a "break-up" of the all-planar *trans* configuration of the methylene groups of the chains. The PMR spectrum provides an impression of any molecular motion, i.e., rotation, translation, or diffusion, of the phospholipid molecule occurring within a much longer time period, some 10^{-7} sec. The fact that the phase transition is concerned primarily with the hydrocarbon chains of the phospholipid is confirmed by the X-ray data.

When phospholipids contain shorter chains, or unsaturated bonds, those marked endothermic phase transitions occur at lower temperatures. The changes in the temperatures at which these transitions occur parallel the trends in the melting points of the related fatty acids. This trend in the transition temperatures also confirms that this phase transition is primarily associated with a melting of the hydrocarbon chains of the phospholipid, while this in turn is related to the dispersion forces between the chains.

2. *Liquid Crystals Formed in the Presence of Water (Lyotropic Mesomorphism of Lipids)*

Small amounts of water can have important effects upon the mesomorphic behavior. Thus, the diacyl phosphatidylcholines (lecithins) exhibit an additional liquid crystalline form between the first transition temperature and the capillary melting point.[11] The intermediate liquid crystalline form is found to exhibit X-ray spacings consistent with a cubic phase organization.

When phospholipids are examined in the presence of increasing amounts of water, the various physical techniques, such as microscopy, PMR spectroscopy, or DTA, show that as the amount of water increases, the marked endothermic transition temperature for a given phospholipid falls. The transition temperature does not fall indefinitely; it reaches a limiting

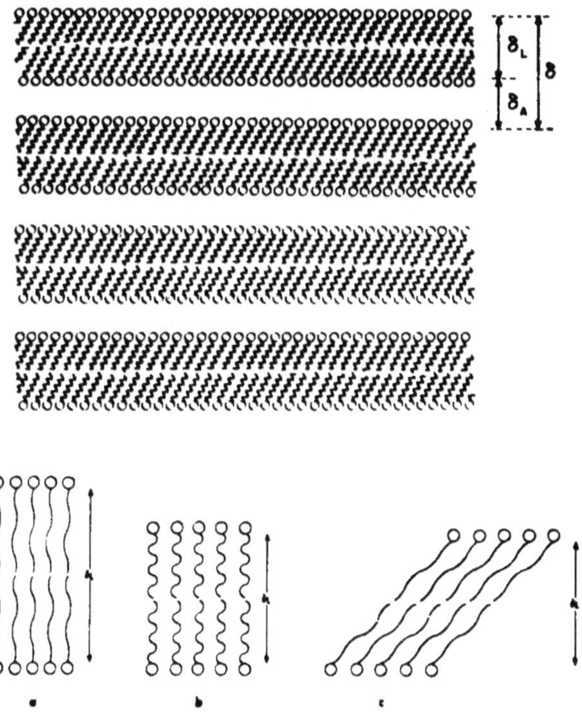

FIGURE 6. Lipid bilayers which spontaneously form in water. The chain organizations showing tilted, vertical, and "melted" arrangements are indicated.

value. Calorimetric heating curves for dimyristoyl lecithin at various concentrations of water are shown in Figure 4.

III. PROPERTIES OF HYDRATED LIPIDS

A. Phase Diagrams

The importance of the thermotropic phase transition temperature can be seen when we appreciate that when water diffuses into the lattice it does so into the polar (ionic) region only when the temperature is reached at which the hydrocarbon chains "melt" (the T transition temperature). If the temperature is higher than this, there is a simultaneous dissociation of the ionic lattice by the penetration of water and melting of the hydrocarbon chain region. The temperature of transition (T_c line) depends on the nature of the hydrocarbon chains and of the polar region of the molecule, the amount of water present, and on any solutes dissolved in the water.

Once the water has penetrated into the lattice of the amphiphile and the sample is then cooled to below the T_c transition temperature, the hydrocarbon chains arrange themselves into an orderly crystalline lattice, but the water is not necessarily expelled from the system. These phases containing crystalline paraffin chain regions are sometimes called gels; these gels may or may not be metastable. If they are metastable they transform over a period of time to a suspension of microcrystals of the amphiphile in water, to coagel state. An understanding of the phase diagram is important for understanding the formation of the

Table 1
HEATS AND TEMPERATURES OF TRANSITION FOR 1,2-DIACYL-L-PHOSPHATIDYLCHOLINE-WATER SYSTEMS

1,2-Diacyl-phosphatidylcholine	Limiting transition temperature (T_c) (°C)	Heat absorbed at T_c(kcal/mol) phospholipid	Entropy change T_c(cal/mol)
Dibehenoyl	75	14.90	42.8
Distearoyl	58	10.70	32.4
Dipalmitoyl	41	8.65	27.6
Dimyristoyl	23	6.65	22.4
Dilauroyl	0 (under the ice peak)	—	—
Dioleoyl	−22	7.6	30.3
Egg yolk[a]	−15/−7	—	—

[a] The egg yolk lecithin transition is broad in water due to the heterogeneous fatty acid composition. The lower temperature is from the heating curve, the higher from the cooling curve.

From Chapman, D., Williams, R. M., and Ladbrooke, B. D., *Chem. Phys. Lipids*, 1, 445, 1967. With permission.

liposome structures. Figure 5 is a simplified phase diagram of the 1,2-dipalmitoyl-L-phosphatidylcholine-water system.[12] We see that upon addition of water, the transition temperature T of the phospholipid is lowered to a limiting value, T_c. This transition temperature is the minimum temperature required for water to penetrate between the layers of the lipid molecules. The multilamellar structure created spontaneously in this way is shown in Figure 6. Above the T_c line the PC-water system exists in a mesomorphic lamellar phase, in which the hydrocarbon chains are in a liquid-like state. The composition of the system at maximum hydration is about 40 wt% water. Addition of more than 50 wt% water gives rise to a two-phase system consisting of fragments of the lamellar phase at maximum hydration dispersed in the excess water.

When the lecithin-water system is cooled below the T_c line, the hydrocarbon chains adopt an ordered packing. The structure of this phase, the gel, is lamellar with the hydrocarbon chains packed in a hexagonal subcell with the chain axes inclined at 58° to the lipid-water interface.

The phase diagrams of the different chain length lecithin-water systems are essentially equivalent and are disposed along the temperature axis according to the melting temperature T_c of the hydrocarbon chains. Differential scanning calorimetric curves of 1,2-diacyl-L-phosphatidylcholine-water system obtained over a range of water concentrations show, in the range $1.0 > c > 0.8$, that the temperature of the endothermic transition T decreases steadily to a limiting value T_c.

Each phospholipid series is characterized by a phase transition temperature T_c below which its fatty acyl chains are in a quasicrystalline array and above which the chains are in a more fluid state. Other changes accompanying chain melting are decreased bilayer thickness and increased area per molecule. These changes have been studied extensively by a variety of physical techniques including X-ray diffraction, nuclear magnetic resonance (NMR), electron spin resonance (ESR), differential scanning calorimetry, and fluorescence depolarization.[7] In general T_c is lowered by decreased chain length by unsaturation of the acyl chains, and by bulky side groups (e.g., cyclopropane rings).

All the different lecithins studied behave similarly (Table 1). The heat absorbed at T_c for lecithin-water systems at high water content is seen to be chain-length dependent. For the

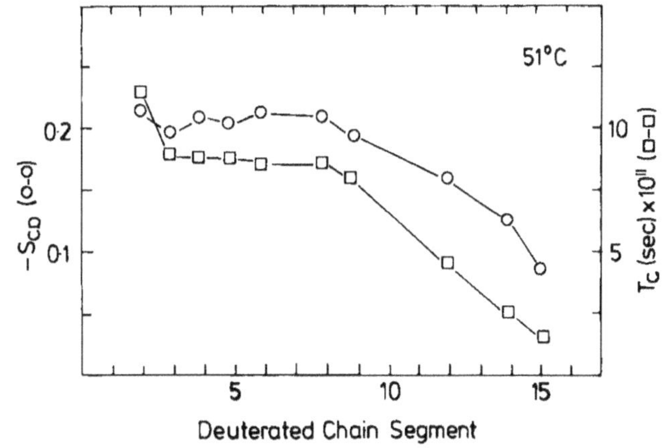

FIGURE 7. A comparison of the rotational correlation times (T_0; □) determined from the T_1 data and the deuterium order parameter ($-S_{CD}$;) as a function of segment position. (From Brown, M. F., Seelig, J., and Haberlen, U., *J. Chem. Phys.*, 20, 5045, 1979. With permission.)

series of fully saturated lecithins a difference of 2 kcal/mol occurs for a chain-length difference of two methylene groups. Extrapolation to zero at this transition shows that the C_7 and shorter chain-length lecithins have a negative heat of transition.

A pretransition observed in the calorimetric studies of pure lecithin-water systems has been shown to indicate the occurrence of a $P\beta^1$ phase corresponding to the rippled phase observed in freeze-fracture pictures.

A complete binary phase diagram of the egg yolk lecithin-water system has been described by Small.[13] The phase diagram is essentially the same as those found for the lecithins of discrete chain lengths. The T_c line is broad and ill-defined, and this observation is supported by other work on phospholipids derived from an heterogeneous mixture of fatty acids.

Liposome structures, as noted earlier, are formed spontaneously in water by phospholipid molecules. Electron micrographs of such liposomes can be found in many recent scientific publications. Liposomes can be formed from many different phospholipids, and the composition most frequently used has been the natural phospholipid egg lecithin extracted from egg yolk (sometimes with or without cholesterol), and with or without anionic lipids such as phosphatidic acid or cationic lipids such as stearylamine. They can also be formed from whole lipid extracts of cell membranes. Under normal conditions cholesterol can be incorporated into liposomes up to a maximum of one cholesterol molecule per phospholipid molecule, although some workers have reported up to 2:1 cholesterol per phospholipid molar ratio in highly sonicated liposome preparations. (Where higher proportions of cholesterol than this are reported there is considerable doubt as to the presence of all the cholesterol in liposomal form, since much of it may be present as dispersed fine crystals of cholesterol.)

The separation between the liposome membranes is determined primarily by the balance between the repulsive forces between the membranes — probably mainly electrostatic interactions between head groups and hydration forces of the head groups — and attractive forces which are consistent with the expected van der Waals forces between the layers. The equilibrium distance between bilayers of pure egg PC in water is 2.75 nm. This separation, and thus the volume of aqueous phase within the liposome, can be increased by the addition of up to 10 mol% of charged lipid (either anionic or cationic) into the phospholipid.

The entrapped volume can be estimated by measuring the degree of entrapment of a nonionic, nonpermeable solute, say glucose, within the liposomes. The entrapped volume is then the fraction of solute entrapped times the total volume of the preparation.[14]

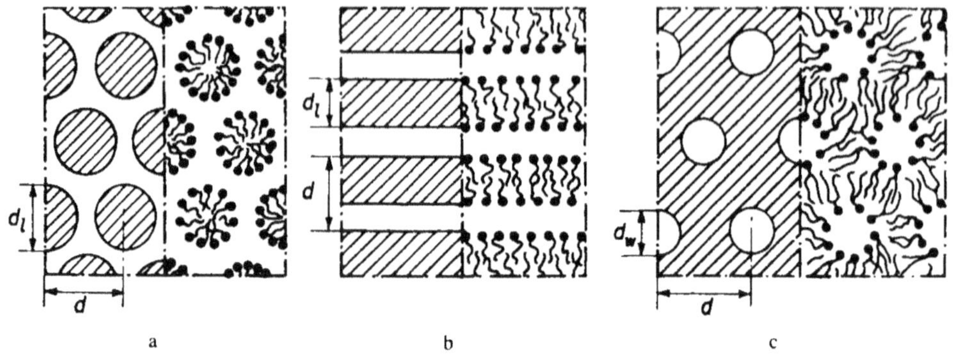

FIGURE 8. Structures of some high-temperature phases of lipid-water systems; schematic representation of a cross-section of the rods and of the lamellae. The hydrophobic group is represented by a dot, the paraffin chains by a wriggle; (a) hexagonal I; (b) lamellar; (c) hexagonal II. (From Luzzati, V., in *Biological Membranes*, Vol. 1, Chapman, D., Ed., Academic Press, New York, 1968, 7. With permission.)

B. Lipid Order Parameters

Early suggestions that the lipid chains in the fluid bilayers were in a random chaotic state are not correct. The liquid-like character of the bilayer interior is different from that of a simple paraffinic melt. The ΔH values associated with the gel-to-liquid crystal transition are lower than the ΔH values for the melting of the pure hydrocarbons. The same holds true for the entropy change in the process. The incremental ΔS per CH group is only about 1 e.u. for the gel-to-liquid crystal transition of bilayers, but is almost twice as large for the melting of simple paraffins.[15] These thermodynamic results showed that the hydrocarbon chains in the bilayer core are not as disordered as they are in a pure liquid hydrocarbon.

The introduction of nonperturbing deuterium probes by Oldfield et al.[16] has been exploited by Seelig and co-workers,[17,18] to examine the "order parameter" of the lipid chains and to show that in the region of constant order parameter gauche conformations can occur only in complementary pairs leaving the hydrocarbon chains essentially parallel to each other. The deuterium probe results differ from those obtained using spin-labeled molecules,[19] e.g., the spin labels detect a continuous decrease of the order parameter while the deuterium probe shows that the order parameter remains approximately constant for the first nine segments (see Figure 7). The number of order profiles reported in the literature is still small.

The deuterium magnetic resonance technique has also been applied to natural biomembranes. This was first carried out by Oldfield et al.[20] with *Acholeplasma laidawii* cells and later by Stockton et al.[21] with selectively deuterated palmitic acid. Quite recently, similar studies have also been made with *Escherichia coli* membranes. The fatty acyl chains in elaidate 9-10-d_2 enriched *E. coli* inner and outer membranes are less ordered by 10 to 20% than in the corresponding model systems.[22]

This characteristic order parameter signature of model membranes is carried over into biological membranes. The agreement between the order profile of *A. laidlawii* and those of the pure phospholipid membranes is striking. The incorporation of a *cis*-double bond promotes larger changes in the order profile than, for example, the introduction of a net negative charge in the polar head group (DPPS) or the incorporation of proteins into the membrane. The divergence of the POPC order profile between carbon atoms 5-9 can be explained by a specific stiffening effect of the *cis*-double bond.[23]

C. Types of Lyotropic Phase

Above the first transition temperature, when sufficient water is present, phospholipids can form several different types of phase organization, e.g., some phospholipids give lamellar- and hexagonal-type structure. Recent studies have shown that the lecithins appear to exhibit only a lamellar-type structure over a large range of water concentrations. Other phospholipids, such as phosphatidylethanolamines and samples of brain lipid, appear to be able to exist in both the hexagonal and lamellar types of organization,[24] depending upon the concentration in water (see Figure 8). Studies by Gulik-Kryzwicki et al.[25] have shown that four phases can be distinguished with a mitochondrial lipid extract. The composition of the extract from beef heart mitochondria was 34% lecithin, 29% PE, 10% PI, 20% cardiolipin, 2% cholesterol (Chol), and 5% neutral lipids. Hexagonal phases were observed, as well as various lamellar phases. As yet, no relationship between these phase properties and their biological function has been made. Nevertheless, these phase properties have to be considered and held in mind in discussions of lipid behavior.

IV. PREPARATION OF LIPOSOMES

An important feature of liposome preparation which is often neglected is that a variety of shapes and sizes of structures can occur. Some workers apparently consider that the only type of structure formed is the spherical or onion-shaped liposome. However, in the light microscope even a casual inspection shows shapes ranging from long tubules to the more usual spheres. These liposomes vary in dimension from several hundred angstroms to fractions of a millimeter, so that a whole range of sizes and shapes can coexist. Regardless, however, of the overall shape, the liposomes are organized as closed concentric "membranes" with an aqueous compartment separating each such "membrane" from its neighbor. By a variety of techniques phospholipids may be dispersed to form multimembrane liposomes and also to form liposomes (250 Å to microns in diameter) in which a single "membrane" encloses the aqueous compartment. The term liposome is frequently used to encompass this entire range of multimembrane or single-membrane structures.

A. Multimembrane Liposomes

Liposomes are sometimes divided into three classes, based on (1) their size, (2) their single membrane, and (3) their multimembrane structures. The spacing between the membrane is determined by a balance of the van der Waals' forces of attraction and the electrostatic and "hydration" forces of repulsion between adjacent membranes and can be varied by adding other charged molecules. These large multimembrane liposomes range in size up to fractions of a millimeter. Within any single preparation, there is considerable size heterogeneity, with the range determined by details of the method of production.

The large multimembrane liposomes are usually formed essentially as described by Bangham et al.[26] in 1965. Appropriate lipids are dissolved in an organic solvent which is subsequently evaporated leaving behind a thin film of dried lipid on the surface of a flask. An aqueous phase containing any water-soluble materials to be trapped is then added, and the solution is swirled by hand to free the lipid from the sides of the flask and to disperse large lipid aggregates. Glass beads are sometimes added to facilitate dispersion. The resulting milky suspension is usually allowed to equilibrate for several hours before use.

Nontrapped solutes in the external aqueous phase can be separated by repeated centrifugation and washing, by dialysis, or by molecular sieve chromatography. In the last method, the liposomes are excluded from the internal volume of the gel and appear in the void volume. The final product contains solute trapped within its internal aqueous space with little solute in the external bathing medium.

The large multimembrane liposomes respond to osmotic gradients by shrinking or swelling, that is, they behave as osmometers, whereas the small unimembrane liposomes do not. The permeability of large liposomes to various solutes has been studied extensively but precise quantitation is difficult because of the size heterogeneity and multicompartmental nature of these liposomes.[14]

B. Small Single-Membrane Liposomes

Small single-membrane liposomes range in diameter from approximately 200 to 500 Å and consist of a single lipid bimolecular layer surrounding an aqueous compartment.[27] These differ from the large multimembrane liposomes in several respects in addition to size: (1) they are osmotically insensitive, (2) a disproportionate amount (about 70%) of the total lipid is located in the outer leaflet of the vesicle in the smallest vesicle, and (3) the small radius of curvature imposes strains in packing of the lipid molecules (these small liposomes are particularly useful for NMR spectroscopic studies).

Ultrasonic dispersion is the method often used for the preparation of the small vesicles. In one approach, phospholipids are sonicated by immersing a metal probe directly into a suspension of the large liposomes. In a second method the lipid dispersion is sealed in a glass vial, which is then suspended in an ultrasonic cleaning bath. Each method has its distinct advantages and disadvantages, depending on the application at hand.

Several other methods have been used to produce small liposomes. In one method an ethanolic solution of phospholipid is injected through a small-bore needle into a rapidly stirred aqueous solution, forming small unilamellar vesicles similar to those prepared by sonication.[28] This technique minimizes the possibility of lipid degradation and eliminates the problem of aerosols. Its disadvantages are a low efficiency of trapping for dissolved solutes and the requirement that the lipids used must be soluble in ethanol.

The small liposomes can also be prepared by injecting an aqueous suspension of phospholipids under high pressure through the small orifice of a "French Press". This approach appears to combine some of the favorable aspects of both the sonication and injection methods.

Finally, single-membrane liposomes of relatively small size can be generated by detergent dilution, a procedure in which detergent is used to solubilize the phospholipid and then is removed by dialysis. This method has been used to "reconstitute" membrane proteins into liposomes and may be useful in future studies of liposome-cell interactions involving protein components.

Preparation of small single-membrane liposomes can be fractionated according to size by either prolonged centrifugation or molecular sieve chromatography. It is thus possible to obtain quite homogeneous preparations of vesicles about 200 Å in diameter. (This value is close to the minimum radius of curvature obtainable in a vesicle.)

C. Large Single-Membrane Liposomes

Large liposomes that consist of only a single lipid membrane have been prepared. Several techniques have been developed for the preparation of these liposomes ranging in diameter from about 600 Å to several microns. Reeves and Dowben described the preparation of phospholipid vesicles 0.5 to 10 μm in diameter, bounded by one (or a few) phospholipid bilayers.[29] These liposomes were formed by slowly hydrating a thin layer of phospholipid in distilled water or an aqueous solution of nonelectrolyte. The vesicles which form can be centrifuged and resuspended in another medium for permeability studies.

Another approach to large "unimembrane" liposome formation makes use of ether injection. In this method, an ether solution of lipid is injected into warm aqueous medium.[30] The resulting liposomes are osmotically active and mostly unimembrane. Their method of formation results in a high trapping efficiency.

Both hydrophobic and polar drugs can be entrapped in the liposomes. To accomplish this the hydrophobic drugs are dissolved in chloroform along with the lipid and cholesterol. Subsequent to evaporation, the drug is embedded in the thin lipid film deposited on the wall of the round-bottomed flask. Polar drugs are dissolved in aqueous buffers used in the liposome formation from the thin film of phospholipid deposited on the wall of the flask. The liposome-entrapped drug can be readily separated from the excess free drug by gel filtration through

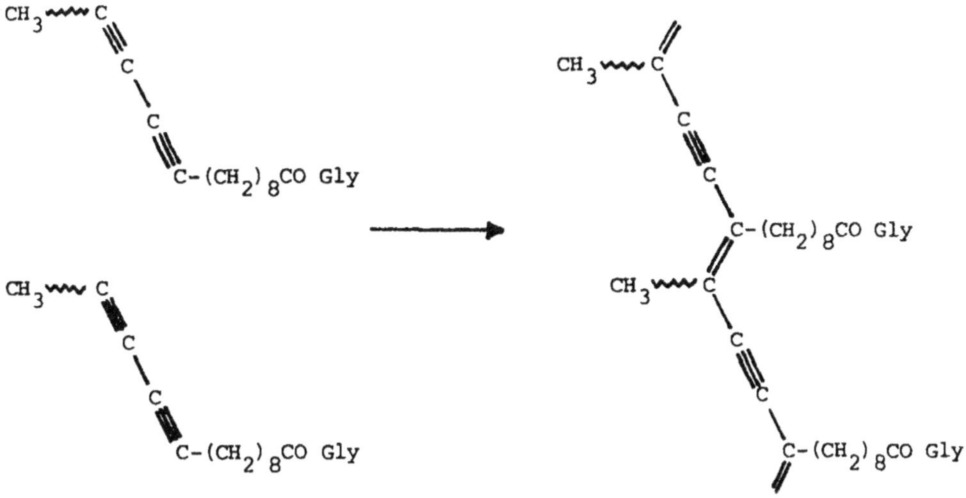

FIGURE 9. Polymerization process triggered by UV radiation of diacetylenic groupings producing cross-linking of phospholipid within liposomes structures. (From Johnston, D. S., Sanghera, S., Pons, M., and Chapman, D., *Biochim. Biophys. Acta*, 602, 57, 1980. With permission.)

a Sephadex® G-50 column. Once the drug is entrapped in the liposome, it can be liberated only by leakage through the bilayer or by destruction of the vesicle. (Conversely, drug solubilization by micelles is an equilibrium situation. There is dynamic equilibrium between monomers and micelles and between micelles and solubilizates.)

V. MODIFICATION OF LIPIDS WITHIN LIPOSOMES

A. Hydrogenation of Double Bonds Within Liposomes

It has previously been shown that the hydrogenation of the double bonds of the lipids in liposome systems can be accomplished using a homogeneous catalyst $RhCl(PPh_3)_3$. This homogeneous catalyst is a hydrophobic complex which has to be introduced into the liposome or biomembrane using a solvent vector. One disadvantage of this catalyst is that once it has partitioned itself into the hydrophobic region of the lipid it can only be removed by disruption of the lipid or biomembrane structure.[31]

In principle, a homogeneous catalyst which is water soluble has a number of advantages. These are (1) the elimination of the solvent vector, e.g., tetrahydrofuran or dimethylsulphoxide, (2) the possibility of more easily removing the catalyst on completion of the reaction, and (3) the possibility of achieving improved selectivity due to the positioning of the active site at particular positions within the hydrocarbon chains of the lipid structures. Recently a water-soluble catalyst has been developed for this purpose and experiments reported.[32] These show that hydrogenation of the double bonds in liposomes using a water-soluble catalyst can be readily accomplished.

There are a number of potential applications of the water-soluble catalyst:

1. The improved selectivity of hydrogenation of the unsaturated bonds of lipids in liposome structures by tailoring the distance at which the active catalytic site is situated from the aqueous interface.
2. An application to the hydrogenation of the unsaturated groups of the lipids within serum lipoprotein structures.
3. Catalytic reactions in addition to hydrogenation reactions, e.g., hydroformylation, using the liposome as an orientating medium, i.e., saturated liposomes can be used

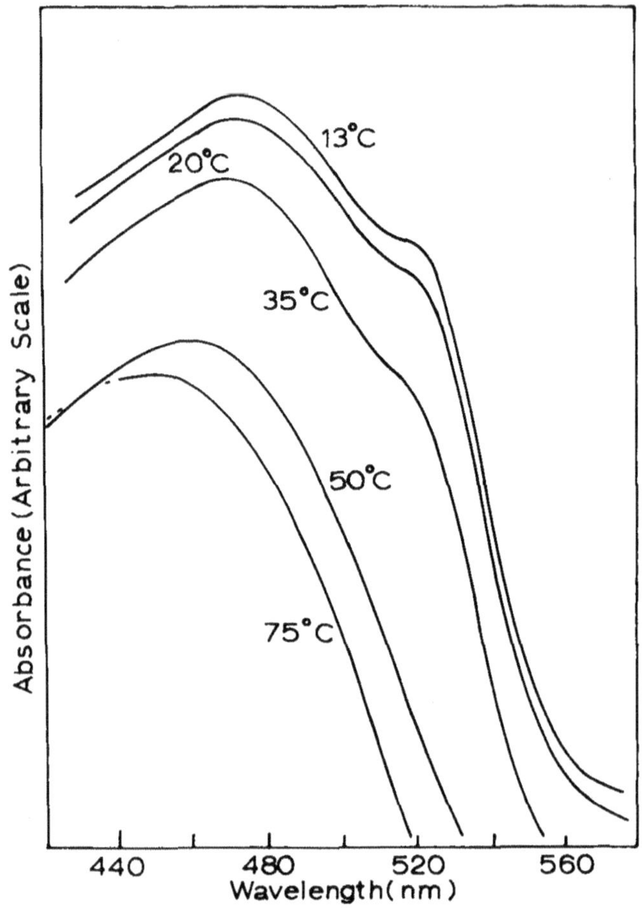

FIGURE 10. Visible spectra of partially polymerized liposomes at different temperatures. (From Johnston, D. S., Sanghera, S., Pons, M., and Chapman, D., *Biochim. Biophys. Acta,* 602, 57, 1980. With permission.)

above the T_c transition temperature so that they are in their fluid state. Molecules (not necessarily lipid molecules) can be introduced into these saturated lipid liposomes, and by means of the homogeneous catalyst highly specific reactions can be brought about.

B. Polymerization Processes

Recent studies[33,34] have included a new approach to the study of liposomes as well as other model membranes, e.g., monolayer and multilayer structures. We have also studied natural biomembrane structures. This new approach causes extensive cross-linking of the lipid chains to occur via a polymerization process triggered via UV radiation. We have achieved this with liposomes by synthesizing phospholipids containing fatty acid residues which have diacetylenic groupings within each lipid chain. The process of polymerization is indicated in Figure 9. It can be seen that the resultant polymer contains a conjugated linkage thereby producing a colored material. The process can be followed by the absorption spectrum in the visible region. The visible spectral absorption of these liposomes in water at different temperatures is shown in Figure 10.

Biosynthetic incorporation of the fatty acids into various cells such as *A. laidlawii* microorganisms has also been achieved and successful polymerization indicated by the same color changes as occur with the liposomes.

VI. FURTHER PROPERTIES OF LIPOSOMES

A. Phase Separation Within Liposomes

Phase separation of lipids can occur within liposome structures. The first studies on phase separation of lipid-water systems were discussed by Ladbrooke and Chapman, who reported studies of binary mixtures of lecithins using calorimetry.[35] These authors examined mixtures of distearoyl and dipalmitoyl lecithin (DSL-DPL) and also distearoyl lecithin and dimyristoyl lecithin (DSL-DML). With the DSL-DPL mixtures the phase diagram shows that a continuous series of solid solutions is formed below the T_c line. It was concluded that compound formation does not occur and that with this pair of molecules having only a small difference in chain length co-crystallization occurs.

With the system DSL-DML monotectic behavior was observed with limited solid solution formation. Here the difference in chain length is already too great for co-crystallization to occur so that as the system is cooled migration of lecithin molecules occurs within the bilayer to give crystalline regions corresponding to the two compounds.[35]

Examination of a series of fully saturated lecithins with dioleoyl lecithin gave similar results with phase separation of the individual components taking place.[36] Later calorimetric studies were reported by Clowes et al.[37] on mixed lecithin-cerebroside systems and on lecithin-phosphatidylethanolamine mixtures.[38] The lecithin-PE systems of the same chain length give a wide melting range with some separation of the different lipid classes.

The use of spin labels such as TEMPO to prove phase separation of mixed lipid systems has been reported by Shimshick and McConnell[39] on similar lipid mixtures. Other phase separation properties have been observed by Ito and Ohnishi and have shown that lipid phase separation can occur in phosphatidic acid-lecithin membranes due to the effects of Ca^{2+}. Butler et al.[40] have shown that the probe stearic acid spin label tends to migrate to the more fluid lipid phase in multiphase systems. This confirms the earlier conclusions of Oldfield et al.[41] and shows that measurements of membrane fluidity in heterogeneous systems are not necessarily representative of the entire membrane.

B. Metal Ion Effects

Metal ion interactions have been known for some years to affect the thermotropic phase transition of soap systems. The thermotropic phase transition of stearic acid occurs at 114°C with the sodium salt and at 170°C with the potassium salt. These phase transitions can be linked to the monolayer characteristics.

Early studies of stearic acid monolayers[42] showed that interaction with Ca^{2+} ions caused an increase in surface pressure (i.e., condensation) and also decreased the permeability to water. The same effect has been observed with PS monolayers[43] Na^+ and K^+ addition gave no such condensation. Later, more extensive studies showed that various acidic phospholipid monolayers undergo an increase in surface potential and decrease in surface pressure on addition of Ca^{2+} and other bivalent cations.[44] PS is found to be more selective than phosphatidic acid but for both systems the order of cation effectiveness is

$$Ca^{2+} > Ba^{2+} > Mg^{2+}$$

The formation of linear polymeric complexes was proposed to account for these findings.

Cationic charge has been observed to be important in some bilayer studies of PS.[45] The results obtained agree well with those obtained previously,[46] with the exception of uranyl cation UO^{2+}. Recent studies with this cation[41] indicate that this ion causes the thermotropic phase transition of lecithins to increase. Two main phase transitions were observed corresponding to the presence of complexed and uncomplexed lipid. When the titration is complete only the higher melting transition remains. The studies by Chapman et al.[41] indicate that the interaction between cation and PS causes greater shifts of transition temperature than is observed with lecithin molecules. All the cations studied were found to shift the phase transition temperature of the phospholipids to higher values.

The precise nature of the interaction between ions and phospholipids is still open to debate. There is some evidence that charge neutralization is the prime interaction of charged phospholipids with divalent cations.[47] Divalent cations were found to increase the transition temperature and the monovalent cations to fluidize the bilayer. Some authors believe that the primary effect of the cation on lecithins may be on the aqueous portion of the lipid bilayer.[48] A recent study of an extensive range of salts with lecithin bilayers[49] indicates that the anion present has a very large effect in determining the state of fluidity of the bilayer, the results obtained being explained best by a thermodynamic treatment based on relative association constants. Träuble[50] has shown that pH can affect lipid transition temperatures, particularly lipids such as the PEs. Verkleij et al.[47] have shown that the thermotropic phase transition of a synthetic phosphatidylglycerol is influenced by pH, Ca^{2+}, and a basic protein of myelin.

C. Permeability Studies

Liposomes have been shown to be permeable to water, ions, and nonelectrolytes, although the permeabilities depend on the chemical composition of the liposomes. Positively charged liposomes (e.g., lecithin plus a positively charged lipid such as stearylamine) are impermeable to cations, whereas negatively charged liposomes (e.g., those containing phosphatidic acid) are permeable to cations; however, the permeability of all liposomes to protons is low. Anions diffuse rapidly through negatively and positively charged lipid membranes. In general, it has been found that increasing the degree of saturation or the length of the phospholipid fatty acyl chains produces a decrease in the permeability of the liposomes to all solutes.[51] This is due to the increase in the T_c transition temperatures resulting from these modifications, which leads to a decrease in fluidity of the membrane interior. A decrease in permeability can also be produced by the addition of ionophores such as valinomycin. The addition of some proteins such as lysozyme, myelin basic protein, and immunoglobulins to a liposome suspension increases the leakage of anions and glucose from the liposomes, while general anesthetics, such as ether and chloroform, produce an increase in cation permeability of the bilayers but not increase in glucose permeability.[14]

Liposomes therefore are capable of sequestering aqueous solutions and the rate at which the solutes leak out of the liposomes is dependent on both the nature of the solute and composition of the liposome. By modification of the composition of the lipid bilayers it is possible to reduce the leakage of particular molecules, although any small molecule not having too many hydrogen bonds can leak rapidly through these structures.

VII. LIPOSOMES CONTAINING INTRINSIC PROTEINS

An important advance in model biomembrane system construction is to include membrane proteins into the system. Some hydrophobic polypeptides, e.g., gramicidin A, can be incorporated by means of simple cosolution into an organic solvent with the appropriate lipid. Many membrane proteins, however, are difficult to manipulate and require a detergent reconstitution method. The basic[52] method involves solubilization of the biomembrane by the detergent (e.g., deoxycholate), thereby keeping the protein in a hydrophobic environment and then replacing the natural lipid by the chosen pure lipid material. There are some problems involving the use of detergents, nevertheless, good reconstituted systems can be obtained.

Experiments with various protein-lipid water systems, i.e., liposomes containing intrinsic proteins, have been carried out both above and below the lipid T_c temperature. Below the T_c transition temperature protein aggregation can occur.[53] The aggregation process produces two regions. This is revealed very clearly in freeze-fracture electron microscopy pictures. As yet more protein is included to the lipid system, the aggregated patch increases in size at the expense of the remaining crystalline region. This is why lipid-water systems show a broadening of the main melting transition.

REFERENCES

1. **Teitel, B. A.,** Note sur les figures myeleniques et leur interpretation physico-chimique, *Arch. Roum Pathol Exp.*, 14, 53, 1947.
2. **Chapman, D., Byrne, P., and Shipley, G. G.,** The physical properties of phospholipids. I. Solid state and mesomorphic properties of some 2,3 diacyl phosphatidyl ethanolamines, *Proc. R. Soc. London*, 290 A, 115, 1966.
3. **Bangham, A. D., Standish, M. M., and Watkins, J. C.,** Diffusion of univalent ions across the lamellae of swollen phospholipids, *J. Mol. Biol.*, 13, 238, 1965.
4. **Chapman, D.,** Phase transitions and fluidity characteristics of lipids and cell membranes, *Q. Rev. Biophys.*, 8, 185, 1975.
5. **Gregoriadis, G., Leathwood, P. D., and Ryman, B. E.,** Enzyme entrapment in liposomes, *FEBS Lett.*, 14, 95, 1971.
6. **Sessa, G. and Weismann, G.,** Incorporation of lysozyme into liposomes A mode for structure-linked latency, *J. Biol. Chem.*, 245, 3295, 1970.
7. **Gregoriadis, G.,** Enzyme entrapment in liposomes, *Meth. Enzymol.*, 44, 218, 1976.
8. **Pearson, R. H. and Pascher, I.,** The molecular structure of lecithin di-hydrate, *Nature (London)*, 281, 499, 1979.
9. **Chapman, D. and Collin, D. T.,** Differential thermal analysis of phospholipids, *Nature (London)*, 206, 189, 1965.
10. **Chapman, D. and Salsbury, N. J.,** Physical studies of phospholipids. V. Proton magnetic resonance studies of molecular motion in some 2,3 diacyl DL phosphatidylethanolamines, *Trans. Farad. Soc.*, 62, 2607, 1966.
11. **Chapman, D., Williams, R. M., and Ladbrooke, B. D.,** Physical studies of phospholipids. VI. Thermotropic and lyotropic mesomorphism of some 1,2 diacylphosphatidylcholines (lecithins), *Chem. Phys. Lipids*, 1, 445, 1967.
12. **Williams, R. M. and Chapman, D.,** Phospholipids liquid crystals and cell membranes, *Progress in the Chemistry of Fats and Lipids*, Pergamon Press, Oxford, 1970.
13. **Small, D. M.,** Phase equilibrium and structure of dry and hydrated egg lecithin, *J. Lipid Res.*, 8, 551, 1967.
14. **Bangham, A. D., Hill, M. W., and Miller, N. G. A.,** Preparation and use of liposomes as models of biological membranes, in *Methods in Membrane Biology* Vol. 1, Plenum Press, New York, 1974, 1.
15. **Phillips, M. C., Williams, R. M., and Chapman, D.,** On the nature of hydrocarbon chain motions in lipid liquid crystals, *Chem. Phys. Lipids*, 3, 234, 1969.
16. **Oldfield, E., Chapman, D., and Derbyshire, W.,** Deuteron resonance. A novel approach to the study of hydrocarbon chain mobility in membrane systems, *FEBS Lett.*, 10, 102, 1971.
17. **Seelig, A. and Seelig, J.,** The dynamic structure of fatty acyl chains in a phospholipid bilayer measured by deuterium magnetic resonance, *Biochemistry*, 13, 4839, 1974.
18. **Seelig, J. and Seelig, A.,** Lipid conformation in model membranes and biological membranes, *Q. Rev. Biophys.*, 13, 19, 1980.
19. **Brown, M. F., Seelig, J., and Haberlen, U.,** Structural dynamics in phospholipid bilayers from deuterium spin lattice relaxation time measurements, *J. Chem. Phys.*, 70, 5045, 1979.
20. **Oldfield, E., Chapman, D., and Derbyshire, W.,** Lipid mobility in Acholeplasma membranes using deuteron magnetic resonance, *Chem. Phys. Lipids*, 9, 69, 1972.
21. **Stockton, B. W., Johnson, K. G., Butler, K. W., Polnasek, C. F., Cyr, R., and Smith, I. C. P.,** Molecular order in Acholeplasma laidlawii membranes as determined by deuteron magnetic resonance of biosynthetically incorporated specifically labelled lipids, *Biochim. Biophys. Acta*, 401, 535, 1975.
22. **Gally, H. V., Pluschke, G., Overath, P., and Seelig, J.,** Structure of *Eshcherichia coli* membranes. Fatty acyl chain order parameters of inner and outer membranes, *Biochemistry*, 20, 1826, 1981
23. **Seelig, A. and Seelig, J.,** Effect of a single cis-double bond on the structure of a phospholipid bilayer, *Biochemistry*, 16, 45, 1977.

24. **Luzzati, V.,** X-ray diffraction studies of lipid-water systems, in *Biological Membranes*, Vol. 1, Chapman, D., Ed., Academic Press, New York, 1968, 7.
25. **Gulik-Krzywicki, T., Sheeter, E., Luzzati, V., and Faure, M.,** Interactions of proteins and lipid structure and polymerisation of protein-lipid-water phases, *Nature (London)*, 223, 1116, 1969.
26. **Bangham, A. D., Standish, M. M., and Watkins, J. C.,** Diffusion of univalent ions across the lamellae of swollen phospholipids, *J. Mol. Biol.*, 13, 238, 1965.
27. **Huang, C.,** Studies on phosphatidylcholine vesicles, formation and physical characteristics, *Biochemistry*, 8, 344, 1969.
28. **Batzri, S. and Korn, E. D.,** Single bilayer liposomes prepared without sonication, *Biochim. Biophys. Acta*, 298, 1015, 1973.
29. **Reeves, J. P. and Dowben, R. M.,** Water permeability of phospholipid vesicles, *J. Membr. Biol.*, 3, 123, 1970.
30. **Deamer, D. and Bangham, A. D.,** Large volume liposomes by an ether vaporization method, *Biochim. Biophys. Acta*, 443, 629, 1976.
31. **Chapman, D. and Quinn, P. J.,** A method to modulate membrane fluidity. The homogeneous catalytic hydrogenation of phospholipids and phospholipid water model membranes, *Proc. Natl. Acad. Sci. U.S.A.*, 73, 3971, 1976.
32. **Madden, T. D., Peel, W. E., Quinn, P. J., and Chapman, D.,** The modulation of membrane fluidity by hydrogenation processes. IV. Homogeneous catalysis of liposomes using a water soluble catalyst, *J. Biochim. Biophys. Meth.*, 2, 19, 1980.
33. **Johnston, D. S., Sanghera, S., Pons, M., and Chapman, D.,** Phospholipid polymers synthesis and spectral characteristics, *Biochim. Biophys. Acta*, 602, 57, 1980.
34. **Albrecht, O., Johnston, D. S., Villaverde, C., and Chapman, D.,** Stable biomembrane surfaces formed by phospholipid polymers, *Biochim. Biophys. Acta*, 687, 165, 1982.
35. **Ladbrooke, B. D. and Chapman, D.,** Thermal analysis of lipids, proteins and biological membranes, *Chem. Phys. Lipids*, 3, 304, 1969.
36. **Phillips, M. C., Ladbrooke, B. D., and Chapman, D.,** Molecular interactions in mixed lecithin systems, *Biochim. Biophys. Acta*, 211, 409, 1970.
37. **Clowes, A., Cherry, R. J., and Chapman, D.,** Physical properties of lecithin-cerebroside films, *Biochim. Biophys. Acta*, 249, 301, 1971.
38. **Chapman, D., Urbina, J., and Keough, K. M.,** Biomembrane phase transitions studies of lipid-water systems using differential scanning calorimetry, *J. Biol. Chem.*, 249, 2512, 1974.
39. **Shimshick, E. J. and McConnell, H. M.,** Lateral phase separation in phospholipid membranes, *Biochemistry*, 12, 2351, 1973.
40. **Butler, K. W., Tattrie, N. H., and Smith, I. C. P.,** The location of spin probes in two phase mixed lipid systems, *Biochim. Biophys. Acta*, 363, 351, 1974.
41. **Oldfield, E., Keough, K. M., and Chapman, D.,** The study of hydrocarbon chain mobility in membrane systems using spin label probes, *FEBS Lett.*, 20, 344, 1972.
42. **Harkins, W. D. and Anderson, T. F.,** *J. Am. Chem. Soc.*, 59, 2189, 1937.
43. **Rojas, E. and Tobias, J. M.,** Membrane model association of inorganic cations with phospholipid monolayers, *Biochim. Biophys. Acta*, 94, 394, 1965.
44. **Papahadjopoulos, D.,** Surface properties of acidic phospholipids: interaction of monolayers and hydrated liquid crystals with uni and bivalent ions, *Biochim. Biophys. Acta*, 163, 240, 1968.
45. **Ohki, S.,** The electrical capicatance of phospholipid membranes, *Biophys. J.*, 9, 1195, 1969.
46. **Blaustein, M. P.,** Phospholipids as ion exchangers: implications for a possible role in biological membrane and excitability and anaesthesia, *Biochim. Biophys. Acta*, 135, 653, 1967.
47. **Verkleij, A. J., de Kruyff, B., Ververgaaert, P. H. J., Tocanne, J. F., and van Deenen, L. L. M.,** The influence of pH, Ca^{2+} and protein on the thermotropic behaviour of the negatively charged phospholipid phosphatidylglycero, *Biochim. Biophys. Acta*, 339, 432, 1974.
48. **Godin, D. V. and Ng, T. W.,** Studies on the membrane perturbational effects of drugs and divalent ions utilising trinitrobenzene sulfonic acid, *Mol. Pharmacol.*, 9, 802, 1974.
49. **Chapman, D., Peel, W. E., Kingston, B., and Lilley, T. H.,** Lipid phase transitions in model biomembranes. The effect of ions, *Biochim. Biophys. Acta*, 464, 260, 1977.
50. **Träuble, H.,** 34th Nobel foundation symposium, Sweden, *Structure of Biological Membranes*, Abrahamsson, S. and Pascher, I., Eds., Plenum Press, New York, 1976.
51. **de Gier, J., Mandersloot, J. G., and van Deenen, L. L. M.,** Lipid composition and permeability of liposomes, *Biochim. Biophys. Acta*, 150, 666, 1968.
52. **Racker, E.,** A new procedure for the reconstruction of biologically active phospholipid vesicles, *Biochem. Biophys. Res. Commun.*, 55, 224, 1973.
53. **Hoffmann, W., Sarzala, M. G., Gomez-Fernandez, J. C., Goni, F. M., Restall, C. J., Chapman, D., Heppeler, G., and Kreutz, P.,** Protein rotational diffusion and lipid structure of reconstituted systems of Ca^{2+} activated ATPase, *J. Mol. Biol.*, 141, 119, 1980.

Chapter 2

A SIMPLE PROCEDURE FOR PREPARING LIPOSOMES CAPABLE OF HIGH ENCAPSULATION EFFICIENCY UNDER MILD CONDITIONS

Christopher J. Kirby and Gregory Gregoriadis

TABLE OF CONTENTS

I. INTRODUCTION

The widespread interest in the use of liposomes as carriers of drugs and other biologically active materials, requires the availability of convenient procedures for the entrapment of such materials into liposomes. From the pharmaceutical standpoint, several criteria must be fulfilled. The procedure must be economical in terms of production time and encapsulation efficiency, particularly in the case of more expensive drugs and materials, and must be capable of being scaled up to meet any demands required of it. It must also be as uncomplicated as possible and avoid any conditions such as sonication and the use of organic solvents and detergents, that are likely to be detrimental to the activity of the material being entrapped, bearing in mind that some candidates for encapsulation might be extremely labile substances.

Numerous procedures for liposome preparation have been developed but only a few of these are capable of entrapping large proportions of water-soluble materials. These methods, which all form large uni- or oligolamellar vesicles include the reverse-phase evaporation technique,[1] ether injection methods,[2,3] and the freeze/thaw method.[4] The first three of these procedures suffer from the fact that they require exposure of the material to be entrapped, to organic solvents and/or sonication and consequent risk of damage to denaturation of sensitive material including many proteins.

The freeze/thaw method requires only brief sonication but has a maximal encapsulation efficiency of 30% and may also have limited usefulness for entrapping high concentrations of proteins.

The original method of Bangham et al.[5] is still the simplest procedure for liposome formation but is of limited usefulness because of its low encapsulation efficiency. As the lipid becomes hydrated and starts to form into closed vesicles, only a small proportion of the total solution volume becomes entrapped within the vesicles. We reasoned that if the whole of the solute is brought into the immediate proximity of the lipid before hydration actually commences, a much higher level of solute loading might be attained. On the basis of this, we have developed a technique which is simple to use and avoids the use of potentially damaging conditions but which allows high concentrations of a variety of materials to be loaded into liposomes of varying lipid compositions. It must be emphasized, however, that these findings are still of a preliminary nature and further investigations are continuing.

II. METHODOLOGY

A. Materials

Egg phosphatidylcholine (PC), phosphatidylserine (PS), and phosphatidic acid (PA) were from Lipid Products, S. Nutfield, Surrey, U.K.; stearylamine (SA) and distearoyl phosphatidylcholine (DSPC) from Sigma Chemical Co., Poole, U.K., and cholesterol (Chol) and Triton® X-100 from BDH Ltd., Poole, U.K. Phospholipon 100 soybean lecithin was a gift from Nattermann, Koln. [^{125}I]-labeled human serum albumin and [^{3}H]-sucrose were purchased from Amersham International. [^{3}H]-DNA *(E. coli)* was a generous gift from Dr. B. Cramp, Hammersmith Hospital, London. Unlabeled human serum albumin was obtained from Blood Products Laboratories, Elstree. Tris/saline buffer was 0.06 *M* Tris, 0.09 *M* NaCl, pH 7.5. Sources of any other materials not given above, and any undescribed assay techniques, have been given elsewhere in the book [6]

B. Preparation of Liposomes

Thirty-three μmoles of total phospholipid, usually together with 33 μmoles of Chol, was rotary evaporated from a chloroform:methanol (2:1, v/v) solution, to form a thin film on the walls of a 50 mℓ round-bottomed flask. After the last obvious traces of solvent had been removed, rotary evaporation was continued for 15 min, followed by drying for a further 15 min under a stream of nitrogen. The lipid was suspended in 2 mℓ of distilled water by shaking with a few glass beads or by briefly immersing in a bath sonicator. In some experiments the dispersed lipid was used as such, leaving it to hydrate for 30 min before proceeding to the next stage of the procedure. In most cases the dispersion was sonicated with a 0.75-in probe to form SUV using the conditions already described for SUV formation.[6] However, it was frequently found that a shortened period of sonication, though insufficient for complete conversion of MLV to SUV, was not detrimental to the efficiency of encapsulation of material during the final stages of the procedure being described. For preparations containing DSPC, initial hydration and sonication were carried out at about 60°C (except for Factor VIII; see later) and liposomes were left to anneal at the same temperature for 40 min before proceeding further.

In one experiment using soybean lecithin (Phospholipon 100), the appropriate amount of lipid, which is provided in paste form, was weighed directly into a glass beaker and homogenized in distilled water for 5 min at full speed using a Virtis 45 homogenizer. The resultant lipid dispersion, which had avoided any use of organic solvents, was then treated in the same way as MLV produced by conventional means (see above) and as SUV.

In the majority of experiments, 2 mℓ of dispersed lipid, whether in the form of SUV or MLV produced from a lipid film or from a mechanically homogenized lipid suspension (Phospholipon 100), was mixed in a 50 mℓ round-bottomed flask with 2 mℓ of a solution of the material to be entrapped, and was then flash-frozen by swirling the flask contents in a freezing mixture of cardice and isopropanol. The flask contents were then lyophilized overnight on a Hetosicc freeze-drier operating at a vacuum of 0.1 torr. Freeze-drying was normally complete within 2 to 3 hr and its cessation after this time was not found to be detrimental to the final level of encapsulation achieved. After lyophilization, the material had a fine lacy appearance which tended to become more solid looking when higher concentrations of solute were used. 0.2 mℓ of distilled water was added to the dried preparation and the flask rotated to wet and suspend the whole of the material. This could be hastened by lightly vortexing the flask. Rehydration of DSPC-containing liposomes was carried out at 60°C, heating the flask before adding distilled water at this temperature. Use of higher temperatures was avoided because of the susceptibility of many solutes, especially proteins, to thermal denaturation. When larger amounts of materials were used during scaled-up experiments, this process of rehydration was much easier (see later). The suspension was left to stand at room temperature (60°C for DSPC liposomes) for 30 min to 1 hr and the liposomes separated from nonentrapped material by diluting to 30 mℓ with Tris/saline buffer and centrifuging for 15 min at 10,000 × *g* or in some cases where dialyzable solutes were entrapped, by diluting to 2 mℓ with distilled water and then dialyzing overnight against 2 ℓ of Tris/saline buffer. In some later experiments, 0.5 *M* NaCl was used for diluting prior to centrifugation, to decrease losses of entrapped material due to possible osmotic stresses on the liposomes. It is suggested that a buffer of appropriate osmolarity be used at this stage after performing preliminary experiments to ascertain the expected level of solute entrapment. After separation of liposomes from nonentrapped material, entrapment efficiencies of [^{125}I]-albumin, [^{3}H]-sucrose and [^{3}H]-DNA were determined by measuring radioactivity in the liposome pellet and supernatant (or dialysate and dialysis buffer). Carboxyfluorescein fluorescence was determined[6,7] using a Perkin Elmer fluorimeter, and Factor VIII: C activity by the two-stage clotting assay.[6] Carboxyfluorescein had to be released from DSPC-containing liposomes prior to measuring, by briefly boiling in the presence of Triton® X-100.[7] Factor VIII was released from the same liposomes by lyophilizing and resuspending in distilled water. Re-entrapment of Factor VIII was discouraged by adding 100 times more water per micromole of phospholipid than was normally used for liposome formation, and by dissolving well below the transition temperature, at 2°C.

For increased-scale experiments, 660 μmoles of egg PC or Phospholipon 100 and 660 μmoles of Chol, were dried to a film on the walls of a 50 mℓ round-bottomed flask and then suspended in 30 mℓ distilled water. After standing for 1 hr the dispersions were cooled to 4°C, submerged in an ice bath, and probe-sonicated for 5 2-min intervals with 30 sec cooling periods in between. Due to the inefficiency of probe sonication when confronted with volumes much larger than 2 mℓ, no significant decrease in opacity of the dispersion was noted, but this was not found to be detrimental to the efficiency of encapsulation. It might, therefore, be possible to achieve the same degree of entrapment with reduced sonication times. After brief centrifugation to remove any particles of titanium from the lipid dispersion, it was transferred to a 500 mℓ round-bottomed flask, mixed with 30 mℓ of 17 m*M* carboxyfluorescein and rapidly frozen by swirling in a cardice/isopropanol bath. The freeze-dried material was rehydrated by mixing and swirling for 2 min with 2 mℓ of distilled water, left for 2 min and mixed with a further 2 mℓ of distilled water. The preparation was then left for 1 hr, diluted to 60 mℓ with Tris/saline buffer and centrifuged at 10,000 g for 30 min. The pellets were washed by suspending in a further 60 mℓ of buffer and recentrifuged using the same conditions. The final liposome pellets were suspended in Tris/saline buffer and the fluorescence intensity in the absence and presence of 1% Triton® × 100 was measured[7] along with that in the supernatant.

III. RESULTS

The liposomal lipid composition most widely used in these studies was equimolar PC:Chol, and levels of entrapment for the various materials so far tested with these liposomes are given in Table 1. The entrapment values represent percent encapsulation of the solute in the total liposomal preparation. Entrapments of each solute are given for both liposomes prepared by the present method (test liposomes) and for MLV prepared by the classical method (control MLV). The former are all derived from 2 mℓ of sonicated lipid mixed with 2 mℓ of solute solution and lyophilized and rehydrated with distilled water as described in the previous section. Control MLV were formed by suspending a dried film of the corresponding amount of lipid in 2 mℓ of the same concentration of solute and shaking vigorously. For all materials tested, entrapment in test liposomes was very much higher than in control MLV. It should be noted that when distilled water is added to lyophilized material during preparation of liposomes, the volume added is only one tenth of the original volume of solute solution used and so the solute concentration at this stage (ignoring the effect of partitioning of the solute by the lipid bilayers) is ten times greater than initially. Control MLV in which the dried lipid film was suspended using 0.2 mℓ of 170 m*M* carboxyfluorescein rather than 2 mℓ of 17 m*M* carboxyfluorescein did produce an increased level of entrapment of 9.3%. Similarly, dispersion of two identical dried lipid films with 2.0 mℓ and 0.2 mℓ of Tris/saline buffer containing identical trace quantities of [^{125}I] albumin, resulted in control MLV with entrapment levels of 3.1% and 11.4%, respectively. Thus, although entrapment by the classical method can be improved by manipulation of the lipid:aqueous phase volume ratio, it is still far less than that obtained with the test liposomes.

Data in addition to that given in Table 1 indicated that the encapsulation efficiency of several of the materials tested was related to the solute concentrations used, this relationship varying from solute to solute. Entrapment of sucrose appears to be independent of its concentration up to a concentration of about 90 m*M* above which there is a gradual fall in entrapment. This might be primarily due to a cryoprotective effect of the sugar, which reduces damage to SUV during the freeze-drying process. Thus, after electron microscopy of SUV mixed with 0.1, 0.3, and 0.5 *M* [^{3}H]-sucrose, freeze-dried and then rehydrated as above, the proportion of SUV increased with increasing sucrose concentration while there was a parallel decrease in incorporation of [^{3}H]-sucrose. With carboxyfluorescein in increasing concentrations there is no evidence of any tendency to retention of SUV structure

Table 1
INCORPORATION OF VARIOUS MATERIALS TO PC:Chol LIPOSOMES

		% Encapsulation	
Entrapped material	**Concn.**	**Test liposomes**	**Control MLV**
Carboxyfluorescein	17 m*M*	30.0 ± 1.5	1.8
	50 m*M*	7.0	
	17 m*M**	43.0	
	50 m*M**	27.0	
Sucrose	50 m*M*	38	1.6
	50 m*M**	44	
	100 m*M*	28	
	100 m*M**	22.4	
	500 m*M*	6	
DNA, *(E. coli)*	0.1 mg/mℓ	52	8.5
		72	
Albumin	0.01%	39	3.1
	0.10%	53.7	
	1.0%	40.6	
Factor VIII	145 units/mℓ	30.4 (0.92)	4.0 (0.07)
	142 units/mℓ	26 (0.79)	

Note: PC:Chol liposomes made by the present method (test liposomes) and MLV by the classical method (control MLV) were prepared from 66 μmole of total lipid as described in the text. Entrapment values represent the percentage of the total material used that becomes incorporated into the liposomes. Preparations marked * were diluted with 0.5 *M* NaCl rather than with Tris/saline buffer after the initial rehydration step. The data for entrapment of 17 m*M* carboxyfluorescein are derived from 10 separate experiments and are expressed as mean entrapment ± standard error. All other data are from individual experiments. Units of Factor VIII are defined in Reference 6. Numbers in parentheses denote percent entrapment per micromole phospholipid used and are given for comparison with values in Reference 6.

after rehydration and the main factor responsible for decreased uptake (at higher concentrations) may be an osmotic effect. Hydration of freeze-dried materials was normally followed, after standing for 1 hr, by diluting with isotonic Tris/saline buffer to a total volume of 30 mℓ prior to separation by centrifugation. In some experiments, however, 0.5 *M* NaCl was used instead of Tris/saline buffer. Table 1 shows that in the case of carboxyfluorescein, use of 0.5 *M* NaCl led to a marked improvement in entrapment efficiency, particularly at the higher solute concentration where any osmotic damage would be expected to be greater. Some variable improvement was also seen with the lower concentrations of sucrose. Since 0.5 *M* NaCl was an arbitrary choice, it is conceivable that further improvements might be made by appropriate manipulation of the osmolarity and possibly the composition of the diluent. When entrapping sucrose or carboxyfluorescein, use of varying concentrations of NaCl or of the appropriate solute solution rather than water in the initial rehydration stage, gave no improvement in encapsulation efficiency and in some cases this was actually reduced.

In some studies the initial sonication stage was omitted and MLV were freeze-dried together with the solute solution. In the case of 17 m*M* CF, this gave a mean entrapment level of 21.8 ± 0.4 in four separate experiments, i.e., not as high as when starting with SUV but still much higher than control MLV. It may even be possible to achieve enhancement of entrapment by simply adding the solute solution to the dried lipid film and freeze-drying without dispersing the lipid. For instance, solutions containing trace amounts of [^{125}I]-albumin or [^{3}H]-DNA freeze-dried together with lipid films and then rehydrated in the usual way, led to 24% and 31% of the respective labels being incorporated. This has not yet been tried with larger solute concentrations or with other small molecular weight markers such as carboxyfluorescein.

In a single experiment the use of organic solvents was completely avoided (not shown). Solid soybean lecithin (Phospholipon 100), was weighed directly into a beaker, mechanically homogenized with distilled water, and then treated in the same way as lipid dispersions prepared by sonication or agitation of solvent-evaporated lipid films. In this way, 30% of a 17 m*M* CF solution was encapsulated compared with entrapments of 38% and 30%, respectively for sonicated and hand-shaken lipid film dispersions (test liposomes) prepared in concurrent experiments.

The volume of distilled water added for the initial rehydration of the lyophilized material has an important effect on the encapsulation efficiency of the liposomes. Whereas smaller volumes can be used at low solute concentrations, 0.2 mℓ was found to be the minimum volume for the amounts of lipid and range of solute concentrations used here. As the volume is increased, entrapment is progressively reduced. Thus, using 17 m*M* carboxyfluorescein, entrapment decreased from 30% with a hydration volume of 0.2 mℓ, to 24% and 21% with hydration volumes of 0.5 mℓ and 1.0 mℓ, respectively. However, addition of Tris/saline buffer 15 min after hydrating with 0.2 mℓ of distilled water, had no effect on entrapment. Addition of hydration volumes of less than 0.2 mℓ to preparations of low solute concentration gave encapsulation efficiencies greater than those shown in Table 1, but for higher solute concentrations, these volumes may not be sufficient to wet the lyophilized material. However this is partly a limitation of the scale at which these experiments are carried out. When a larger-scale experiment (Table 2) was performed in which all amounts of lipids were increased 20-fold (to 660 μmol phospholipid and 660 μmol Chol), it was found possible to rehydrate the lyophilized material with half (2 mℓ) the relative volume needed in the small-scale procedure. This resulted in a total entrapment of 42% compared to a value of 34% for a parallel, small-scale control experiment. This was despite the fact that prior to freeze-drying, most of the lipid was probably in the form of MLV rather than SUV since the efficiency of sonication is greatly reduced in the presence of such large volumes. Entrapment might have been increased further if after rehydration the liposomes had been diluted with 0.5 *M* NaCl rather than with Tris/saline buffer.

Other attempts at producing greater levels of entrapment by optimizing the rehydration process included slow rehydration using a stream of nitrogen saturated with water vapor or a jet of steam, but neither process produced any encapsulation advantage. Increased encapsulation could, however, be obtained by increasing the liposomal lipid concentration. The phospholipid concentration of 33 μmole per 2 mℓ of water, used throughout these studies, was quite arbitrary. Thus in an experiment where phospholipid concentrations of 16.5, 33, and 66 μmoles per 2 mℓ solute were used, final entrapments of 16.6, 29.1, and 40.6% of total carboxyfluorescein were obtained. Again, further increases might have been used if 0.5 *M* NaCl had been used to dilute the liposomes prior to centrifugation.

Table 2 shows entrapment levels of 17 m*M* carboxyfluorescein by a variety of different liposomal lipid compositions. With the exception of the increased-scale experiments, all had the same molar quantities of total phospholipid and in two types, PC and DSPC, no Chol was present. Both showed marked reductions in carboxyfluorescein entrapment compared to when Chol is present. This may have been due to the decreased amounts of total lipid present and a consequent reduction in total liposome volume. However, the increased membrane permeability associated with the absence of Chol, might have given rise to increased osmotic fragility of the liposomes leading to the decreased entrapment levels. All of these liposome types were diluted with Tris/saline buffer rather than with a hypertonic solution prior to centrifugation.

Table 2
ENTRAPMENT OF CARBOXYFLUORESCEIN BY LIPOSOMES OF VARYING LIPID COMPOSITION

Liposome composition	% Encapsulation of 17 m*M* carboxyfluorescein
PC	14
	12.8
PC:Chol	30.0 ± 1.5
PC:Chol (increased scale)	42%
PC:SA:Chol (+ ve)	18.7
PC:PA:Chol	30.8
(− ve)	22
PL:Chol	38
DSPC	7.8
DSPC:Chol	26.2 ± 2.5

Note: Liposomes were prepared as described in the text using 33 μmoles (or 660 μmol; increased scale) of total phospholipid. Cholesterol was used in amounts equimolar with total phospholipid and PA and SA were incorporated as 10% (on a molar basis) of total phospholipid. Data for DSPC:Chol is derived from 4 separate experiments and is given as mean entrapment ± standard error. All other details are as in the legend to Table 1.

The presence of a negative charge as seen in the PC:PA:Chol liposomes, had no effect on CF entrapment whereas positively charged PC:SA:Chol showed a decreased entrapment in the single case studied. Phospholipon:Chol gave an entrapment of 38% which is higher than that of PC:Chol, and even when the sonication stage was omitted and hand-shaken MLV were used (not shown) a high carboxyfluorescein entrapment of 30% was found. Encapsulation of carboxyfluorescein by DSPC:Chol liposomes, which had to be hydrated at an increased temperature, was not markedly different from PC:Chol liposomes. When Factor VIII was entrapped in the former (at 40°C to preserve the factor's activity) an encapsulation efficiency of 37% was achieved.

IV. DISCUSSION

While this work is still in its early stages of development and assessment, these preliminary findings suggest that rehydration of lyophilized or otherwise dried lipid/solute dispersions produces liposomes capable of achieving encapsulation efficiencies which are easily reproducible and comparable to those of any other high entrapment technique so far reported. The liposomes so produced appear to be as stable and impermeable as those produced by more conventional means. Thus, carboxyfluorescein-loaded liposomes remained highly latent during prolonged (24 hr) incubation at 37°C with mouse plasma (not shown). The limited electron microscopic examination so far carried out suggests that mainly multilamellar vesicles are formed, but the possibility of other liposome types cannot be discounted and the nature of the material being entrapped might have some influence on the eventual morphology. An increased number of lamellar would be expected to make the liposomes less permeable than large uni- and oligolamellar vesicles, particularly when the latter are prepared using organic solvents and detergents, lingering traces of which might contaminate the liposomes and affect their stability and permeability (see Chapter 8, Volume I). Where a more homogeneous preparation of smaller liposomes is required, techniques such as polycarbonate membrane filtration[8] should be applicable, and the smaller liposomes formed should still contain acceptably high solute concentrations because of the high loading capability of the initial preparation.

The procedure described here is extremely mild, requiring the presence of neither solvents nor detergents, and though sonication was used routinely throughout these studies, this takes place in the absence of the material being entrapped, so avoiding any possibility of damage to the latter. If it is desired that the use of sonication be completely abolished, this can easily be replaced by procedures such as French pressure cell filtration[9,10] to produce the "empty" SUV. In fact, high levels of entrapment are often attained without starting with SUV or even MLV (i.e., dried lipid film) and it is expected that such levels will be further improved by minor manipulations such as optimization of the lipid:solute ratio and the avoidance of osmotic damage to the newly formed liposomes.

The mildness of the conditions involved make this a particularly attractive technique for the encapsulation of labile materials such as enzymes, DNA, cofactors, and other biologically active substances. The procedure actually evolved while searching for an efficient means of entrapping Factor VIII, a markedly labile glycoprotein, inside liposomes. The level of entrapment of Factor VIII in PC:Chol rehydration liposomes (Table 1) was, in terms of total units incorporated per micromole of phospholipid, 8.5-fold higher than in any liposome types described in Reference 6. In addition, loss of VIII:C activity during liposome preparation was generally much lower than in previous efforts.[6] Another important aspect was the high entrapment of Factor VIII in DSPC:Chol liposomes (see above). No other liposome preparation technique so far investigated allowed incorporation of active Factor VIII into DSPC-containing liposomes. This is mainly because the biological activity of the protein is very sensitive to detergents and is rapidly destroyed by any solvent tested that was found to be capable of dissolving DSPC.

The chief virtue of the rehydration procedure[11] is its great simplicity. Even if it is decided to commence with SUV, these can if necessary be prepared in batches and stored for long periods until required. Also, scaling up of the procedure presents no problems. For these reasons the technique lends itself readily to the pharmaceutical exploitation of liposomes as drug carriers.

REFERENCES

1. **Szoka, F. and Papahadjopoulos, D.,** Procedure for preparation of liposomes with large internal aqueous space and high capture by reverse-phase evaporation, *Proc. Natl. Acad. Sci. USA.*, 75, 4194, 1978.
2. **Deamer, D. W. and Bangham, A. D.,** Large volume liposomes by an ether vaporization method, *Biochim. Biophys. Acta*, 443, 629, 1976.
3. **Schieren, H., Rudolph, S., Finkelstein, M., Coleman, P., and Weissmann, G.,** Comparison of large unilamellar vesicles prepared by a petroleum ether vaporization method, with multilamellar vesicles, *Biochim. Biophys. Acta*, 542, 137, 1978.
4. **Pick, U.,** Liposomes with a large trapping capacity prepared by freezing and thawing of sonicated phospholipid mixtures, *Arch. Biochem. Biophys.*, 212, 1986, 1981.

5. **Bangham, A. D., Standish, M. M., and Watkins, J. C.,** The action of steroids and streptolysin S on the permeability of phospholipid structures to cations, *J. Mol. Biol.*, 13, 138, 1965.
6. **Kirby, C. J. and Gregoriadis, G.,** Incorporation of Factor VIII into liposomes, in *Liposome Technology*, Vol. I, Gregoriadis, G., Ed., CRC Press, Boca Raton, Fla., 1984, in press.
7. **Senior, J. and Gregoriadis, G.,** Methodology in assessing liposomal stability in the presence of blood, clearance from the circulation of infected animals and uptake by tissues, in *Liposome Technology*, Vol. III, Gregoriadis, G., Ed., CRC Press, Boca Raton, Fla., 1984, in press.
8. **Olson, F., Hunt, C. A., Szoka, F. C., Vail, W. J., and Papahadjopoulos, D.,** Preparation of liposomes of defined size distribution by extrusion through polycarbonate membranes, *Biochim. Biophys. Acta*, 557, 9, 1979.
9. **Hamilton, R. L., Goerke, J., Guo, L. S., Williams, M. C., and Havel, R. J.,** Unilamellar liposomes made with the French pressure cell: a preparative and semi-quantitative technique, *J. Lip. Res.*, 21, 981, 1980.
10. **Barenholzt, Y., Amselem, S., and Lichtenberg, D.,** A new method for preparation of phospholipid vesicles (liposomes) — French press, *FEBS Lett.*, 99, 210, 1979.
11. **Kirby, C. J. and Gregoriadis, G.,** U.K. patent application 8321012, 1983.

Chapter 3

PREPARATION OF SOLVENT VAPORIZATION LIPOSOMES

David W. Deamer

TABLE OF CONTENTS

I. INTRODUCTION

The solvent vaporization method for producing liposomes was introduced by Deamer and Bangham[3] and grew out of a perceived need for unilamellar liposome preparations that were capable of encapsulating significantly greater volumes than the sonicated preparations available at that time. The original procedure used diethyl ether as the solvent and was termed the ether injection method, in that a lipid solution in ether was injected into an aqueous phase. However, other solvents can be used as well, and this term is no longer precise. The term ''solvent vaporization liposomes'' is therefore proposed. This also serves to distinguish the preparation method from the alcohol injection method of Batzri and Korn,[1] in which lipid solutions in ethyl alcohol are injected, but the solvent is not vaporized.

The rationale is that lipids can be dissolved in solvents that are relatively water insoluble and then injected into aqueous phases at temperature and pressure ranges that cause the solvent to vaporize and form bubbles. The lipid presumably spreads to fill the vapor/water interface with a thin multilayer of lipid or a monolayer in equilibrium with a multilayer. As vapor bubbles rise through the aqueous phase, uni- and oligolamellar liposomes form at the interface and remain behind in the solution.

It is instructive to calculate some of the numbers associated with such a process. Suppose that 1 μℓ of solvent containing 5 m*M* lipid is injected into an aqueous phase at 60°C and at atmospheric pressure. The solvent vaporizes to form a gas bubble of 0.24 mℓ volume, about 0.8 cm in diameter, with a surface area of 1.9 cm^2. Since the lipid available in the bubble would form a monolayer of 21 cm^2 (assuming 70 A^2 per molecule) it follows that each such bubble would have a film equivalent to about ten lipid layers (or five bilayers) available at the vapor/water interface. In essence, then, the method is similar to preparing liposomes by agitating an aqueous solution above a lipid film spread on a glass surface. However, the film is only a few lamellae thick, is spread at a gas-liquid interface, and forms primarily unilamellar, rather than multilamellar vesicles.

A significant improvement in the method was reported by Schieren et al.,[15] who used petroleum ether, rather than diethyl ether. The latter solvent has two disadvantages. First, it can accumulate peroxides that may damage lipids sensitive to peroxidation. Second, diethyl ether is soluble in water to the extent of about 7.5% by volume, and its final concentration may approach molar ranges. Alkanes such as those composing petroleum ether do not have significant water solubility and would not be expected to promote peroxidation. Pentane is a convenient alternative to petroleum ether and is used in the procedure described here.

Solvent vaporization liposomes have been used by Nichols and Deamer[9] and Rossignol et al.[14] to produce vesicles with buffered pH gradients in order to study proton-hydroxyl flux across lipid bilayer membranes. The lipids in these studies included PC-PA (9:1 mole ratio) and a mixture of phospholipids naturally occurring in plant material. Ostro et al.[10] used the method to encapsulate rabbit globin messenger RNA. The vesicles were interacted with cells in tissue culture, and it was demonstrated that the RNA was delivered to the cells and translated. In a more recent study, Ostro et al.[11] investigated several parameters of the liposome-cell interaction by encapsulating radioactive RNA from *E. coli* and monitoring delivery of the RNA to human cells in culture.

II. METHODS AND MATERIALS

The principle of the method can be demonstrated quite easily by injecting diethyl ether, petroleum ether, or pentane solutions of phospholipid into an aqueous phase warmed to 60°C. This is also sometimes useful to determine whether a new lipid mixture will be suitable for the preparation method. Approximately 2 mℓ of the lipid solution (1 to 10 m*M* lipid is useful range) is injected from a 5 mℓ glass syringe into 10 mℓ of the aqueous phase. The

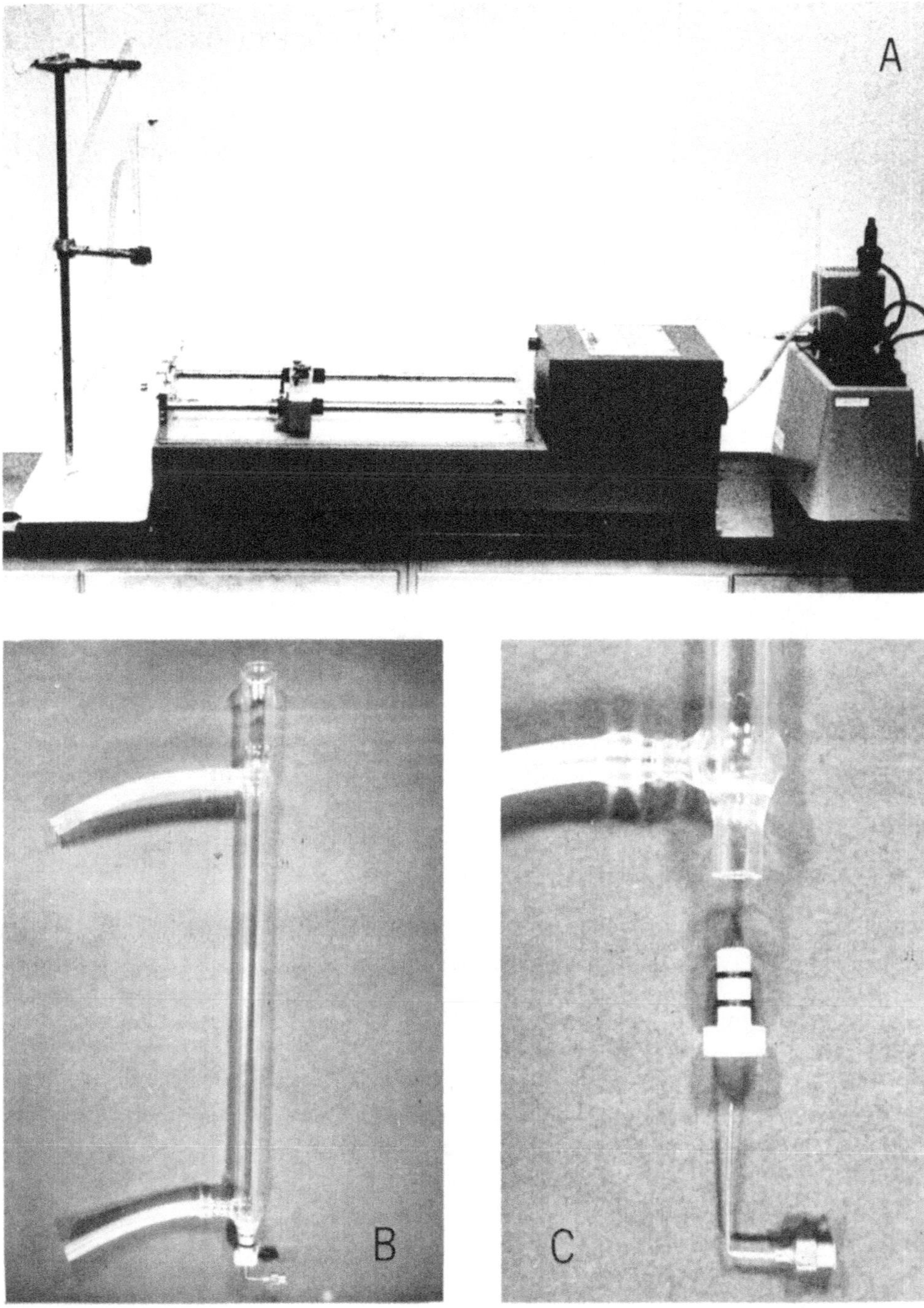

FIGURE 1. Apparatus for the preparation of liposomes (A) assembly, (B) column, (C) Teflon® plug sealing the end of the column and bent needle.

injection rate should be about 1 mℓ/min, using a 20-gauge hypodermic needle, and the procedure should be carried out in a hood to remove vapor fumes. The resulting turbid suspension can be filtered through a Millipore or Nucleopore polycarbonate filter to disperse the lipid and remove aggregates. If diethyl ether is used, the resulting liposomes should be passed through a Sephadex® G-50 column to remove the remaining ether.

Table 1
PROPERTIES OF LARGE UNILAMELLAR VESICLE PREPARATIONS

Lipid composition[a]	Vesicle diameter (μm)	Captured volume (ℓ/mol)	Encapsulation (% original volume)	Ref.
PC:PA 90:10 (1 m*M*, diethyl ether solvent vaporization)	0.05—0.25	17 ± 4	1.7	3
PC:DCP:Chol 7:2:1 (15 m*M*, pet ether solvent vaporization)	0.1—0.4	23—31	38—46	15
PG:PC:Chol 1:4:5 (66 m*M*, REV method)	0.2—1.0	11.7	35—65	16
PS (20 m*M*, Ca + EDTA method)	0.2—1.0	7	15	12
PC (50 m*M*, DOC + dialysis)	0.1	2.4	12	5
PC (16 m*M*, octyl glucoside + dialysis)	0.2—0.3	6—7.4	22	8
Soybean phospholipid (20 mg/mℓ, freeze-thaw-sonicate method) method)	—	8	—	13
Single chain amphiphile (50 m*M*, dodecyl sulfate-dodecanol)	1—10	—	42	7

[a] PC, phosphatidylcholine; PA, phosphatidic acid; DCP, dicetylphosphate; PG, phosphatidyl glycerol; Chol, cholesterol; PS, phosphatidylserine; REV, reverse phase evaporation; DOC, deoxycholate.

To prepare liposomes under more controlled conditions, we have developed the apparatus shown in Figure 1. The following equipment is required:

- Syringe perfusion pump for 1 to 10 mℓ volume range
- 10-mℓ glass syringes (2)
- 20-gauge hypodermic needle, approximately 5 cm in length
- Stainless steel tubing (~ 1 mm O.D.) attached to nitrogen source
- Leibig condenser column, 1 cm × 40 cm with water jacket
- Teflon® plug milled to fit bottom of column
- Circulating water bath
- Hood or water vacuum
- Nucleopore polycarbonate filter (1" diameter, 0.2 to 1.0 μm pore size) and mount for syringe

The equipment is assembled so that the perfusion pump drives the injection of lipid solution from the syringe into the condenser column which contains the aqueous phase. The needle of the syringe is crimped at one end and carefully ground to produce a fine orifice approximately 0.1 mm × 0.5 mm. This serves to prevent larger droplets of solvent from entering the aqueous phase before being driven out by pressure from the perfusion pump and is particularly important if the column is placed under vacuum, as will be described later. The needle is then bent at right angles and passed through a Teflon® plug which seals the end

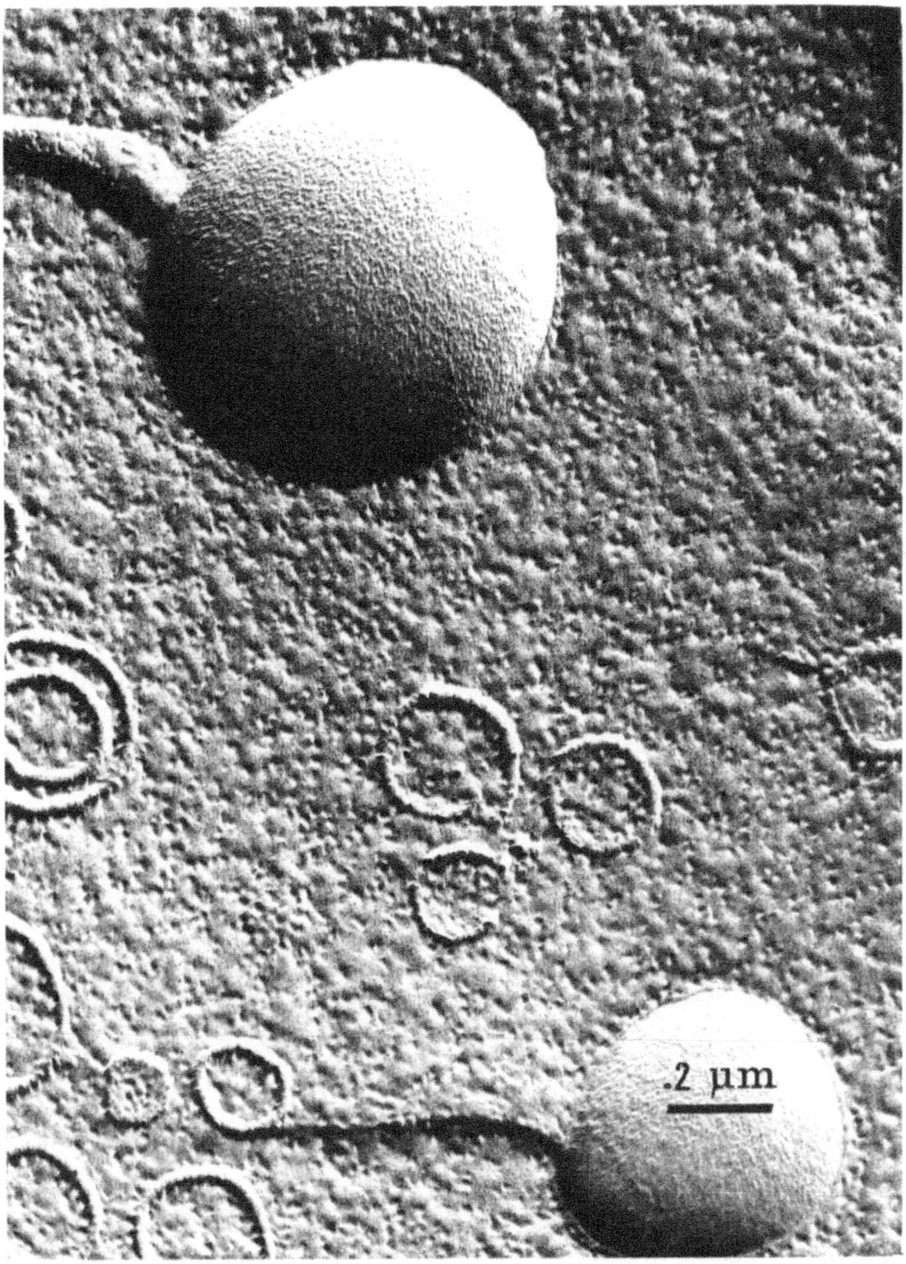

FIGURE 2. Freeze-fracture micrograph of liposomes produced by injection of phosphatidylcholine-phosphatidic acid (9:1 mole ratio) into 10 m*M* sodium chloride.

of the column (Figure 1, B and C). A converted Leibig condenser forms a temperature-controlled chamber, the bottom being cut and ground to receive the plug (Figure 1C). A double O-ring seals the plug against possible leaks. The condenser jacket is attached to a circulating water bath so that the aqueous phase may be maintained at a given temperature. The chamber will contain up to 10 mℓ of solution in the temperature-controlled water jacketed section. It is important that there be an unheated portion at the top of the column so that overflow does not occur during solvent vaporization.

In a typical preparation with the apparatus, 5 mℓ of the lipid solution in pentane (1 to 10 m*M* lipid) is injected into 10 mℓ of the aqueous phase warmed to 60°C. The injection rate is 0.2 mℓ/min, and the aqueous phase is flushed during injection with a gentle stream of nitrogen to promote mixing and provide an inert atmosphere. The nitrogen may also serve to increase the size of the vapor bubbles, thereby increasing the yield of unilamellar vesicles.

A funnel attached to a water vacuum aspirator is placed above the column to draw away flammable vapor. (Alternatively, the procedure can be carried out in a hood.) After the injection phase is completed, the turbid suspension is removed from the chamber by drawing it through a suitably long, thin plastic tube into a second syringe. At this point it is convenient to adjust the volume for any loss occurring during injection. A polycarbonate filter with the desired pore size is then mounted to the syringe, and the liposome preparation is forced through under moderate pressure. This step greatly increases the homogeneity of the preparation, since larger vesicles are sized downward to the approximate diameter of the filter pores. The liposomes may be passed through a gel permeation step if it is desired to exchange the external phase. We use the centrifugal method described by Fry et al.[6] which does not dilute the liposomes.

Occasionally, it is desired to produce liposomes from heat-sensitive materials or to encapsulate molecules such as proteins which may undergo thermal damage. We have found that by attaching the water vacuum aspirator so that it draws directly from the top of the column, it is possible to vaporize the solvent at much lower temperatures.

III. PROPERTIES OF SOLVENT VAPORIZATION LIPOSOMES

Table 1 shows some characteristic parameters of a number of preparations including the solvent vaporization liposomes. Most of the large unilamellar vesicle (LUV) preparations have entrapment values in the range of 2.4 to 31 ℓ/mol and encapsulation efficiencies ranging from 2 to 65%. The latter parameter depends primarily on lipid concentration and can be determined from the relationships described by Deamer and Uster.[4]

Figure 2 shows a freeze-fracture micrograph of liposomes produced by injection of PC-PA (9:1 mole ratio) into 10 m*M* sodium chloride. In this instance, the preparation was etched sufficiently to show cross-fractured vesicles, as well as vesicles fractured along the plane of the membrane. Note that the cross-fractured vesicles are primarily unilamellar, although an occasional bilamellar vesicle can be seen. In this preparation, the unilamellar vesicles composed 82% of the vesicles, bilamellar 15%, and trilamellar 3%.

IV. ADVANTAGES AND LIMITATIONS OF THE SOLVENT VAPORIZATION METHOD

The main advantage of the solvent vaporization method is that it is relatively simple and robust and is applicable to a wide variety of lipid mixtures and aqueous solutions. It should also be convenient to scale up for production of large quantities of liposomes. The preparation can be completed in less than an hour, while other methods, such as detergent dialysis, require significantly longer preparation times. The primary limitations are that the organic solvents used may be deleterious to certain classes of solute and that the method cannot be used to insert integral proteins into liposome membranes. Certain lipids, particularly phospholipids containing saturated fatty acids, are not sufficiently soluble in ether or pentane, but methanol (20% by volume) can be added to increase their solubility.[2] The methanol must be removed afterward by gel permeation chromatography. Finally, in our hands, all LUV preparations composed of lipid carrying no net charge ultimately aggregate. Therefore, small amounts of a charged lipid such as PA or PS must be added for long-term stability of the preparation.

REFERENCES

1. **Batzri, S. and Korn, E. D.,** Single bilayer liposomes prepared without sonication, *Biochim. Biophys. Acta,* 298, 1015, 1973.
2. **Deamer, D. W.,** Preparation and properties of ether-injection liposomes, *Ann. N.Y. Acad. Sci.,* 308, 250, 1978.
3. **Deamer, D. W. and Bangham, A. D.,** Large volume liposomes by an ether injection method, *Biochim. Biophys. Acta,* 443, 629, 1976.
4. **Deamer, D. W. and Uster, P.,** Liposome preparation methods, *The Liposomes,* Ostro, M. J., Ed., Marcel Dekker, New York, 1983, 27.
5. **Enoch, H. G. and Strittmatter, P.,** Formation and properties of 1000-Å-diameter single bilayer phospholipid vesicles, *Proc. Natl. Acad. Sci. U.S.A.,* 76, 145, 1979.
6. **Fry, D. W., White, C., and Goldman, D. J.,** Rapid separation of low molecular weight solutes from liposomes without dilution, *Anal. Biochem.,* 90, 809, 1978.
7. **Hargreaves, W. R. and Deamer, D. W.,** Liposomes from ionic, single chain amphiphiles, *Biochemistry,* 17, 3759, 1978.
8. **Mimms, L. T., Zampighi, G., Nozaki, Y., Tanford, C., and Reynolds, J. A.,** Phospholipid vesicle formation and transmembrane protein incorporation using octylglucoside, *Biochemistry,* 20, 833, 1981.
9. **Nichols, J. W. and Deamer, D. W.,** Net proton-hydroxyl permeability of large, unilamellar liposomes, *Proc. Natl. Acad. Sci. U.S.A.,* 77, 2038, 1980.
10. **Ostro, M. J., Giacomoni, D., Lavelle, D., Paxton, W., and Dray, S.,** Translation of rabbit albin mRNA introduced by liposomes into mouse lymphocytes, *Nature (London),* 274, 921, 1978.
11. **Ostro, M. J., Lavelle, D., Paxton, W., Matthews, B., and Giacomoni, D.,** Parameters affecting the liposome-mediated insertion of RNA into eucaryotic cells *in vitro, Arch. Biochem. Biophys.,* 201, 392, 1980.
12. **Papahadjopoulos, D. and Vail, W. J.,** Incorporation of macromolecules within large unilamellar vesicles, *Ann. N.Y. Acad. Sci.,* 308, 259, 1978.
13. **Pick, U.,** Liposomes with a large volume trapping capacity prepared by freezing and thawing sonicated phospholipid mixtures, *Arch. Biochem. Biophys.,* 212, 186, 1981.
14. **Rossignol, M., Thomas, P., and Grignon, C.,** Proton permeability of liposomes from natural phospholipid mixtures, *Biochim. Biophys. Acta,* 684, 195, 1982.
15. **Schieren, H., Rudolph, S., Finkelstein, M., Coleman, P., and Weissmann, G.,** Comparison of large unilamellar vesicles prepared by a petroleum ether vaporization method with multilamellar vesicles, *Biochim. Biophys. Acta,* 542, 137, 1978.
16. **Szoka, F. and Papahadjopoulos, D.,** Procedure for preparation of liposomes with large internal aqueous space and high capture by reverse phase evaporation, *Proc. Natl. Acad. Sci. U.S.A.,* 75, 4194, 1978.

Chapter 4

FRENCH PRESSURE CELL LIPOSOMES: PREPARATION, PROPERTIES, AND POTENTIAL

Robert L. Hamilton and Luke S. S. Guo

TABLE OF CONTENTS

I. INTRODUCTION

The method described herein evolved from a simple observation: intact Golgi membranes, isolated from rat liver, are transformed into homogeneous vesicles that look like microsomes by passing the subcellular fraction once through a French pressure cell.[1] The pressure cell, first designed (Figure 1) and described by French and associates in 1950, was used to make colloidal dispersions of chloroplasts that retained substantial photochemical activity.[2] A facsimile is now manufactured by SLM-Aminco (Urbana, Ill.) in sizes with capacities of either 3 or 40 mℓ and in designs that tolerate either 20,000 or 40,000 psi. An automatic hydraulic press is also sold by SLM-Aminco designed specifically for these pressure cells, a machine that is probably worth the investment if one plans to use the pressure cell.

The reasons for developing yet another technique for preparing liposomes stem largely from the need to avoid ultrasonic irradiation of lipids and proteins and to completely eliminate detergents and organic solvents from the product.

II. PREPARATION

The method has proved to be both simple and versatile. Aqueous suspensions containing up to 25 mg/mℓ of egg phosphatidylcholine (PC) (prepared in the usual manner of resuspending the lipids with glass beads in a round bottom flask) are poured or pipetted into the cylinder of a French pressure cell. Wetting the cylinder wall and O-rings of the piston with water or the aqueous buffer just before the lipid suspension is introduced makes it easier to insert the piston part way into the cylinder. The sample level is brought just to the relief valve pressure hole by manually adjusting the piston, and the closure lid is then placed into the open end of the cylinder forcing most air from the cylinder. It is probably best to extrude a small volume of the sample manually to ensure that all air has been removed. This small amount of sample can be processed in the next cycle and therefore is not lost. The relief valve is closed snugly and the pressure cell is placed into position in the automatic press which is operated by following the manufacturer's directions, including presetting the dial to the desired psi. The optimum pressure for producing mostly small unilamellar liposomes is 20,000 psi.[1,3] (We have no experience with the pressure cell that can generate up to 40,000 psi.) The press quickly generates 20,000 psi in the cell and the operator manually opens the relief valve causing rapid release of the sample into a test tube. The process is more efficient if the flow rate is kept high (i.e., ~20 to 40 mℓ/min.) while still maintaining the selected 20,000 psi. A slow flow rate transforms somewhat less of the lipid into unilamellar liposomes (Figure 2). A second reason for releasing the sample rapidly is to keep the time under high pressure reduced to a minimum. Prolonged exposure to high pressure can cause progressive loss of biological activity if proteins are present.[2] The milky parent suspension of PC is transformed within seconds into a transparent opalescent dispersion (Figure 3) of mostly (~70%) unilamellar liposomes of about 200 Å diameter (Figures 4, 5). The sample may be recycled one or more times thereby transforming the small amount of lipid left in the cylinder and that which was bled off with the air. Two to three cycles transforms 90 to 95% of the lipid into unilamellar vesicles that enter the included volume of 2 or 4% agarose (BioRad®, Richmond, Calif.) gel columns. Ten consecutive cycles of egg PC at 20,000 psi did not cause any detectable degradation of lipid as measured by thin layer chromatography (TLC) and gas liquid chromatography (GLC).[1]

III. PROPERTIES

By agarose (2 or 4%) gel chromatography and electron microscopy (Figure 5), we found that only unilamellar vesicles were in the included volume after preparation of liposomes

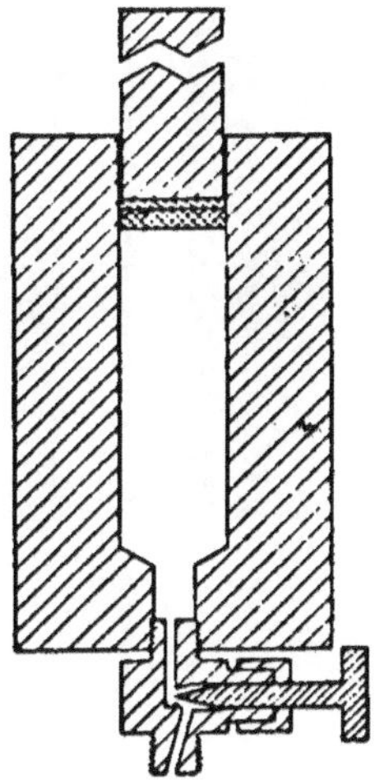

FIGURE 1. "Dispersion Unit" is the term used to describe the first pressure cell. (From Milner, H. W., et al., *Science*, 3, 633, 1950. With permission.)

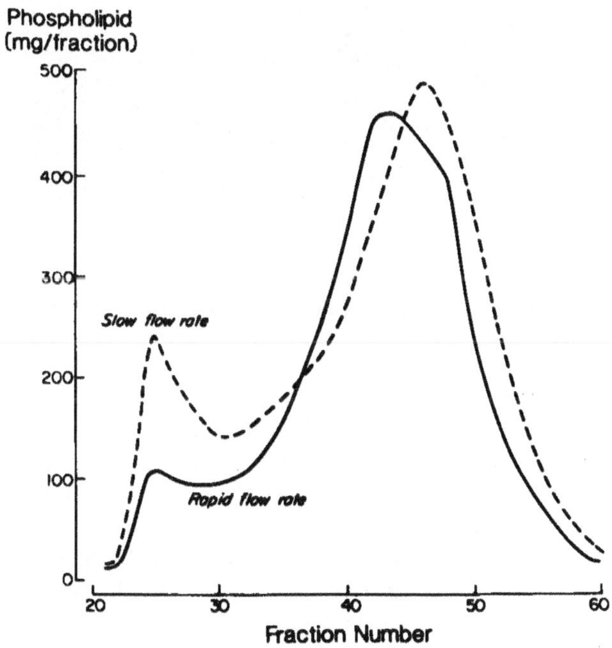

FIGURE 2. Elution profiles from a 2% agarose column (1.2 cm × 90 cm) of egg PC (10 mg/mℓ) liposomes formed by passing lipid suspensions two times through a French pressure cell at 20,000 psi at room temperature. The lipid is routinely suspended in sodium chloride solutions (0.15 to 0.20 *M*) containing 0.04% EDTA and 0.02% NaN_3, referred to as column buffer. Rapid flow rate is about 20 to 40 mℓ/min, slow flow rate is about 1 mℓ/min.

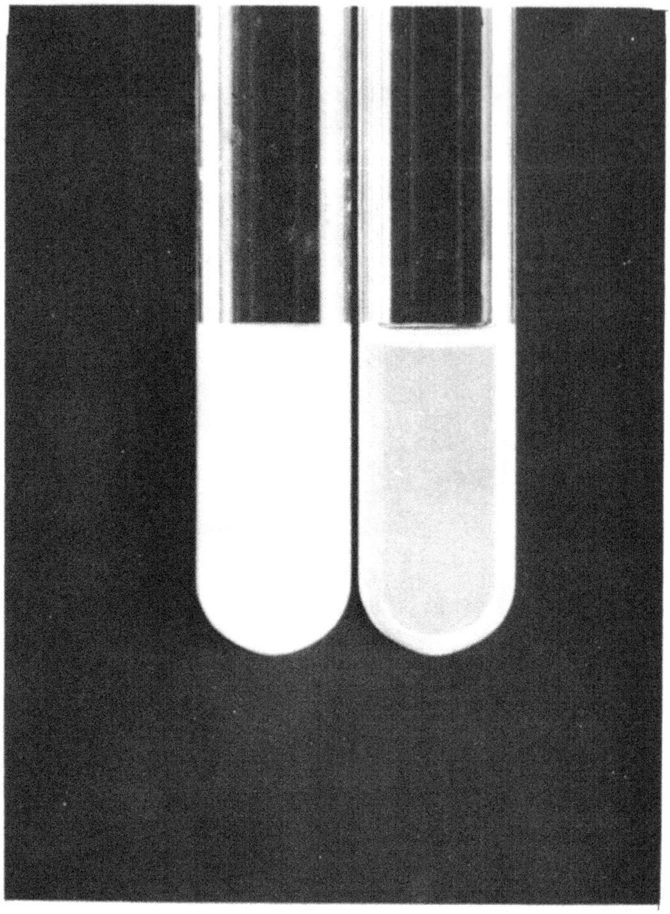

FIGURE 3. Egg PC (10 mg/mℓ) suspension in column buffer before (left) and after (right) a single pass through a French pressure at 20,000 psi at room temperature.

at 20,000 psi; small multilamellar structures were in the void volume.[1] There was little difference in the recovery of PC in multilamellar structures (void volume) as compared to unilamellar liposomes (included volume) under the following conditions: (1) when 5 or 15 mg/mℓ PC was cycled two or six times;[1] (2) when the temperature of the cell and sample was 4, 25, or 37°C; (3) when the unilamellar liposomes were stored in the refrigerator for several weeks; (4) or when they were incubated at 37°C for 4 hr (Figure 6). A small fraction (2 to 8%) of PC was usually in the void volume even after repeated cycles and most of this was pelleted by centrifugation at 100,000 × *g* for 1 hr (Figure 6). Up to 45 mol % cholesterol can be incorporated making the product liposomes larger (~300 to 400 Å diameter).[1] More lipid is in multilayered liposomes when equimolar cholesterol is present in the parent suspension.[1] We have made unilamellar liposomes by this technique from pure dimyristoyl PC (DMPC), pure dipalmitoyl PC (DPPC), soybean PC, and from mixtures of egg PC containing 10 mol % lysolecithin. We have also found that this procedure transforms complex polar lipid suspensions obtained from lipid extracts of organs or from biological fluids into fairly homogeneous, single-walled vesicles. For each type of pure lipid or lipid mixture, the size, homogeneity, and proportion transformed into small vesicles will have to be determined independently. For example, we found that pure DPPC was completely transformed[1] (i.e.,

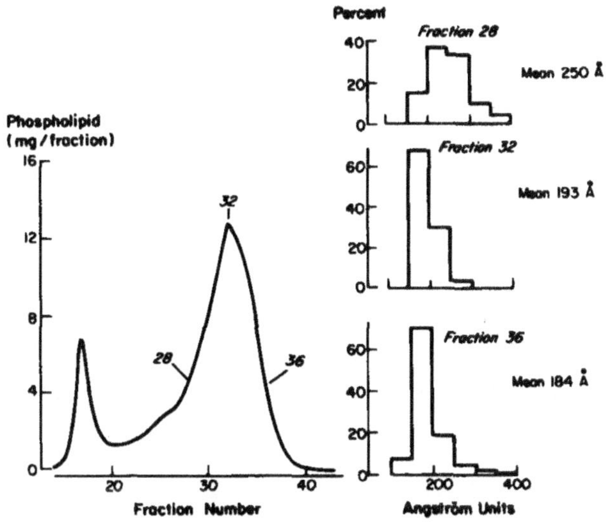

FIGURE 4. Elution profiles (from a 4% agarose column, 2.5 cm × 95 cm) after a single pass of egg PC, 13 mg/mℓ, through a French pressure cell at ~19,000 psi; 70% of the phospholipid recovered from the column is eluted between fractions 28 and 37. Measurements of particle size of three fractions were made from photographs of negatively stained preparations and are shown in the histograms.

no detectable void volume lipid) into very homogeneous unilamellar vesicles (400 to 600 Å) by pressing the lipid suspension at 43 to 45°C three times at 20,000 psi.[4] However, commercially prepared sphingomyelin (SPH) formed a heterogeneous population of unilamellar vesicles with considerable lipid remaining as multilamellar structures in the void volume.[1]

Lower pressure produces some interesting liposome structures. A single pass at about 2000 to 3000 psi formed a widely heterogeneous group of vesicles. Some were multilamellar with the size and appearance of lamellar bodies of lung Type II cells; some were small whereas others were large unilamellar vesicles; many were irregular and elongated multilamellar structures.[1] Two cycles at exactly 3000 psi (with or without cholesterol) transformed about half of the egg PC into fairly homogeneous, small (~500 to 1500 Å) multilayered vesicles eluting in the void volume (Figures 7, 8). In this experiment, the included volume vesicles appeared in negative stains to have double bilayers with an average diameter of about 600 Å (Figure 8).[1]

The method described (i.e., one to four cycles at 20,000 psi, with egg PC at 1 to 20 mg/mℓ at room temp) is especially efficient for trapping small molecules such as carboxyfluorescein (CF) because high initial concentrations of both lipid and dye can be used.[5] Such vesicles containing entrapped CF are also unilamellar although they may be slightly larger (Figure 9) (~250 Å/mean diameter) than those without dye.[1] The reason for this small increase in vesicle size was not pursued but somewhat larger vesicles commonly are formed when the concentration of the suspending medium is made high with either a solute (e.g., CF) or a colloid (e.g., serum albumin). This observation may be worthy of further study. Although only small quantities of macromolecules (i.e., albumin) are captured in the trapped volume of these small unilamellar vesicles, the entrapment of a nondiffusable marker such as radioactive inulin, can be used to estimate the size of vesicles prepared with the French pressure cell.[1]

Unilamellar liposomes made with a French pressure cell leak entrapped CF quite slowly (hours) at 4°C and somewhat more rapidly at 37°C.[5] This slow release of solute (probably

A
B
C

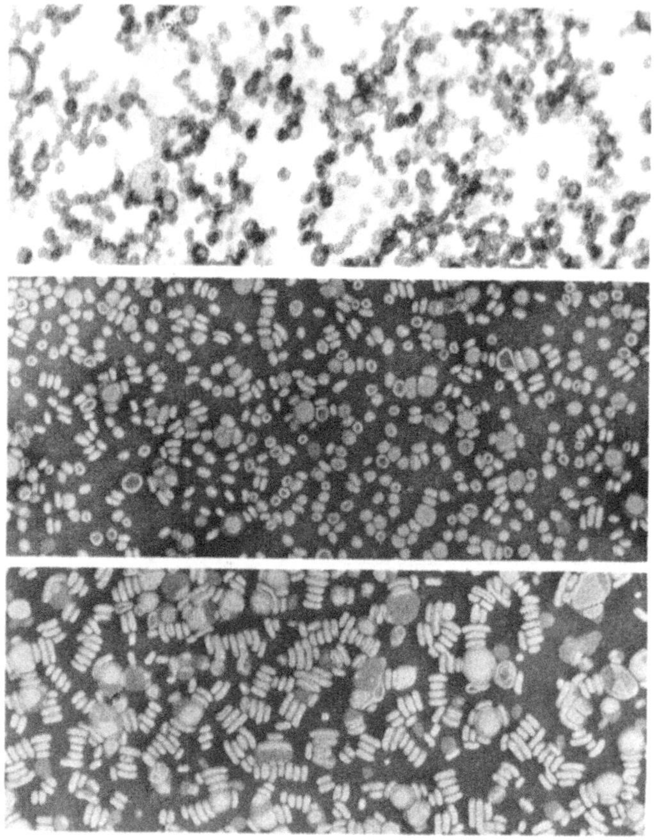

FIGURE 5. Examples of electron microscopic images of unilamellar liposomes used to measure particle size (Magnification × 90,000). (A) Thin section of homogeneous unilamellar liposomes prepared by passing egg PC four times through the French pressure cell at 20,000 psi. This image illustrates the trilaminar staining of the head group regions of phospholipids, the spherical shape of the liposomes, and the trapped volume compartment. The few larger liposomes can be removed by ultracentrifugation or gel chromatography. (B) Negative stain of same sample shown in top image. (C) Negative stain of egg PC liposomes containing 45 mol % cholesterol. Although the mean particle size is much larger, (~375 Å), some particles as small as those shown above for egg PC alone are also present.

by diffusion) is further retarded if cholesterol is present in the bilayer (Figure 10). When egg PC vesicles of the same size are recovered from agarose gel column fractions, no difference in the rate of dye release is obtained between vesicles made by this technique or by ultrasonic methods (Figure 10). This finding conflicts with a report by others that vesicles made by ultrasonic methods are far more leaky than the ones made by the French pressure cell method.[3] These authors also reported that unilamellar liposomes of egg PC made with the French pressure cell method were twice the size (~420 Å diameter) that we found[1] and that cholesterol had no effect on size.[3] The reasons for these different results are not established. However, we sonicated very briefly in short bursts[1] whereas the commonly used procedure[3] described by Huang and Thompson is that of continuous ultrasonic irradiation of lipid suspension for 2 hr.[6] There appear to be subtle differences in egg PC liposome structure produced by the standard sonication technique because they reportedly are stable for only about a week,[6] whereas the unilamellar vesicles produced by the French pressure cell appear to be stable for many months.[1]

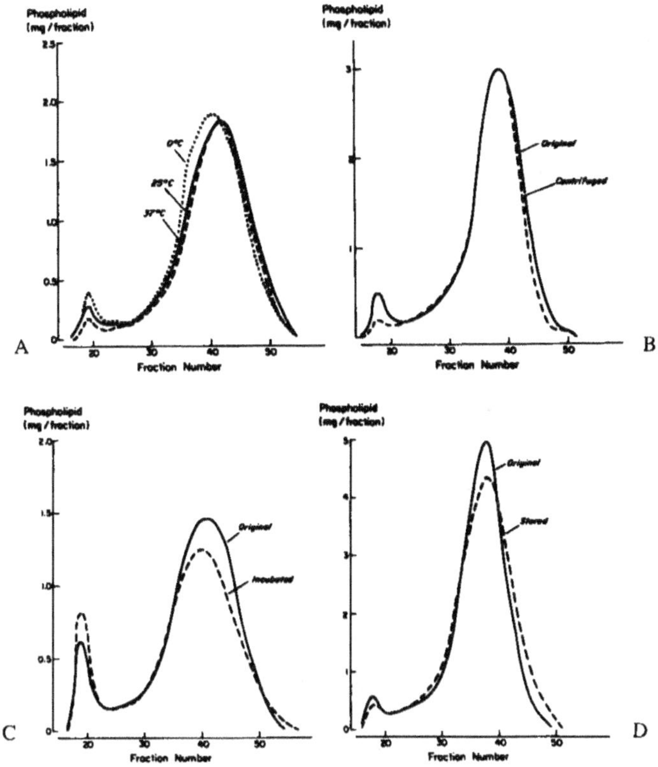

FIGURE 6. Elution profiles (from a 2% agarose column, 1.2 cm × 95 cm). (A) Egg PC liposomes formed at 0, 25, and 37°C by passing each sample twice through a French pressure cell at the temperature shown. Each initial sample contained 25 mg PC in 3 mℓ of column buffer. (B) Egg PC liposomes (solid line) and the supernatant fraction after ultracentrifugation (dashed line) of liposomes at 100,000 × *g* for 1 hr. The liposomes were formed at 20,000 psi from a 10 mg/mℓ suspension in column buffer. (C) Freshly prepared egg PC liposomes (solid line) and after liposomes were incubated at 37°C for 4 hr (dashed line). The liposomes were formed at 20,000 psi from a 20 mg/mℓ suspension of egg PC in 0.2 *M* aqueous 6-carboxyfluorescein at pH 7.7 (see Guo et al.[5]) (D) Freshly prepared egg PC liposomes (solid line) and following storage (dashed line) at 0 to 4°C for 14 days. The liposomes were formed at 20,000 psi from a 10 mg/mℓ suspension of egg PC in column buffer.

The unilamellar liposomes made by the described method are rapidly disrupted (seconds) by whole serum, isolated serum lipoproteins, and pure apolipoproteins.[5] The incorporation of cholesterol progressively inhibits this disruption and almost completely blocks the disruptive effect at 37 mol % or more (Figure 11). This disruptive effect of isolated apoproteins (apo A-I and apo E) on egg PC liposomes results from the transformation of bilayer vesicles into bilayer discoidal particles (Figure 12) which have an annulus of apoprotein at the edge.[7] These recombinants formed by apo E and apo A-I are similar to nascent discoidal HDL particles from perfused rat liver and to the discoidal HDL in the plasma of humans with the genetic disease of lecithin:cholesterol acyltransferase deficiency.[8] The recombinant that forms when either apo A-I or apo E is added to liposomes that contain greater than 37 mol % cholesterol remains as an intact bilayer vesicle (Figure 12)[5] similar to the abnormal plasma lipoprotein of cholestasis, commonly called LP-X.[9]

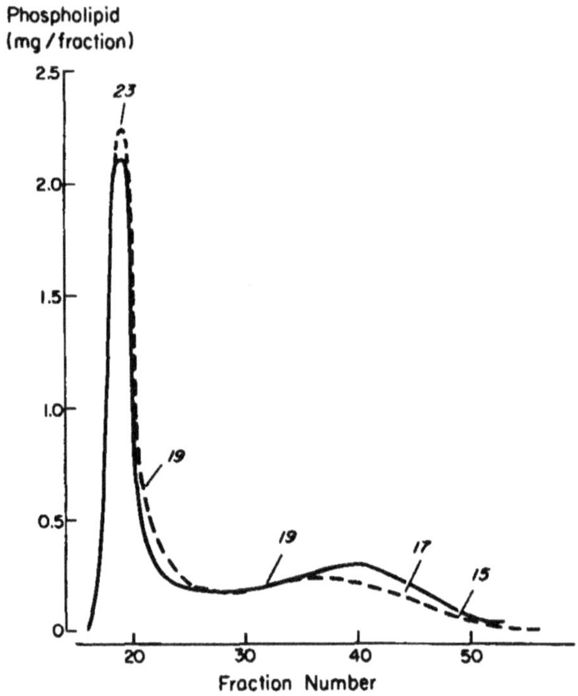

FIGURE 7. Elution profiles of liposomes (from a 2% agarose column, 1.2 cm × 95 cm) of egg PC alone (solid line), and egg PC plus cholesterol (50 mol %) (dashed line) after passing each sample twice through a French pressure cell at the low pressure of 3000 psi. Each sample contained 9 mg/mℓ of egg PC. The numbers in the figure are the measured mol % of cholesterol of the fractions indicated.

The French pressure cell method is also efficient in incorporating vitamin E into liposomes.[10] Up to 33 mol % of α-tocopherol (α-T) can easily be incorporated into unilamellar vesicles of egg PC. Tocopherol changes liposomes in a manner similar to cholesterol, making them larger, less permeable, and even more resistant to disruption by serum, even at lower molar ratios. Thus, vitamin E containing liposomes may become useful vehicles for tissue culture studies or even in vivo for targeting in body fluids.[10]

In summary, unilamellar vesicles prepared by the described French pressure cell technique appear to be similar in many properties to those made by commonly used sonication methods. Thus, they are also similar to the unilamellar vesicles prepared by the method of Batzri and Korn.[11] However, there are important differences. French pressure cell liposomes are far more stable than those prepared by sonication,[12] thereby permitting storage under refrigeration. In addition, the small amounts of organic solvent or detergent that resists removal from vesicles, even after extensive dialysis,[13] is not present in liposomes made with the French pressure cell.

IV. ADVANTAGES

There are a number of advantages to the technique of preparing liposomes with the French pressure cell.

Time saving — It routinely takes only a few minutes to transform multilamellar lipid suspensions into mostly (90 to 95%) unilamellar vesicles of about 200 Å diameter.

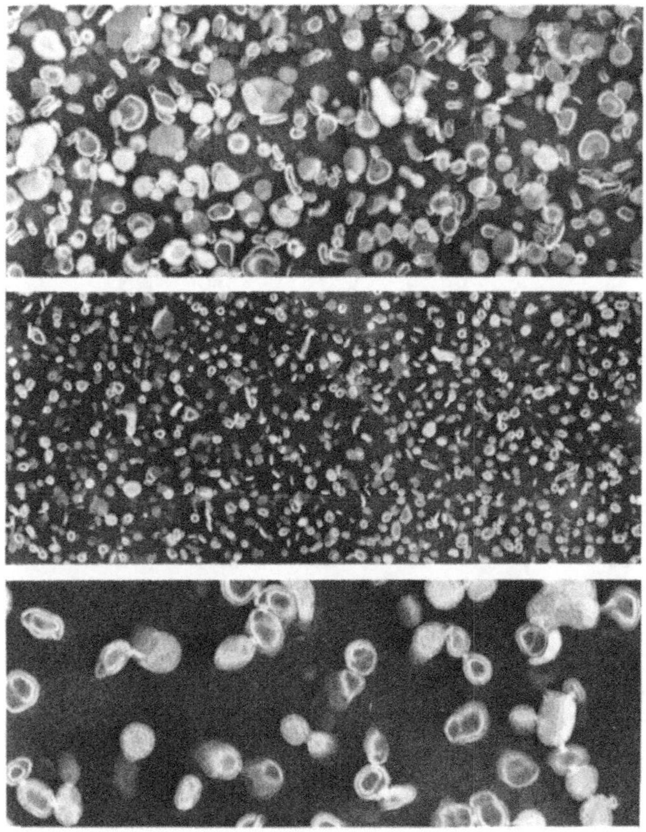

A

B

C

FIGURE 8. Small multilamellar liposomes made with the French pressure cell. (A) The void volume fraction of egg PC liposomes following two passes through the French pressure cell at 3000 psi (see Figure 7). Diameters range between 500 to 1500 Å (mean 980 Å). The same type of image is obtained from negative stains of the small amount of lipid in the void volume after 20,000 psi (see Figure 6B) (Magnification × 30,000). (B) The included volume fraction from egg PC liposomes following two passes through the French pressure cell at 3000 psi (see Figure 7). Diameters range from 300 to 900 Å (mean 570 Å) and the majority of the liposomes appear to contain two lamellae. (Magnification × 30,000.) (C) This higher magnification of the above sample illustrates the double lamellae of these liposomes. (Magnification × 90,000.)

Multiple samples — Because of the rapidity of preparation, one can readily prepare a series of different vesicles of differing lipid proportions or compositions within a short period.

Reproducibility — With a given lipid, such as egg yolk PC, the same particle size and agarose gel column distribution of lipid is obtained routinely when the same pressure, number of cycles, and rate of flow are used.

Quantity — The larger pressure cell has a capacity of 40 mℓ and, because the technique is effective with concentrations up to about 20 to 25 mg/mℓ PC, large quantities of liposomes can be obtained.

Degradation — No detectable degradation of egg yolk PC was observed following ten consecutive cycles through the pressure cell at 20,000 psi as measured by TLC and GLC.[1]

Denaturation — Several enzymes retain substantial biological activity after processing through a French pressure cell at 20,000 psi.[2] Furthermore, apoprotein C-11 that is bound

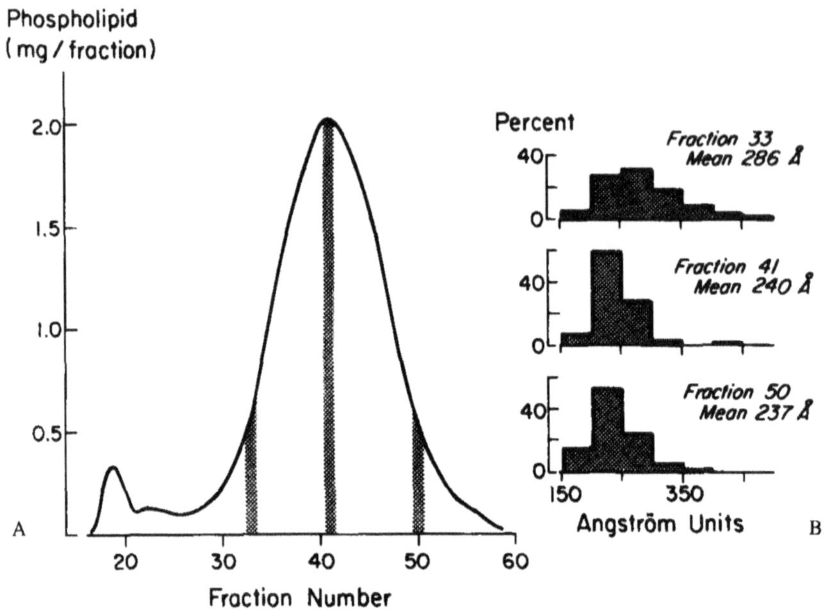

FIGURE 9. (A) Elution (from a 2% agarose column, 1.2 cm × 95 cm) of PC liposomes containing CF. (B) Mean particle diameter and distribution of diameters of unilamellar liposomes of different elution volumes. Particle diameters were measured from negatively stained images.[1]

to very low density lipoproteins continues to activate lipoprotein lipase after processing these serum lipoproteins through a pressure cell.[14]

Stability — Unilamellar vesicles of egg yolk PC (with or without cholesterol) appear to remain intact during storage in the refrigerator at ~4.0°C for more than a year (if under N_2 gas), as determined by the absence of precipitation and the appearance of vesicles in negative stains. This is true only if the vesicles are centrifuged at 100,000 × *g* for ~1 hr prior to storage to pellet the small mass of multilamellar structures which otherwise form a precipitate during prolonged storage.[1]

Versatility —

1. The procedure works well with different pure lipids and with biological mixtures of polar lipids.
2. A wide variety of volumes may be used (~1 to 40 mℓ).
3. A wide range of lipid concentrations may be used <1 to ~25 mg/mℓ PC.
4. A wide range of temperatures may be used from 0 to ~50°C by icing the pressure cell or placing it in an oven for ~1 hr before use. The large mass of metal holds the temperature constant long enough to make liposome preparations at the chosen temperature.
5. A variety of vesicular structures may be produced, including homogeneous small (~200 Å) unilamellar vesicles, small (500 to 1500 Å) multilamellar vesicles, and even small (~500 to 600 Å) double bilayer vesicles (Figures 5, 8).
6. Trapping small solutes is very efficient because concentration can be made quite high thereby saving on the cost of materials or precious reagents.

Omissions — Perhaps one of the most favorable features of the technique is the fact that many materials and procedures that necessarily accompany other techniques are not needed or are optional when liposomes are prepared with the French pressure cell method.

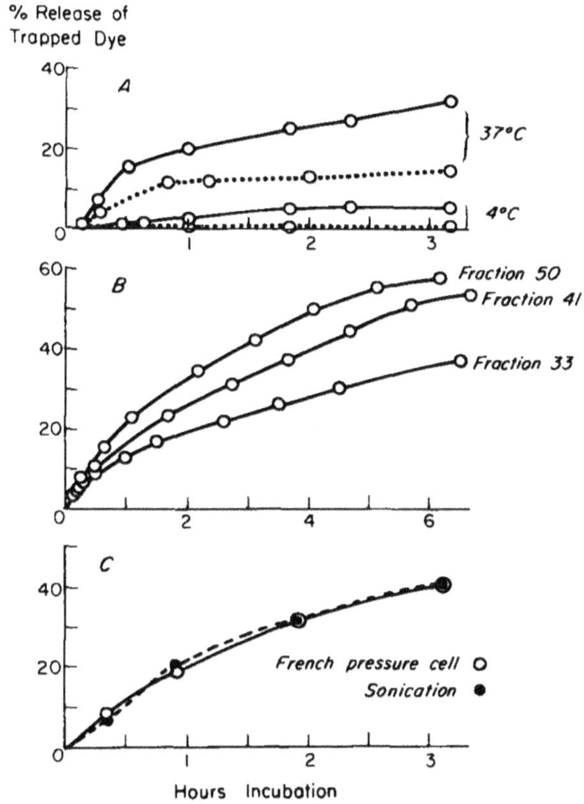

FIGURE 10. Slow (hours) release of CF from liposomes. (A) Release of dye from PC liposomes (solid line) and PC-cholesterol liposomes (43 mol % cholesterol, dotted line) at shown temperatures. (B) Release of dye (at 37°C) from selected fractions of PC liposomes shown in Figure 9. (C) Release of dye (at 37°C) from PC liposomes prepared by French pressure cell (○) or sonication (●). Liposomes were from the fraction corresponding to fraction 41 in Figure 9.

1. Dialysis is not needed because neither detergents nor organic solvents are used. In addition, many electrolytes, buffers, antimicrobial agents, EDTA, etc., may be used without altering the results.
2. Concentrating procedures are not needed because high initial concentrations of PC can be used and sample dilution by column chromatography is often unnecessary.
3. Column chromatography can be omitted for many studies because ultracentrifugation for 1 hr at 100,000 × *g* pellets the small (2 to 8%) mass of multilamellar vesicles. For each new lipid, however, it is advisable to establish the characteristics of the elution profile.
4. Centrifugation can be omitted for many studies because most (~90 to 95%) of the lipid is in homogeneous unilamellar vesicles.
5. Inert gases such as nitrogen and argon are unnecessary for most PC commonly used because the method does not induce lipid degradation in air.
6. Titanium particles are not generated.
7. Aerosols are not generated.
8. Cooling jacket is not required because the temperature change in the sample during transformation is only a few degrees.

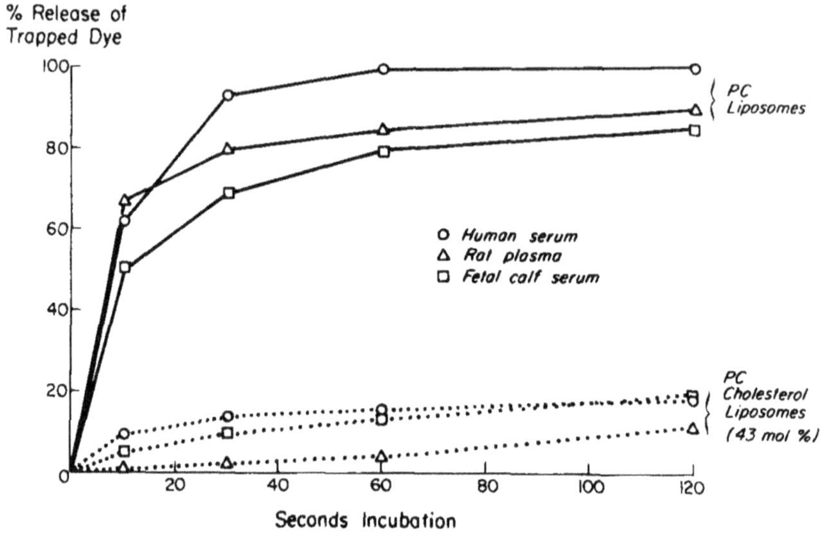

FIGURE 11. Fast (seconds) release of CF from liposomes. Induced release of CF from liposomes by whole serum or plasma at 37°C occurs within seconds and is largely inhibited by cholesterol. Similar results occur with isolated plasma lipoproteins or their pure apoproteins.[5]

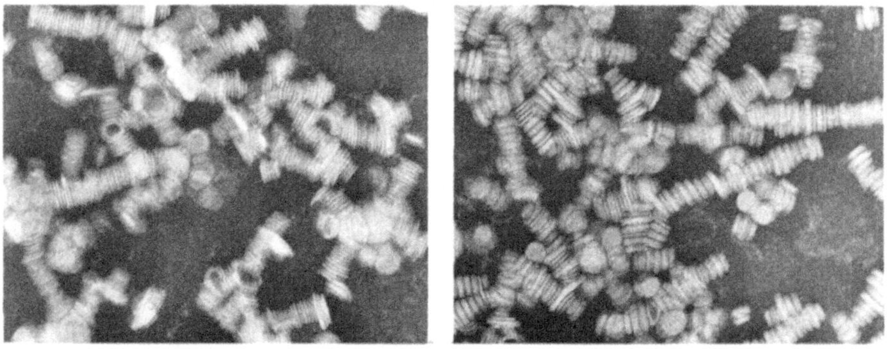

A B

FIGURE 12. Images obtained by negative staining mixtures of liposomes and rat apo HDL (lipid-protein 2:1 w/w) incubated at 37°C for 5 min. (A) PC liposomes appear almost completely transformed into discoidal particles with an edge thickness of about 35 Å, appropriate for a single bilayer. (The stacking in rouleaux is an artifact of negative staining that occurs more in the presence of apoproteins.) (B) PC-cholesterol liposomes (43 mol % cholesterol) appear to remain intact as shown by the thickness of the edges of individual particles which slightly exceed 100 Å. A few particles in the rouleaux, however, appear to be discs consistent with the data in Figure 11. (Magnification × 90,000.)

V. DISADVANTAGES

Cost — The initial investment in the purchase of the pressure cell and the automatic hydraulic press is considerable. However, we find that a single press is adequate for several research groups. Therefore, each group need purchase only a single pressure cell.

Maintenance — Care must be taken to properly clean and dry the pressure cell between each use. The O-rings and the Teflon® ball must be replaced when wear begins to show.

VI. POTENTIAL

Because of the speed of preparation, versatility, reproducibility, and simplicity of the technique, the method described here for preparing small liposomes with the French pressure cell appears to have a large potential. Once the established investigators in the field of "liposomology" discover the method, we believe that it will become the method of choice.

Perhaps the greatest application of this technique must await the successful development of tissue and cellular targeting of small unilamellar liposomes containing drugs, etc. If and when successful technology in targeting becomes a reality and drug companies enter the scene, the great simplicity, versatility, and quantitative advantages of the French pressure cell procedure, together with remarkable stability of the product vesicles, will likely make it the most cost-effective method of small liposome production.

ACKNOWLEDGMENTS

This research was supported by the National Institutes of Health Arteriosclerosis SLOR Grant HL-14237 at University of California San Francisco.

REFERENCES

1. **Hamilton, R. L., Goerke, J., Guo, L. S. S., Williams, M. C., and Havel, R. J.,** Unilamellar liposomes made with the French pressure cell: a simple preparative and semiquantitative technique, *J. Lipid Res.*, 21, 981, 1980.
2. **Milner, H. W., Lawrence, N. S., and French, C. S.,** Colloidal dispersion of chloroplast material, *Science*, 3, 633, 1950.
3. **Barenholz, Y., Amselem, S., and Lichtenberg, D.,** A new method for preparation of phospholipid vesicles (liposomes). French Press, *FEBS Lett.*, 99, 210, 1979.
4. **Oyarzun, M. J., Clements, J. A., and Baritussio, A.,** Ventilation enhances pulmonary alveolar clearance of radioactive dipalmitoyl phosphatidylcholine in liposomes, *Am. Rev. Respir. Dis.*, 121, 709, 1980.
5. **Guo, L. S. S., Hamilton, R. L., Goerke, J., Weinstein, J. N., and Havel, R. J.,** Interaction of unilamellar liposomes with serum lipoproteins and apolipoproteins, *J. Lipid Res.*, 21, 993, 1980.
6. **Huang, C. and Thompson, T. E.,** Preparation of homogeneous, single-walled phosphatidylcholine vesicles, *Meth. Enzymol.*, 32, 485, 1974.
7. **Tall, A. R., Small, D. M., Deckelbaum, R. J., and Shipley, G. G.,** Structure and thermodynamic properties of high density lipoprotein recombinants, *J. Biol. Chem.*, 252, 4701, 1977.
8. **Hamilton, R. L., Williams, M. C., Fielding, C. J., and Havel, R. J.,** Discoidal bilayer structure of nascent high density lipoproteins from perfused rat liver, *J. Clin. Invest.*, 58, 667, 1976.
9. **Felker, T. E., Hamilton, R. L., and Havel, R. J.,** Secretion of lipoprotein-X by perfused livers of rats with cholestasis, *Proc. Natl. Acad. Sci. U.S.A.*, 75, 3459, 1978.
10. **Halks-Miller, M., Guo, L. S. S., and Hamilton, R. L.,** Tocopherol-containing unilamellar liposomes produced with a French pressure cell, submitted.
11. **Batzri, S. and Korn, E. D.,** Single bilayer liposomes prepared without sonication, *Biochem. Biophys. Acta*, 298, 1015, 1973.
12. **Lichtenberg, D., Freire, E., Schmidt, C. F., Barenholz, Y., Felgner, P. L., and Thompson, T. E.,** Effect of surface curvature on stability, thermodynamic behavior, and osmotic activity of dipalmitoyl phosphatidyl choline single lamellar vesicles, *Biochemistry*, 20, 3462, 1981.
13. **Allen, T. M., Romans, A. Y., Kercret, A., and Segrest, J. P.,** Detergent removal during membrane reconstitution, *Biochem. Biophys. Acta*, 601, 328, 1980.
14. **Hamilton, R. L. and Havel, R. J.,** unpublished studies.

Chapter 5

THE USE OF FRENCH PRESSED VESICLES FOR EFFICIENT INCORPORATION OF BIOACTIVE MACROMOLECULES AND AS DRUG CARRIERS IN VITRO AND IN VIVO

Peter I. Lelkes

TABLE OF CONTENTS

I. INTRODUCTION

The French press has been known for a long time as a useful tool in cell biology, especially in botany, to disrupt whole cells and to obtain organelles, membranes, etc., efficiently with high purity.[1] The principles of a French press, are briefly as follows: a constant pressure (adjustable between 35 to 1400 atm (500 to 20,000 psi) is applied hydraulically via a piston onto a liquid contained in a pressure-tight chamber. Upon opening of the pressure relief valve, the pressure gradient between inside and outside, created across the tiny outlet hole, generates shearing forces, which are sufficient to disrupt fragile, particulate matter, contained in the extruded suspension. The size of these submicroscopic fractions depends primarily on the applied pressure (i.e., on the size of the shearing forces acting as a consequence of the pressure gradient across the opening outlet).

Recently, Barenholzt and co-workers[2] adapted this method to produce, by extrusion of handshaken multilamellar liposome suspensions at 20,000 psi, a very homogeneous population of small unilamellar vesicles (SUV). This technique had been in laboratory use previously elsewhere, but had not been reported.[3] High-pressure extrusion in a French press is a simple, efficient technique, which allows the encapsulation into small uni- and/or oligolamellar liposomes of numerous labile bioactive macromolecules, which are deactivated by other preparational methods. A question, not resolved as yet, concerns the size of the resulting liposomes. While Hamilton et al.[3] claim, that upon extrusion at 20,000 psi very small vesicle (about 200 Å) are obtained, i.e., even smaller than sonicated SUV,[4] other laboratories found, that the resulting liposomes are somewhat larger (about 400 to 800 Å) than SUV obtained by extensive sonication.[2,5-8] In this chapter, the methodology will be described, as used in our laboratory, to produce by extrusion in a French pressure cell homogeneous populations of unilamellar vesicles (FPV) of small size (≤800 Å). We shall discuss the dependence of the size on several experimental parameters (e.g., lipid composition of the vesicles, pressure, extrusion velocity, etc.). In addition some applications of FPV as drug carriers in vitro and in vivo will be demonstrated.

II. PREPARATION OF LIPOSOMES BY HIGH-PRESSURE EXTRUSION IN A FRENCH PRESS

Extrusion of liposome suspensions in a French press is one possible technology[9] to convert heterogeneous preparations of handshaken and/or vortexed large multilamellar liposomes into a more homogeneous population of either smaller multilamellar vesicles (about 1000 to 3000 Å) and/or into oligo- and unilamellar vesicles of varying sizes between 300 to 1000 Å in diameter. The characteristic features of the resulting liposome suspensions depend on a combination of factors like phospholipid composition, cholesterol content, temperature of operation, and speed of extrusion. As a general rule, the higher the pressure the smaller the resulting liposome size, provided all operations are performed above the phase transition temperature of the phospholipids. On the other hand, the final liposome population will be more homogeneous, the slower the extrusion process is performed. For a typical experiment, 10 mg chromatographically pure egg-phosphatidylcholine (PC) (Lipid Products, Nutfield, U.K.) and 5 mg cholesterol (CHOL) (Merck, Darmstadt, F.R.G.), dissolved in 0.5 mℓ chloroform: methanol (1:1, v/v) are dried onto the wall of the glass bottle, by evaporating the organic solvent under a stream of N_2 (for 60 min) and further evaporation under vacuum in a dessicator for an additional hour; 2 mℓ aqueous buffer (150 m*M* NaCl, 10 m*M* Tris, pH = 7.5) are added and the lipid film is completely removed from the glass wall by vigorous vortex mixing. Prior to adding the liposomes, the French pressure cell (Figure 1) is kept at 4°C, and carefully rinsed with ice-cold buffer. The milky white multilamellar large vesicles (MLV) suspension is transferred into the French pressure cell, which is held upside

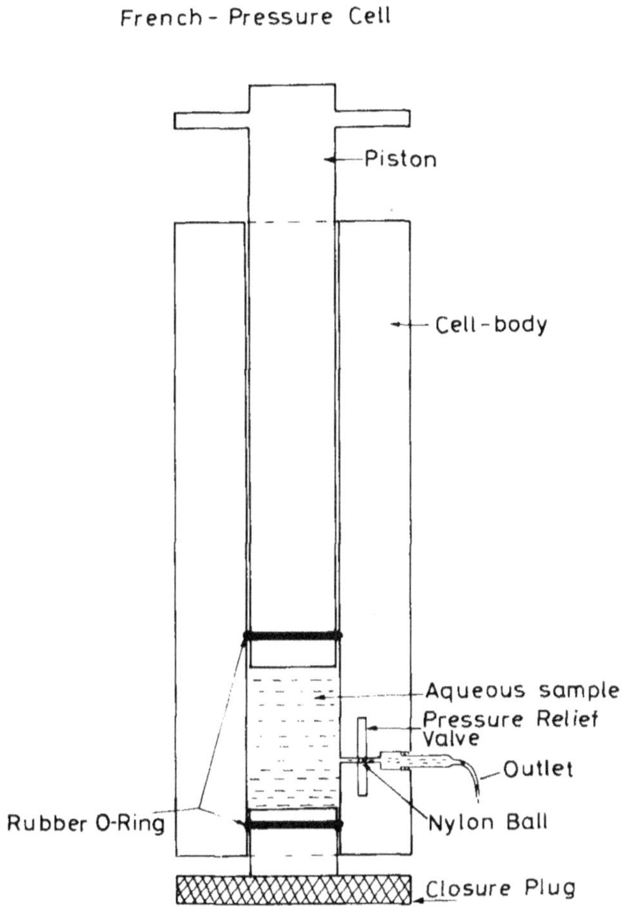

FIGURE 1. French pressure cell, schematic cross-section.

down, the shaft being tightly closed by the piston. Before closing the cell the piston is carefully raised until the surface of the aqueous suspension reaches just below the outlet hole. Occasionally this is rather tedious to observe, since either the outlet hole is not visible under unfavorable light conditions and/or the upper level of the aqueous suspension is reflected in the piston shaft wall and the mirror image of the suspension surface seems to be in place, rather than the surface itself. However, it is of importance to align the aqueous surface and to extrude the liquid free of air. The compressed air, when released, upon extrusion, will cause unnecessary splashing of the sample, which potentially increases the biohazards of the procedure, especially when working with radioactive material, etc.

Upon proper alignment of the upper level of the fluid in the piston shaft and insertion of the closure plug, the pressure relief valve is firmly closed, and the pressure cell is turned right-side up. When closing the valve, care must be taken, that (1) the tiny nylon ball is in place and in working shape and (2) that the valve is not squeezed too much, to prevent damage of the outlet hole. The pressure cell is inserted into the hydraulic press, and the desired pressure is applied, according to the operating instructions. As a precaution, the outlet tubing should be held into a long-necked reservoir before applying the pressure to collect fluid which may be extruded prematurely due to incomplete closure of the outlet valve and/or damage of the nylon ball. If the closure of the pressure cell is performed properly, the cell will hold the set pressure and can now be emptied carefully. Very slow

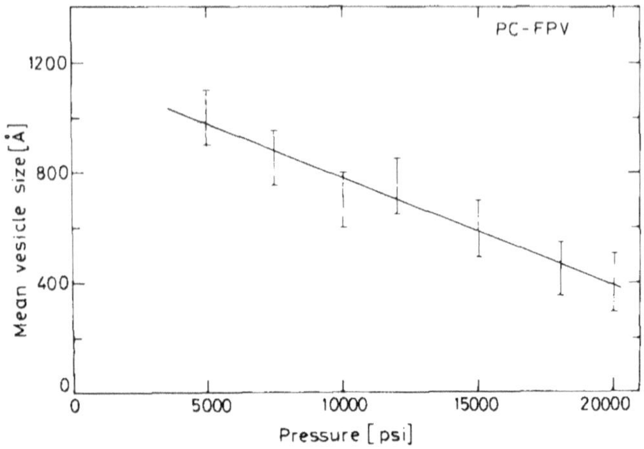

FIGURE 2. Dependence of the mean vesicle size (i.e., outer diameter) of French pressed egg PC vesicles on the extrusion pressure. Freeze-fracture electromicrographs of FPV, extruded at various pressures were analyzed, and histograms of the size-distribution for 400 to 500 vesicles in each preparation were plotted in 100-Å increments. The error bars indicate ± 100 Å.

extrusion is a crucial step in the preparation procedure and must be carried out with utmost care and great patience. In order to prevent a rapid drop of the pressure, the fluid must be extruded slowly, drop by drop, at a flow rate not exceeding 1 mℓ/min. Such slow flow rates are absolutely essential to obtain homogeneous vesicle populations, especially of small unilamellar liposomes. In addition, careful extrusion, drop by drop, enables the use of small sample volumes (0.6 to 0.8 mℓ) when using the small French pressure cell (3/8-in.-diameter piston). To assure complete conversion of MLV into French pressed vesicles (FPV), the filling and extrusion steps are repeated at least three times, although, more than 90% of the MLV are transformed into FPV during the first extrusion, provided this has been performed carefully as described. Upon completion of the preparation, FPV extruded at >10,000 psi are centrifuged at 10,000 × *g* in a microfuge for 10 min to sediment contaminations from the rubber O-ring and remaining larger, nondisrupted liposomes or liposomal aggregates. The FPV remaining in the supernatant, usually contain >95% of the total lipid.

III. SIZE AND PHYSICOCHEMICAL CHARACTERISTICS OF FPV

The size of the FPV depends on three main parameters: (1) their phospholipid composition, (2) whether or not they contain cholesterol, and (3) the temperature of preparation.

If the lipids, comprising the liposome suspension, are in a fluid state, i.e., above their phase-transition temperature, a good correlation can be obtained between the pressure at which the liposome suspension is extruded and the resulting FPV size (Figure 2). Between 5000 and 20,000 psi applied pressure, the mean outer diameter of the largely unilamellar FPV obtained by extruding a preparation of PC-MLV decreases linearly from ≈1000 Å to ≈400 Å, as determined by negative staining and freeze-fracture electromicroscopic observation. Including 20% and 50% Chol (molar ratio) into the lipid mixture will yield vesicles with an outer diameter of ≈550 Å, and of ≈700 Å, respectively, when the MLV are extruded at 20,000 psi (Figure 3). Similarly, at lower pressures the vesicle size increases by approximately 30 to 50% upon inclusion of up to 50 mol % cholesterol. In addition to increasing the mean vesicle diameter, inclusion of Chol also results in a more widespread size distribution of the resulting FPV, as can be seen in the histograms in Figure 3. When applying

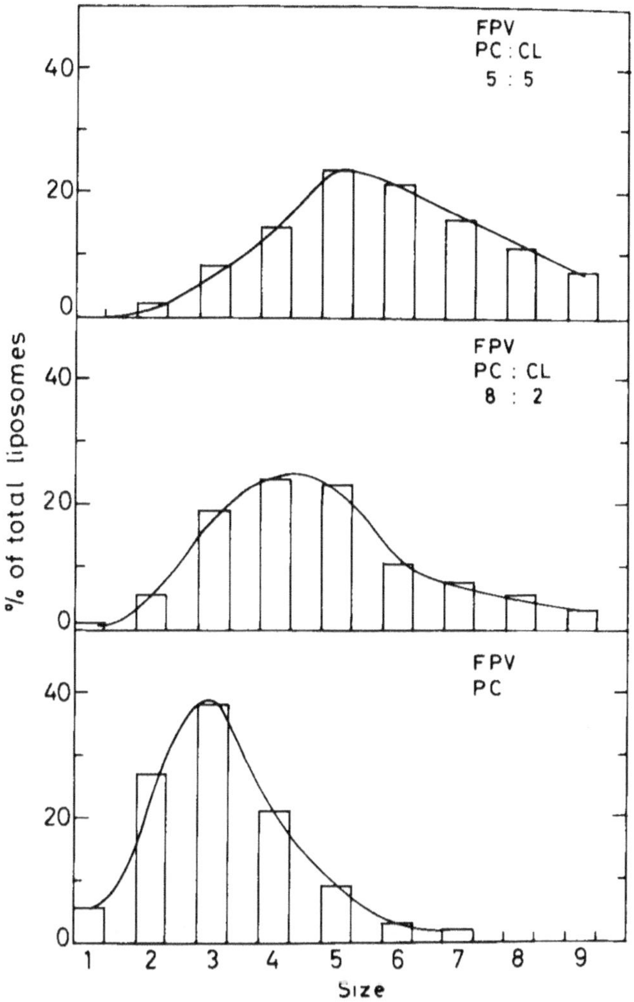

FIGURE 3. Effect of increasing Chol content on the average vesicle size (outer diameter in Å) of French pressed vesicles prepared by high-pressure extrusion of egg PC MLV containing 0, 20, and 50 mol % Chol. The histograms were obtained as described in the legend to Figure 2. The size groups are (1) 200 to 300 Å; (2) 300 to 400 Å; (3) 400 to 500 Å; (4) 500 to 600 Å; (5) 600 to 700 Å; (6) 700 to 800 Å; (7) 800 to 900 Å; (8) 900 to 1000 Å; (9) >1000 Å.

lower pressures (500 to 5000 psi), the resulting FPV are mainly multilamellar; their mean outer diameter decreases from ≥5000 Å to ≈1000 Å with increasing pressure.

Small unilamellar FPV composed of phospholipids with fully saturated long-chain fatty acids, such as dipalmitoyl phosphatidylcholine (DPPC) or distearyl phosphatidylcholine (DSPC), are obtained, only if the extrusion is performed above their respective phase-transition temperatures. While this is still feasible for DPPC (42°C), operation at 55°C (DSPC) is inconvenient. If MLV preparations composed of one of these lipids or mixtures of them are extruded at 20,000 psi at ambient temperature, the resulting FPV preparation will be a mixture of oligo- and multilamellar vesicles of ≈1500 Å in diameter.

Assuming that the resulting FPV are uni- or oligolamellar ($n \leq 3$) vesicles, the number of phospholipids per liposome, n, can easily be calculated according to:[10]

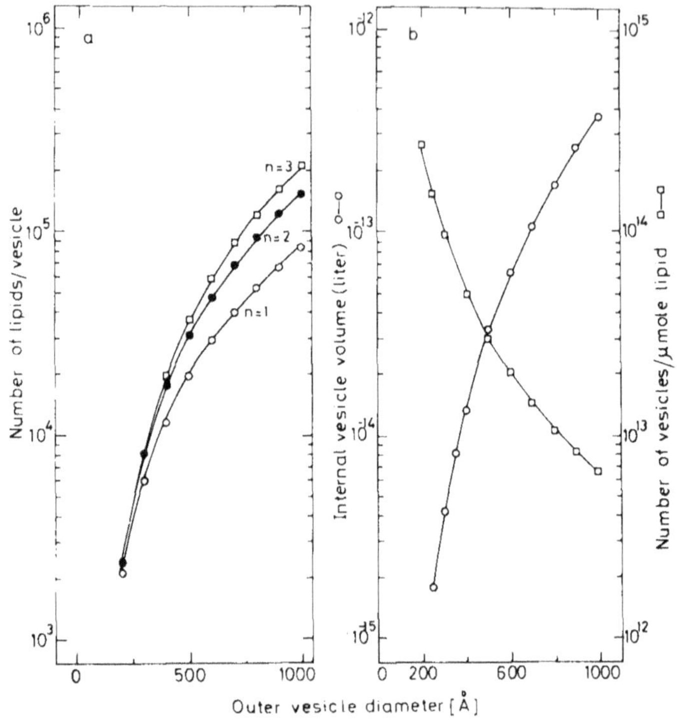

FIGURE 4. Correlation between the outer vesicle diameter and (a) the number of lipid molecules per liposome and (b) the number of liposomes per micromole lipid (□-□-□) as well as their internal vesicle volume per micromole lipid (○-○-○) for unilamellar liposomes. In (a) the number of lipids per vesicle is also plotted for oligolamellar vesicles, containing n (n = 2 or n = 3) lipid bilayers, separated by a 50-Å aqueous space from each other.[11] The curves in this graph are purely theoretical, calculated according to Reference 10, and serve as an estimate of some fundamental properties of the liposome preparations.

$$n = \frac{2\pi}{65}(D-50)^2 \tag{1}$$

where D is outer diameter (in Ångstrom), (DC-50) takes into account the average width of one bilayer, and $65(\text{Å})^2$ the average area of one phospholipid molecule. For fully hydrated oligolamellar vesicles, the distance between two adjacent lipid bilayers can be assumed to be ≈50 Å,[11] i.e., for each layer the value for D has to be modified accordingly. This above calculation is very approximate and does not take into account the variety of phospholipids and the effects of cholesterol, as manifested in the subtle differences for each species in their average headgroup areas, the width of the lipid-bilayers, and the inter-bilayer distances.[12] However, the above formula can be used to obtain a rough assessment of the number of phospholipid molecules per vesicle and hence for obtaining an estimate of the number of vesicles per micromole lipid as well as of their internal volume, as shown in Figure 4.

When comparing FPV and SUV prepared from PC, with their average diameters of 450 Å and 210 Å, respectively (cf. Figure 3 and Reference 4), two striking differences exist between FPV and SUV: (1) the number of lipid vesicles per FPV (15,500) is ≅ six times larger than that of SUV (2500); (2) due to their larger size, the internal aqueous space of FPV is approximately 30 times larger than that of SUV (cf. Figure 4). These two features are reflected in some of the physicochemical properties of FPV, as discussed on the next page.

Table 1
HALF-TIME OF CF RELEASE ($T^{1}/_{2}$) FROM PC LIPOSOMES PREPARED BY SONICATION (SUV) AND BY HIGH-PRESSURE EXTRUSION (FPV)

Liposomes	Temperature (°C)	$T^{1}/_{2}$ (hr) Buffer	$T^{1}/_{2}$ (hr) Buffer + 10% human serum
SUV	4	29.8 ± 3.1	ND[a]
FPV	4	312.2 ± 36.2	ND
SUV	37	12.8 ± 2.4	1.2 ± 0.4
FPV	37	71.3 ± 9.5	6.4 ± 1.3

Note: Large multilamellar liposomes (2 μmol) were prepared from egg PC in 100 m*M* CF, 10 m*M* Tris, pH = 7.5, and divided into two samples. One was sonicated in an ice-water bath under N_2 for 30 min, the other extruded four times at 20,000 psi in a French press. The vesicle suspensions were centrifuged for 10 min at 10,000 × *g*, and the nonentrapped CF was removed by the Sephadex® G-50 minicolumn centrifugation technique. Leakage of CF was assayed in PBS or PBS + 10% human serum, as described.[27] The results shown are the mean ± SD from three independent experiments, performed in quadruplicates.

[a] ND, not determined.

1. The larger external diameter of FPV also means an augmented internal space; therefore the entrapment efficiency of FPV for water-soluble macromolecules is considerably larger than that of SUV.
2. Due to the larger size of FPV, the lipid asymmetry between the inside and outside half of the FPV bilayer (concomitant with defects in the lipid packing) is relieved.[4,9] This is reflected in the augmented stability properties of FPV. Thus FPV, prepared from PC by extrusion at 20,000 psi, retain their aqueous contents up to ten times longer than sonicated SUV, as assessed by the carboxyfluorescein (CF) technique.[2,13]
3. Due to their larger size and/or less imperfect lipid packing characteristics, FPV are more stable towards disruption by serum proteins, than sonicated SUV.[13] Similarly SUV, when incubated with isolated cells in vitro, will loose a considerable part (up to 25% within 30 min) of their aqueous contents, due to cell-induced vesicle destabilization.[14,15] FPV of the same composition, on the other hand, will retain more than 95% of their contents, even after 2 hr of incubation.[15]

Shown in Table 1 are the half-times for CF release from SUV and FPV, prepared from egg PC, under various experimental conditions, when the liposomes are suspended in phosphate-buffered saline (PBS). CF leakage at 4°C is ≈10.5 times and at 37°C: ≈5.5 times slower from FPV than from SUV. Similarly in the presence of 10% human serum, released CF from FPV is approximately five times slower than from SUV. Similar data have been found, when using ^{14}C-sucrose as the marker for the release of the aqueous contents of the liposomes.

IV. ENTRAPMENT OF MACROMOLECULES INTO FPV

A crucial question in encapsulating macromolecules into liposomes is whether the bio-

Table 2
RESIDUAL ACTIVITY[a] OF BIOLOGICAL MACROMOLECULES FOLLOWING SONICATION AND HIGH-PRESSURE EXTRUSION

	Residual activity (%) following	
Compound	Sonication[b]	French press[c] extrusion
Trypsin	45	100
Actin	15	85
DNAse I	0	95
Heavymeromyosin	0	85

[a] Residual activity as compared to initial bioactivity, before the respective treatment; bioassays were carried out as described.[8]

[b] 2-mℓ samples were sonicated for 10 min under N_2 in an ice-water bath with a tip-sonicator (Branson B12 sonifier) at a nominal output of 70 W.

[c] 2-mℓ samples were extruded four times at 20,000 psi at room temperature.

logical activity of these compounds is preserved during the preparation. The strong shearing forces and high local temperatures during sonication are known to disrupt and deactivate many macromolecules, especially those composed of several subunits.[9,16] In the presence of organic solvents, as for instance in the reverse-phase-evaporation (REV) technique,[9] many macromolecules will precipitate or will be deactivated.[17] High temperatures, e.g., >40°C, as required during the ether/ethanol injection method, will also cause deactivation of heat-labile compounds. As seen in Table 2, the biological activity of several biologically important macromolecules is essentially fully retained upon fourfold extrusion at ambient temperature in a French press at 20,000 psi; 10 min of sonication of the same samples in a Branson-type 12-tip sonicator, will lead to nearly complete deactivation, even when the samples are cooled in an ice bath during sonication.

It should be noted, that high-pressure extrusion may not always be the method of choice for encapsulation into liposomes. Thus, molecules that are extremely labile to shearing forces, like human β-interferon, will lose up to 90% of their antiviral activity when extruded even once at 20,000 psi.[18] On the other hand, the encapsulation of magnetic microparticles (≈100 Å) by this technique is not possible, due to the ferromagnetic properties of the French pressure cell.[19]

The entrapment of molecules with >40,000 mol wt into sonicated SUV is severely limited, due to the extremely small internal free space of such liposomes.[9] However, by high-pressure extrusion, even large molecules up to a molecular weight of 350,000 can be entrapped efficiently into FPV, as seen in Figure 5. The dependence of the entrapment efficiency on the molecular weight indicates, that in FPV, too, the internal space is limited, but still sufficient, to accommodate large macromolecules. The deviation from the straight line, i.e., the augmented entrapment, as seen for heavy meromyosin (350,000 mol wt) may be due to a partial association of this molecule with the lipid bilayer.

V. LIPOSOMAL DRUG DELIVERY IN VITRO

Our initial interest in FPV as drug delivery systems in vitro originated from studies into the involvement of the cytoskeleton in the mechanism of stimulus-secretion coupling in

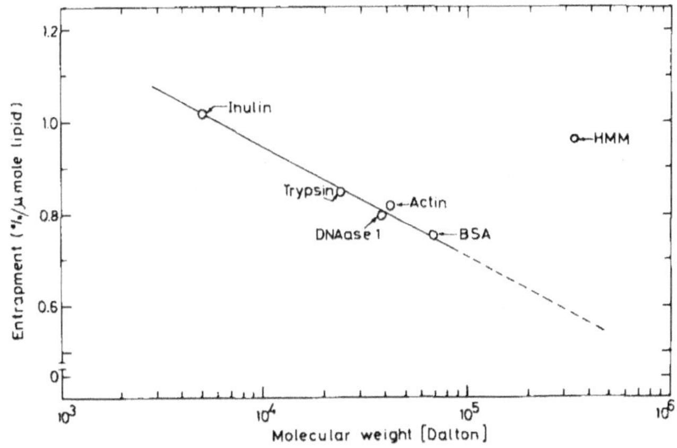

FIGURE 5. Entrapment efficiency of FPV for macromolecules. FPV, from PC:PS:Chol (at a molar ratio 7:1:2, total lipid 35.6 μmol) were prepared in the presence of the various solutes, at a concentration of 5 mg/mℓ. Entrapment was calculated as described.[8]

secretory cells, such as the adrenal chromaffin cell or basophils.[8,13,20] Preliminary experiments showed that only by high-pressure extrusion in the French press we could retain the biological activity of the encapsulated cytoskeleton specific and/or cytoskeletal enzymes and proteins such as DNAse I, actin, heavymeromyosin, etc.[8,13,21,22] The lipid composition of the vesicles we chose was as a first approximation similar to the lipid composition of the plasma-membranes, i.e., PS:PC:Chol at a molar ratio of 7:1:2. FPV were prepared as described in Section II. Nonentrapped macromolecules were separated by column-chromatography on either Sephacryl® S-200 or on Sepharose® 4-B columns (2.5 cm × 40 cm). The liposome fraction eluting in the void volume was analyzed for phosphate and protein contents to determine the entrapment efficiency.[7,8] Prior to the introduction of cytoskeletal proteins into the cells via FPV several types of experiments as described below were performed to determine the extent and mechanism of interaction between the liposomes and the cells, using radiolabeled ^{14}C-DPPC and ^{3}H-inulin, as markers for the lipid phase and the aqueous contents of vesicles, respectively:

1. Dose-dependent (0.01 to 10 μmol lipid/10^6 cells) uptake of both radiolabels was measured in the presence and absence of a tenfold excess of "cold" vesicles of identical composition and in the presence and absence of a combination of metabolic inhibitors (5 m*M* NaN_3 and 50 m*M* deoxyglucose).
2. At several liposome: cell ratios the time course of the uptake of the radiolabels by the cells was followed.
3. For a given liposome: cell ratio the temperature dependence of the uptake of the radiolabels was determined upon incubation with the cells for 60 min. In addition transfer of fluorescent vesicle contents (either CF or rhodamine isothiocyanate-labeled actin) was observed by flow cytometry in a fluorescence activated cell sorter and/or by visualization in a fluorescence light microscope.

In both our systems (the chromaffin cells and the basophils), we found qualitatively similar results,[8,13,20-23] i.e., the uptake of both the lipid and the aqueous space markers is parallel and depends on the ratio liposome: cells. This uptake is only modestly (≤ 30%) effected by either metabolic inhibitors and/or chasing by nonlabeled liposomes. The temperature dependence shows a characteristic break at ≈ 16 to 18°C. All these features indicate that

Table 3
INCREASE IN CATECHOLAMINE (CA) RELEASE FROM ISOLATED BOVINE CHROMAFFIN CELLS FOLLOWING 1 HR INCUBATION WITH LIPOSOMES OF VARIOUS COMPOSITIONS CONTAINING 150 m*M* NaCl[a]

Lipid composition of the liposomes[b]	Increase in CA release[c] (%)
PC : PS : CL 7 : 1 : 2	211.9 ± 24.2
PE : PS : CL 6 : 2 : 2	135.0 ± 18.1
PE : PC : CL 6 : 2 : 2	184.3 ± 5.8
PE : PC : PS : CL 3 : 3 : 2 : 2	162.4 ± 19.4
PC : CL 8 : 2	178.1 ± 17.0

[a] 10^6 cells were incubated with 3 μmol FPV in a total volume of 1 mℓ buffer I. For experimental details see Reference 8.
[b] The molar ratios of the constituent lipids are given.
[c] The basal CA release of nontreated cells is set to 100%. The data thus indicate the % increase in CA secretion over the basal release.

in our system, liposomes are taken up by the cells preferentially by fusion and not by endocytosis. This conclusion is underlined by fluorescence microscopic observations, e.g., by uniform staining of the cytoplasma, with the exception of the nuclear region, upon transfer of the fluorescent vesicle contents via fusion. Endocytosis of the vesicles would have resulted in punctuate fluorescence of the cells.[24,25] The efficacy of drug delivery via FPV into secretory cells is easy to follow, since changes in, for instance, the cytoplasmic ion milieu, the state of the microfilaments, etc., are immediately reflected in the rate of spontaneous and/or stimulated release of catecholamines or histamine in the chromaffin cell and the basophil, respectively.[8,13,20-23] Increasing the internal sodium concentration will result in increased basal release in both systems. Thus we used FPV, containing 150 m*M* NaCl, to examine the lipid requirement for optimal fusion of the vesicles with the cells, as reflected by maximal increase in the basal release. In each experiment increasing amount of liposomes, between 0.01 and 10 μmol total lipid per 10^6 cells were used. The cells were incubated with liposomes of various compositions for 60 min at 37°C. Thereafter the amounts of spontaneously secreted catecholamines and histamine, respectively, were compared to that of control cells, which either had not been incubated with the liposomes, or were incubated with identical concentrations of liposomes containing 150 m*M* KCl, that do not increase the basal catecholamine or histamine secretion. Table 3 shows the results for some of the lipid mixtures with PC:PS:Chol at a molar ratio of 7:1:2 being the optimum composition.

Having established that fusion is the main mechanism of interaction between FPV and the two types of secretory cells, we are currently attempting to optimize the experimental conditions for efficient transfer of the vesicle contents into the cells. We can thus investigate the biological effects caused by the introduction via FPV of cytoskeleton-specific enzymes and cytoskeletal proteins into chromaffin cells and rat leukemic basophils and study more extensively the molecular mechanisms of stimulus-secretion coupling in these cells.

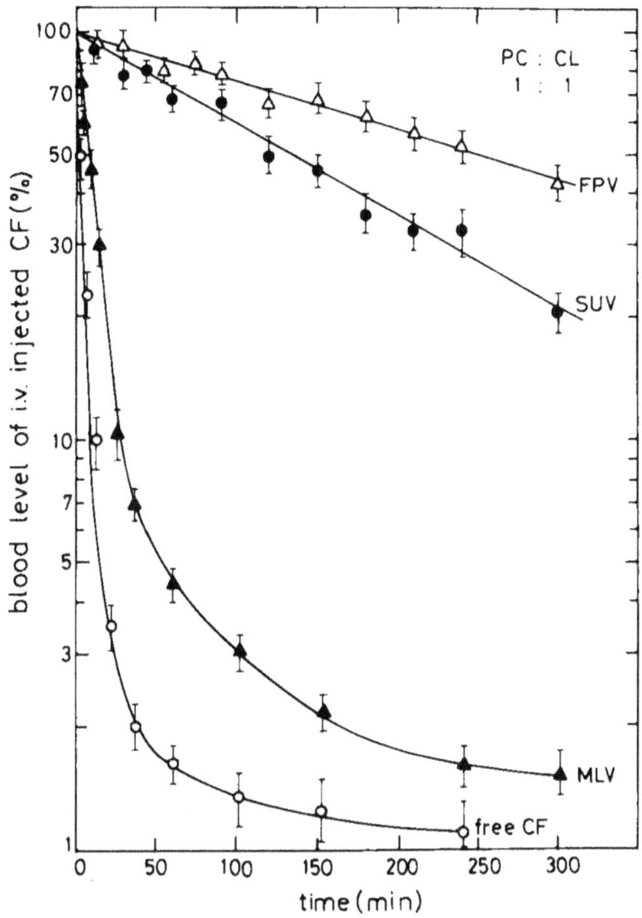

FIGURE 6. Blood levels in female Balb/c mice of intravenously injected PC/Chol-liposomes containing 100 m*M* CF, 10 m*M* Tris, pH = 7.5. Experimental details in text.

VI. FPV AS DRUG CARRIERS IN VIVO

Two inherent features of FPV, offering advantages over sonicated SUV, make them attractive as drug carrier systems in vivo:

1. Their slightly larger size (400 to 800 Å) and hence their increased entrapment efficiency allows the encapsulation of approximately one order of magnitude more hydrophilic drugs and larger macromolecules. FPV are however still small enough to cross capillary barriers, especially the sinusoidal ones.
2. FPV in vitro were shown to be more resistant to destabilization by serum proteins than SUV (cf. Table 1).

These two characteristics, increased carrying capacity and augmented stability, make FPV a good candidate as drug delivery systems in vivo whenever a circulating depot effect is sought. Compared in Figure 6 are the blood levels of intravenously injected CF-containing FPV, SUV, and MLV composed of PC:Chol at a molar ratio 1:1. The liposomes were prepared by standard methods.[9,26] 100-μλ liposome suspension, containing 2 μmol total lipid were injected into the tail veins of female Balb/c mice. At time intervals 10λ blood

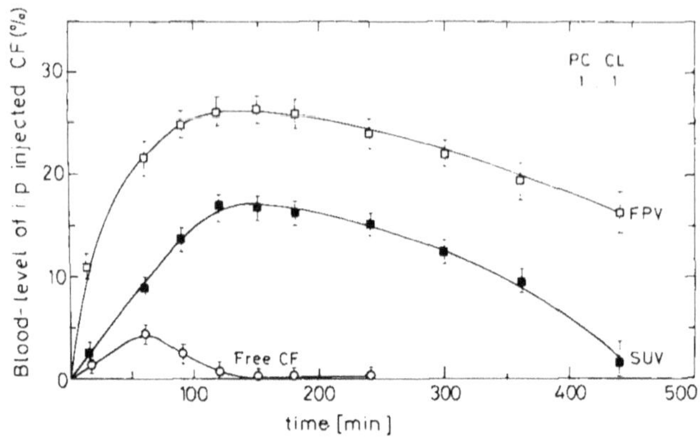

FIGURE 7. Time course of the blood level female Balb/c mice of intraperitoneally injected PC/Chol-liposomes containing 100 m*M* CF, 10 m*M* Tris, pH = 7.5. Experimental details in text.

samples were withdrawn via the tail vein of the animals and transferred into 1 mℓ of PBS. The samples were centrifuged for 10 sec in a Beckmann microfuge. The supernatant was transferred into glass cuvettes and the fluorescence determined before and after the addition of 0.2% (v/v) of sodium deoxycholate. Following the correction for vesicle binding to the red blood cells, serum-induced fluorescence quenching, and centrifugation effects, the half-times ($T_{1/2}$) of the vesicle clearance from the bloodstream were calculated as described elsewhere.[27]

The data are given as mean ± SD from quintuplicate samples of three independent experiments. $T_{1/2}$ for MLV was 8 ± 3 min, whereas $T_{1/2}$ for SUV was in the order of 130 ± 15 min. FPV, however, were cleared significantly ($p < 0.01$) slower; their $T_{1/2}$-value was found to be 248 ± 25 min.

In similar experiments, the blood levels of CF-containing MLV, SUV, and FPV composed of equimolar amounts of PC and Chol were measured upon intraperitoneal injection of 200 μℓ CF-containing liposomes (2 μmol total lipid) into female Balb/c mice. The withdrawal of blood samples and the calculation of the blood levels were identical to those for i.v. injection. As seen in Figure 7, the time course of the blood level upon injection of the various liposome preparations is biphasic and depends on the type of liposome. Of the total amount of injected SUV approximately 2% were detected in the bloodstream within 15 min. This value increased to 16% at 120 min and stayed nearly constant for the next 60 min; thereafter the blood content decreased to ≈1% of the injected dose within 7 hr. The blood level of injected FPV is similar, qualitatively; however there are significant quantitative differences to SUV. Approximately 11% of the administered FPV are found in the bloodstream at 15 min, and this value increases to 26% after 90 min. This blood level stays practically constant for up to 200 min, followed by gradual decrease to 16% at 420 min. The blood level of injected MLV, sized by French press extrusion at 2000 psi to ≈3000 to 5000 Å, is again biphasic, however quantitatively nearly one order of magnitude lower than that of SUV, indicating minimal passage of intact MLV of that size from the peritoneal cavity into the bloodstream (data not shown). The blood level of intraperitoneally injected free CF is also biphasic, however it reaches its maximum (4.5% of the total injected amount) within 60 min and rapidly decreases to <0.5% within 180 min.

The quantitative differences between SUV and FPV may be due to the increased stability of the latter type of vesicles, which is manifested in (1) increased amount of intact FPV reaching the blood circulation within the first 2 hr and (2) prolonged survival of FPV in the circulation.

VII. CONCLUSIONS

The preparation of liposomes by high-pressure extrusion in a French press is a convenient technique, which has several distinct advantages over other techniques:

1. By variation of the applied pressure, the size of the resulting multi-, oligo-, and/or unilamellar vesicles can readily be controlled. In addition the vesicle size can be manipulated by addition of cholesterol.
2. Unilamellar vesicles (FPV) obtained by extrusion at 20,000 psi are larger in size than sonicated SUV of the same lipid composition. As a direct consequence, their internal aqueous space is considerably larger and hence their entrapment efficiency, especially for larger macromolecules. In addition FPV are more stable than SUV, especially towards disintegration by serum proteins (high density lipoproteins, etc.).
3. The technique of high-pressure extrusion is comparatively a gentle one and permits the encapsulation into liposomes of a number of labile macromolecules, which lose their biological activity, if encapsulation is attempted by other techniques.

In spite of all these advantages, we would like to point out some biohazards, which may possibly occur, when using radioactive, carcinogenic, and/or otherwise toxic substances in an open system, like that of the French press. If such biohazardous materials are to be processed, the greatest care has to be taken to prevent contamination of personnel and machinery by accidental spilling of the fluid during the extrusion (cf. Section II). Following the use of such compounds, all parts of the French pressure cell have to be rinsed excessively and even more carefully than for common usage, to remove all traces of the biohazardous materials.

As a last, additional, point in using the French press, we would like to mention two biological applications, which directly relate to the preparation of closed vesicles:

1. Purified plasma membranes of whole cells and/or organelles, when extruded at 20,000 psi will yield small, tightly closed French pressed membrane vesicles (FPMV). During this vesiculation process, hydrophylic substances, e.g., CF, can be entrapped into FPMV, just like into liposomes. FPMV have been used recently to study fluxes across the plasma membrane of bovine chromaffin cells and to assay the fusion between isolated chromaffin granules and such plasma membrane vesicles.[28]
2. As a result of high-pressure extrusion in the French press, membrane-associated proteins can be sheared off the membrane matrix with high efficiency. We have recently observed, that upon fourfold extrusion of isolated bovine chromaffin cell plasma membranes at 20,000 psi, up to 90% of the membrane-associated peripheral proteins (e.g., actin, acetylcholine esterase, etc.) were removed from the vesicles, whereas intrinsic membrane proteins were essentially untouched.[29] In similar experiments Muller and Shinitzky[30] could isolate blood group antigens by high-pressure rigidification of erythrocytes in a French press.

ACKNOWLEDGMENTS

The author would like to acknowledge the contribution of several co-workers that participated in one or another aspect of the experiments described in this chapter: Jonathan E. Friedman, Dalia Goldenberg, Dalia Gurari-Rotman, Avraham Oplatka, Kurt Rosenheck, Ronit Sagi-Eisenberg, and Milton B. Yatvin. The skillful efficiency of Haim Great in preparing the drawings is greatly appreciated. Part of this research has been supported by research grants from the Israel Ministry of Absorption and Immigration and from the Israel Academy of Sciences and Humanities Basic Research Funds.

REFERENCES

1. **Milner, H. W., Lawrence, N. S., and French, C. S.,** Colloidal dispersion of chloroplast material, *Science*, 111, 633, 1950.
2. **Barenholzt, Y., Amselem, S., and Lichtenberg, D.,** A new method for preparation of phospholipid vesicles (liposomes) — French Press, *FEBS Lett.*, 99, 210, 1979.
3. **Hamilton, R. L., Jr., Goerke, J., Guo, L. S. S., Williams, M. C., and Havel, R. J.,** Unilamellar liposomes made with the French pressure cell; a simple preparative and semiquantitative technique, *J. Lipid Res.*, 21, 981, 1980.
4. **Mason, J. T. and Huang, C.,** Hydrodynamic analysis of egg phosphatidylcholine vesicles, *Ann. N.Y. Acad. Sci.*, 308, 29, 1978.
5. **Oyarzun, M. J., Clements, J. A., and Baritussio, A.,** Ventilation enhances pulmonary alveolar clearance of radioactive dipalmitoyl phosphatidylcholine in liposomes, *Am. Rev. Resp. Dis.*, 121, 709, 1980.
6. **Lichtenberg, D., Freire, E., Schmidt, C. F., Barenholz, Y., Felgner, P. L., and Thompson, T. E.,** Effect of surface curvature on stability, thermodynamic behavior, and osmotic activity in dipalmitoyl phosphatidylcholine single lamellar vesicles, *Biochemistry*, 20, 3462, 1981.
7. **Klein, O.,** M.Sc. thesis, The Weizmann Institute, Rehovot, Israel, 1983.
8. **Lelkes, P. I. and Friedman, F. E.,** Interaction of French pressed liposomes with isolated bovine adrenal chromaffin cells. I. characterization of the cell-liposomal interactions, submitted, 1983.
9. **Szoka, F., Jr. and Papahadjopoulos, D.,** Comparative properties and methods of preparation of lipid vesicles (liposomes), *Ann. Rev. Biophys. Bioeng.*, 9, 467, 1980.
10. **Weinstein, J. N., Yoshikami, N., Henkart, P., Blumenthal, R., and Hagins, W. A.,** Liposome-cell interaction. Transfer and intracellular release of a trapped fluorescent marker, *Science*, 195, 489, 1977.
11. **Rand, R. P.,** Interacting phospholipid bilayers: measured forces and induced structural changes, *Ann. Rev. Biophys. Bioeng.*, 10, 277, 1981.
12. **Israelachvili, J. N., Marcelja, S., and Horn, R. G.,** Physical principles of membrane organization, *Q. Rev. Biophys.*, 13, 121, 1980.
13. **Lelkes, P. I., Friedman, J. E., and Neeman, M.,** French Pressed liposomes as drug carrier systems, in *Liposomes, Drugs and Immunocompetent Functions*, Nicolau, C. and Paraf, A., Eds., Academic Press, London, 1981, 174.
14. **Van Renswoude, A. J. B. M. and Hoekstra, D.,** Cell induced leakage of liposome contents, *Biochemistry*, 20, 540, 1981.
15. **Lelkes, P. I., Friedman, J., and Sagi-Eisenberg, R.,** unpublished data, 1982.
16. **Lelkes, P. I.,** unpublished data, 1982.
17. **Lelkes, P. I.,** unpublished data, 1982.
18. **Gurari-Rotman, D. and Lelkes, P. I.,** Encapsulation of human fibroblast interferon activity in liposomes, *Biochem. Biophys. Res. Commun.*, 107, 136, 1982.
19. **Lelkes, P. I. and Yatvin, M. B.,** unpublished data, 1980.
20. **Lelkes, P. I., Sagi-Eisenberg, R., and Pecht, I.,** Fusion of liposomes with rat basophils: modulation of histamine release by internalization of vesicle contents, *Isr. J. Med. Sci.*, 18a, 13, 1982.
21. **Lelkes, P. I., Friedman, J. E., Oplatka, A., and Rosenheck, K.,** Interaction of French pressed liposomes with isolated bovine adrenal chromaffin cells. II. Modulation of calecholamine release and transmembrane potential, submitted, 1983
22. **Lelkes, P. I., Friedman, J. E., Rosenheck, K., and Oplatka, A.,** Liposomes as vectorial drug carriers in modulating stimulus-secretion coupling in bovine adrenal chromaffin cells, 4th Int. Conf. Surf. Coll. Sci. (Abstr.), Jerusalem, Israel, 1981, 396.
23. **Lelkes, P. I., Sagi-Eisenberg, R., and Pecht, I.,** manuscript in preparation, 1983.
24. **Geiger, B., Gitler, C., Calef, E., and Arnon, R.,** Dynamics of antibody and lectin-mediated endocytosis of hapten-containing liposomes by murine macrophages, *Eur. J. Immunol.*, 11, 710, 1981.
25. **Lelkes, P. I. and Tandeter, H. B.,** Studies on the methodology of the carboxyfluorescein assay and on the mechanism of liposome stabilization by red blood cells *in vitro*, *Biochim. Biophys. Acta*, 716, 410, 1982.
26. **Yatvin, M. B. and Lelkes, P. I.,** Clinical prospects for liposomes, *Med. Phys.*, 9, 149, 1982.
27. **Lelkes, P. I.,** Methodological aspects dealing with stability measurements of liposomes *in vitro* using the carboxyfluorescein-assay in *Liposome Technology*, Gregoriadis, G., Ed., CRC Press, Boca Raton, in press.
28. **Lelkes, P. I., Lavie, E., Naquira, D., Schneeweiss, F., Schneider, A., and Rosenheck, K.,** Acetylcholine induced *in vitro* fusion between cell membrane vesicles and chromaffin granules from the bovine adrenal medulla, *FEBS Lett.*, 115, 129, 1980.
29. **Lelkes, P. I., Naquira, D., Friedman, J. E., Rosenheck, K., and Schneider, A. S.,** Plasma membrane vesicles from bovine adrenal chromaffin cells: characterization and fusion with chromaffin granules, in *Advances in the Biosciences*, Vol. 36, Izumi, F., Ed., Perganom Press, New York, 1982, 143.

30. **Muller, C. P. and Shinitzky, M.,** Passive shedding of erythrocyte antigens induced by membrane rigidification, *Exp. Cell Res.*, 136, 53, 1981.

Chapter 6

FORMATION OF UNILAMELLAR LIPOSOMES BY pH ADJUSTMENT

Nigel Gains and Helmut Hauser

TABLE OF CONTENTS

I. SUMMARY

Small vesicles (< 600 Å) can be formed from mixed dispersions of phosphatidylcholine (PC) and phosphatidic acid (PA) provided that the molar proportion of PC is 70% or less. These vesicles are formed when the phospholipid mixtures are dispersed either directly in sodium hydroxide at pH ~10 or in water the pH of which is then rapidly (~ 1 sec) increased. Exposure of the phospholipids to high pH is short (< 2 min) and during this time no degradation is detectable by thin layer chromatography (TLC). The small vesicles can be separated from larger ones by centrifugation, by column chromatography on Sepharose 4B, and by filtration. The vesicles formed are at present being characterized by a combination of centrifugation, column chromatography, isotope encapsulation, NMR, electron microscopy, and quasielastic light scattering. Some of these data have been[1,2] or are to be[3] published. The results discussed here are mainly relevant to the efficient production of small unilamellar vesicles (SUV) from PA and mixtures of PA and PC. However these results also give an insight into how the experimental conditions can be manipulated to produce larger vesicles. Preliminary experiments indicate that vesicle size can be varied and predetermined; vesicle size is dependent on the acidic phospholipid used, the molar ratio of acidic phospholipid to PC, the ratio in the organic phase of counter ion to acidic phospholipid, and the rate and extent of the pH change. So far, PA is the only one of the phospholipids tested (phosphatidylserine, -inositol, -ethanolamine and -glycerol) that on its own and in combination with PC forms vesicles that are not sedimented by centrifugation at 160,000 *g* for 180 min.[4]

It has been shown elsewhere[4] that the vesicles formed from PA and mixtures of PA and PC encapsulate $^{22}Na^+$, $^{35}SO_4^{2-}$ and the fluorochromes harmaline and pyranine.[2] The half-time of efflux of $^{22}Na^+$ and $^{35}SO_4^{2-}$ is of the order of days.[2]

II. MATERIALS

Egg PA and egg PC were purchased from Lipid Products (South Nutfield, U.K.). Dilauroyl PA (disodium salt) was from R. Berchtold, Biochem. Lab. (Berne, Switzerland); Sepharose 4B from Pharmacia Fine Chemicals AB (Zürich, Switzerland); Permablend III from Packard Instrument International S.A. (Zürich, Switzerland). ^{3}H-dipalmitoyl phosphatidylcholine (DPPC) was tritiated by EIR, Würenlingen, Switzerland.

III. METHODS

A. Preparation of Vesicles

Vesicles were made either from PA or a mixture of PA and PC; 10 mg or less of these phospholipids dissolved in 1 to 2 mℓ of chloroform/methanol (2:1) and, where appropriate, containing 25 μg ^{3}H-DPPC as a label were rotary evaporated at room temperature and under reduced pressure in a 50 mℓ round-bottomed flask. The phospholipids were then dispersed in distilled water, a NaCl solution of the indicated concentration, or in 10 m*M* NaOH. Dispersion was achieved by shaking by hand with glass beads. For dispersions of PA alone it was also necessary to freeze thaw about six times from ice/methanol at ~ −15 to ~5°C; this gives a translucent, white dispersion in which some of the particles are just visible. The pH of the pure PA dispersions at 10 mg/mℓ was between 2.5 and 3. For PC-PA dispersions the initial pH is dependent on the amount of PA present. Previously[1] we described the use of a vortex mixer to help suspend dispersions of PA. However, this method should not be used if the dispersion contains hazardous material, as occasionally the glass beads perforate the round-bottomed flask and scatter the contents.

For PA and mixtures of PC and PA dispersed in water the initial pH was raised to a maximum value with NaOH added either slowly (~ 2 min) from an Agla Syringe (Wellcome,

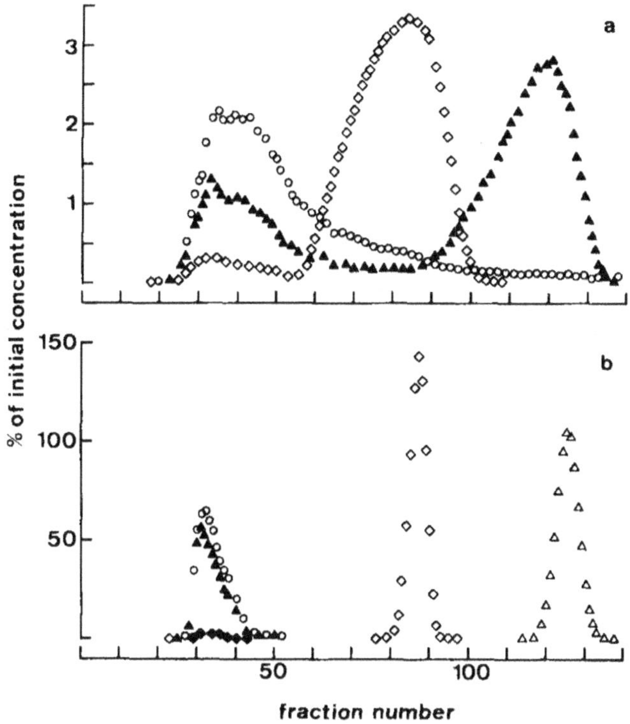

FIGURE 1. The effect of centrifugation on the elution of PC-PA vesicles (○) from Sepharose 4B. The PC to PA ratio was 6:1 (w/w). The samples were dispersed in distilled water containing $^{22}Na^+$ (△, ▲) and $^{35}SO_4^{2-}$ (◇, ◆), the pH was then raised slowly (~ 1 min) to pH 7.6. Sample a (250 μℓ at 10 mg/mℓ) was not centrifuged, Sample b (2.5 mℓ at 1.67 mg/mℓ) was centrifuged at 12,000 × *g* for 3 min before loading onto the column. In b the symbol ▲ corresponds to one tenth of the ordinate scale.

Beckenham, U.K.), or rapidly (~ 1 sec) from a micropipette, both into stirred dispersions. The rate of addition of sodium hydroxide from the Agla Syringe was not controlled. After addition of sodium hydroxide the pH was reduced from the maximum value to 7.6 by addition of hydrochloric acid. The exposure of the phospholipids to high pH was short, less than 2 min, and not sufficient to cause detectable degradation.

B. Column Chromatography: Experimental Details

Phospholipid dispersions were chromatographed at about 22°C on a Sepharose 4B column[5] (40 to 50 cm × 0.9 cm) with a flow rate of 2.5 to 3 mℓ/hr. This was maintained when the column was not in use. The elution buffer contained 0.02% sodium azide, 2 m*M* Hepes (N-2-hydroxyethylpiperazine-N′-2-ethane sulfonic acid) plus sodium hydroxide to pH 7.55. Before loading the samples (usually 1 mℓ) onto the column, they were centrifuged in 1.5-mℓ tubes for 3 min at 12,000 × *g*. This pellets more than 97% of the material in an unsonicated dispersion of PC at 10 mg/mℓ in distilled water. 300-μℓ fractions were collected in an automatic fraction collector. Before reproducible results can be obtained the Sepharose 4B must be saturated with the appropriate phospholipid mixture.

C. Concentration of Eluted Material

In Figures 1 to 3, the concentration of the eluted isotopes is expressed as a percentage (p) of their initial concentration in the sample applied to the column. It should be noted that Σp for all the fractions is equal to the (sample volume × 100) / fraction volume.

D. Justification of the Use of Sepharose 4B for the Chromatography of PC-PA Vesicles

The suitability of Sepharose 4B for separation and characterization of liposomes made by sonication of mixed dispersions of PC and PA has been questioned.[8] It was observed that the second elution peak (at about 0.5 of the total column volume) was not symmetrical but tailed severely. From this it was concluded that the phospholipids were adsorbed onto the Sepharose. The elution pattern of a PC-PA mixture (6:1, w/w; 250 $\mu\ell$ sample at 10 mg/mℓ) dispersed in water containing $^{22}Na^+$ and $^{35}SO_4^{2-}$ is shown in Figure 1a. The phospholipid peak and those of the encapsulated $^{35}SO_4^{2-}$ and $^{22}Na^+$ associated with it are asymmetric and give the impression that the elution conditions are not optimal. This is supported by the observation that the unencapsulated $^{22}Na^+$ and $^{35}SO_4^{2-}$ are eluted over about 0.4 column volumes. In contrast, when a mixture of PC and PA at the same weight ratio (6:1) is dispersed at 1.67 mg/mℓ (sample size 2.5 mℓ) and centrifuged (3 min at 12,000 $\times$ *g*) to remove undispersed material the peaks in the elution profile are well resolved (Figure 1b). The phospholipid peak is asymmetric but shows virtually no tailing, the unencapsulated $^{22}Na^+$ and $^{35}SO_4^{2-}$ are eluted over about 0.12 column volumes, and the three peaks are clearly separated. Adsorption of the phospholipids does occur,[5] but it is not the cause of the ill-resolved peaks in Figure 1a. This results from poorly dispersed material partially blocking the gel or the porous plug at the outlet of the column. The amount of material that is removable by centrifugation (12,000*g* for 3 min) but which can effectively disrupt elution is small ($\sim$ 5% for a 1-mℓ sample at 10 mg/mℓ).

Sepharose 4B absorbs PA and in so doing acquires a negative charge and the characteristics of a cation exchanger. In the absence of an adsorbed negative charge, anions and cations should elute from Sepharose 4B together at the total column volume. However, as can be seen in Figure 1 the unencapsulated $^{35}SO_4^{2-}$ and $^{22}Na^+$ are resolved into two distinct peaks, the former elutes before, and the latter after the total column volume. The cation-exchange properties of the PA equilibrated column are relatively stable; $^{35}SO_4^{2-}$ and $^{22}Na^+$ applied to the column 24 hr after a sample of PA has been eluted are resolved into two distinct peaks.

E. Measurement of Encapsulation

The ability of the vesiscles to encapsulate $^{22}Na^+$ and $^{35}SO_4^{2-}$ was measured. Both isotopes were, unless otherwise stated, included in the initial dispersion medium. Tritiated PC, $^{35}SO_4^{2-}$ and $^{22}Na^+$ were counted simultaneously using a scintillation fluid containing 2ℓ toluol, 1ℓ Triton® X-100, 11 g Permablend III, and 100 mℓ glacial acetic acid[6] in a Beckman LS 7500 scintillation counter. The fractions (300 $\mu\ell$) from column chromatography were collected in scintillation vials and 2.5 mℓ of the scintillation fluid was added.

F. Phosphate Determination

Phosphate was determined by using a modification of the procedure of Chen et al.[7]suitable for small sample volumes containing little organic material ($\sim$ 700 g / mol P_i). 20$\mu\ell$ samples were refluxed for 10 min at 100°C and then evaporated to dryness at 200°C in 1.3 mℓ soda glass tubes of 5-mm internal diameter. 25 $\mu\ell$ of concentrated H_2SO_4/70% $HClO_4$ (10:1, v/v; it is not known if this mixture is potentially more hazardous than its individual components) were added. As soon as the yellow color, if any, had disappeared ($<$ 10 min), the temperature was allowed to cool to 45°C, 0.8 mℓ of a freshly mixed solution containing 1 volume 0.4% ammonium molybdate and 1 volume 1.67% ascorbic acid was then added. The reagent and phosphate containing acid were mixed within 2 min, by sucking into a pasteur pipette. The blue color was allowed to develop at 45°C; this is complete within 45 min and is then stable for several hours. The optical density was read at 820 nm.

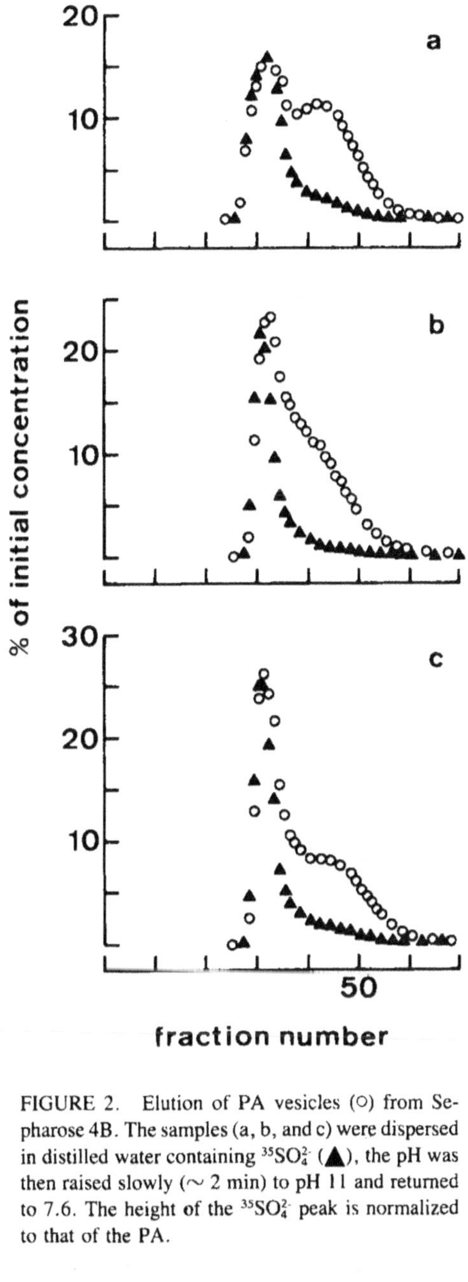

FIGURE 2. Elution of PA vesicles (○) from Sepharose 4B. The samples (a, b, and c) were dispersed in distilled water containing $^{35}SO_4^{2-}$ (▲), the pH was then raised slowly (~ 2 min) to pH 11 and returned to 7.6. The height of the $^{35}SO_4^{2-}$ peak is normalized to that of the PA.

IV. VESICULATION OF PA AND PA-PC DISPERSIONS

A. Effect of the Rate of Addition of Sodium Hydroxide on the Size of PA Vesicles

The elution profiles of three samples dispersed in water are shown in Figure 2. The pH of these dispersions was raised (to pH 11) over a period of 1 to 2 min, by the addition of sodium hydroxide from an Agla Syringe, and then returned to pH 7.6 before gel filtration. Two distinct, but not well-resolved, peaks of PA can be seen in Figure 2a. The peak eluting at or near the void volume (fraction 31 and 32) is referred to as the first vesicle peak, that eluting between fractions 45 and 50 as the second vesicle peak. In Figure 2b and c the PA vesicle peaks are less well resolved. The only major variable in the preparation of the samples in Figure 2 was the uncontrolled rate of sodium hydroxide addition. This would, therefore,

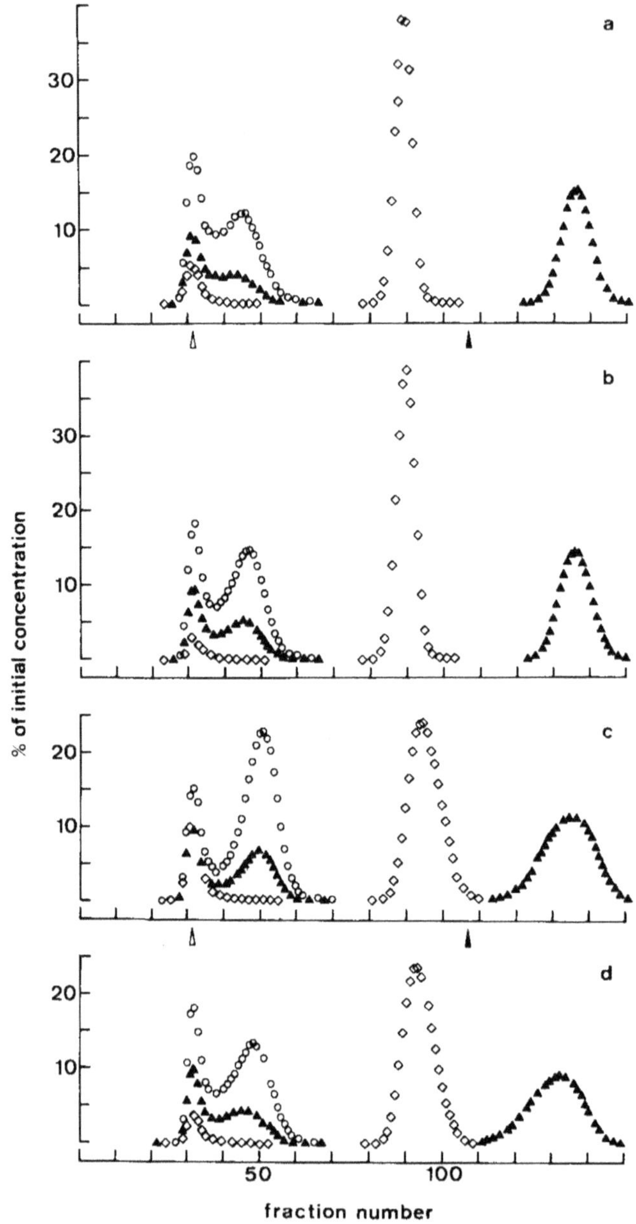

FIGURE 3. Elution of PA vesicles (○), $^{35}SO_4^{2-}$ (◇) and $^{22}Na^+$ (▲) from Sepharose 4B. The samples were dispersed in distilled water containing $^{22}Na^+$ and $^{35}SO_4^{2-}$, the pH was then raised rapidly (~ 1 sec) to 7.9 (+ 16 m*M* NaOH) in a, to 9.0 (+ 17 m*M* NaOH) in b, to 12.2 (+ 30 m*M* NaOH) in c and to 8.6 (+ 16 m*M* NaOH) in d, and returned to 7.6. To Sample d, 14 m*M* sodium chloride was added after the pH had been raised. The void volume (fraction 31 or 32) is indicated by △ and the total column volume (fraction 107) by ▲.

seem to effect either or both the volume at which the smaller vesicles elute and the fraction of the smaller vesicles formed. The latter can be approximately determined. The $^{35}SO_4^{2-}$ peak elutes almost entirely with the first PA peak (Figure 2a). Assuming that these two peaks have the same shape, then it is possible to determine, approximately, the fraction of the PA that is present as small vesicles. This can be done by scaling the $^{35}SO_4^{2-}$ peak to the same height as the first PA peak (Figure 2, a to c) and then subtracting the area of the $^{35}SO_4^{2-}$ peak (A_s) from the area of the total PA elution profile (A_t-A_s) and finally by dividing this by the total area of the PA elution profile (1-A_s/A_t). As these values (a, 55%; b, 50%; c, 57%) are similar for each of the three elution profiles the decrease in the resolution of the two PA peaks must be caused by a decrease in the elution volume of the second peak. Further, as the conditions under which the vesicles were formed and chromatographed were very similar, then the difference in the elution volume indicates that the size of the vesicles is somehow dependent on the rate of addition of the sodium hydroxide. This was tested by the rapid addition of the sodium hydroxide, the mixing time being about 1 sec. Elution profiles of various samples treated in this way are shown in Figure 3. These samples differ from one another in that different quantities of sodium hydroxide were added to give different maximum pH values. The maximim pH value of the samples in Figure 2 was 11 and this value lies between the maximum values of the samples in Figure 3b (pH 9.0) and 3c (pH 12.2). Although, as can be seen in Figure 3 (a to c), the separation of the two peaks increases as the maximum pH value is increased, it can also be seen that the separation achieved after rapidly (~ 1 sec) raising the pH to 9 is better than that achieved after slowly (22 min) raising the pH to 11 (Figure 2, a to c). The percentage vesiculation is also increased from ~55% for the samples in Figure 2 to 75 and 80% for those in Figure 3b and c, respectively. The dependency of vesicle size on rate of sodium hydroxide addition has not yet been investigated in detail.

B. Effect of the Maximum pH on the Size of PA Vesicles

The position of the second vesicle peak in the elution profile is dependent not only on the rate of sodium hydroxide addition but also on the maximum pH value. The fractional elution volume of the small vesicles in Figure 3 (a to c) (expressed as a fraction of the total column volume) is, respectively, 0.42 at a maximum pH value of 7.9, 0.43 at pH 9.0, and 0.47 at pH 12.2. However, in addition to its pH dependence, the fractional elution volume is also dependent on the sodium ion concentration. Although 16 m*M* sodium hydroxide was added to both the sample in Figure 3a and d, 14 m*M* sodium chloride was also added to that in Figure 3d after the pH had been raised. This increases the fractional elution volume from 0.42 (for 16 m*M* NaOH) to 0.45 (for 16 m*M* NaOH plus 14 m*M* NaCl). However, the fractional elution volume (0.47) is larger when the same concentration of sodium ions (30 m*M*) is added directly as sodium hydroxide (Figure 3c). From this it is apparent that as the maximum pH value is increased, the size of the vesicles is decreased. Further, as the sodium ion concentration is increased, the repulsion between the negative charges of the PA vesicles and those of the PA adsorbed onto the Sepharose is decreased.

C. Encapsulation of $^{22}Na^+$ and $^{35}SO_4^{2-}$ and its Relevance to Vesicle Size

The PA samples chromatographed in Figure 3 were dispersed in water containing $^{22}Na^+$ and $^{35}SO_4^{2-}$. From the amount of these isotopes that is encapsulated by the vesicles (and therefore co-elutes with them) it is possible to determine an approximate size for the vesicles. The theoretical relationship between the encapsulated aqueous volume (V), in liters per mole, and the external radius (r), in Å, of the vesicles is given by

$$V = A\,(r - d)^3/30((r - d)^2 + r^2) \qquad (1)$$

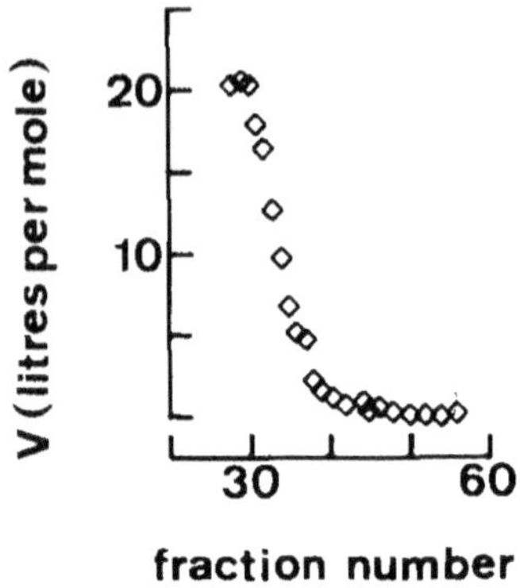

FIGURE 4. Internal aqueous volume of PA vesicles determined by $^{35}SO_4^{2-}$ encapsulation. The data from Figure 3a are plotted as the percentage of the initial $^{35}SO_4^{2-} \times [PA]^{-1}(V)$, in liters per mole, in each fraction against the fraction number.

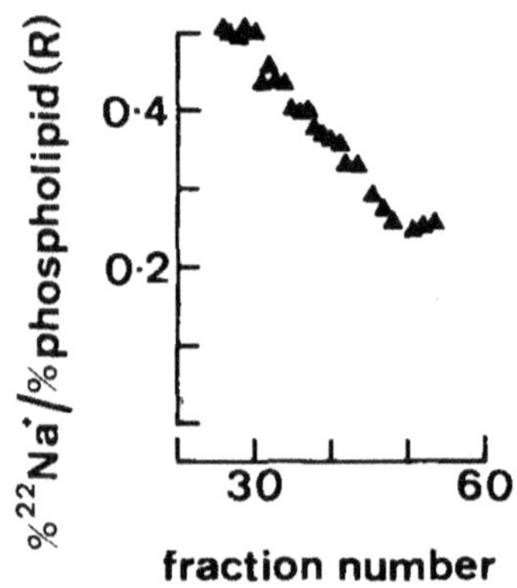

FIGURE 5. Encapsulation of sodium ions by PA vesicles. The data from Figure 3a are plotted as the ratio (R) of the percent of the total $^{22}Na^{+}$ to that of PA (derived from that of the ^{3}H-PC marker) in each fraction against fraction number.

where A is 0.42 km^2, the area occupied by 1 mol of phospholipid packed at 70 Å^2 per molecule and d is the bilayer thickness, taken as 40 Å.[8,9] For this calculation it is necessary to assume that the molecular packing density is equal in both halves of the vesicle bilayer. This assumption and these values of A and d are used in Equations 2 and 3.

If it is assumed that the percent encapsulation of water is the same as that of $^{35}SO_4^{2-}$, then the encapsulated aqueous volume (V) can be calculated. In any given fraction this volume, in liters per mole, is equal to the ratio of the percent of the total $^{35}SO_4^{2-}$ present to that of ^{3}H-PC divided by the initial PA concentration. This has been done for the ^{3}H-PC-containing fractions in Figure 3a and is plotted against fraction number in Figure 4. The encapsulated volume (V) decreases with increasing fraction number from a value of 17 at the first peak to a limiting value of 0.23 l (of $^{35}SO_4^{2-}$ containing water per mole of PA) for the second

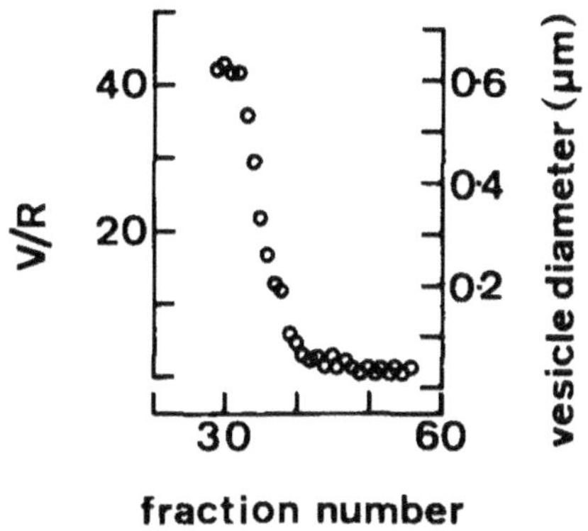

FIGURE 6. Variation of average vesicle diameter with elution volume. The ratio V/R is calculated from the values plotted in Figures 4 and 5. The relationship between this ratio and vesicle diameter is given in the text.

vesicle peak. From these and other such data and by using Equation 1 to generate a standard curve an external diameter of 400 to 1700 nm can be calculated for the vesicles in the first peak and one of about 300 Å for the vesicles in the second peak. These calculations require the assumption that the vesicles are unilamellar. This assumption is supported by the $^{22}Na^+$ encapsulation data which is discussed below.

The theoretical relationship between the ratio (R) of the inner to the total surface area of the vesicles and their external radius (r) is given by

$$R = (r - d)^2/((r - d)^2 + r^2) \tag{2}$$

The high efficiency of $^{22}Na^+$ encapsulation, about 40 times that of $^{35}SO_4^{2-}$, indicates that in the initial dispersion virtually all the $^{22}Na^+$ (~ 3 n*M*) is bound to the phosphatidic acid (15 m*M*) and remains so during vesiculation. If this is assumed, then an experimental value for the internal to the total surface area of the vesicles can be derived from the ratio (R) of the percent $^{22}Na^+$ to percent ^{3}H-PC present in any given fraction. This has been done for the ^{3}H-PC containing fractions in Figure 3a and is plotted against fraction number in Figure 5. The ratio for the vesicles in the first PA peak is about 0.5 and indicates that the vesicles have about half the total PA molecules on their inner surface and that they are unilameller. A ratio of 0.5 was found provided that the maximum pH value was between 8 and 10. At higher maximum pH values the ratio for the vesicles at the void volume was ~ 0.7; this indicates that a proportion of the vesicles were not unilamellar. In Figure 5 the initial ratio decreases to a limiting value of 0.25. From this and other similar data and by using Equation 2 to generate a standard curve, an external diameter of 200 Å can be calculated for the smaller vesicles.

By combining Equations 1 and 2 it can be shown that the external vesicle radius (r), in Å, is directly proportional to V/R, provided that the vesicles are homogeneous in size.

$$r = (30/A) \cdot (V/R) + d \tag{3}$$

The experimentally determined ratio V/R is plotted in Figure 6 against fraction number. The ordinate scale on the right-hand side gives the theoretical external vesicle diameter calculated from Equation 3. The ratio at or near the void volume indicates an average vesicle size of about 600 nm, which decreases to a limiting value of 220 Å for the vesicles in the second peak.

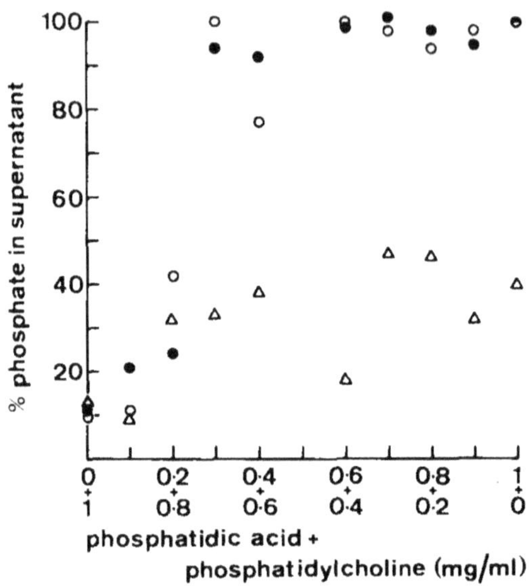

FIGURE 7. Centrifugation of PC-PA.dispersions (160,000 × *g* for 180 min). ●, 1 mℓ of 10 m*M* sodium hydroxide was added directly to 1 mg of the phospholipid mixture. ○, after dispersion of the phospholipid mixture (1 mg) in water (1 mℓ) sodium hydroxide was rapidly added (~ 1 sec) to give a pH of 10.5 to 11.0. For both of the above, the pH was then adjusted to 7.5. △, after dispersion of the phospholipids (1 mg) in water (1 mℓ) the pH was raised slowly (~ 1 min) to pH 7.5. After centrifugation the phosphate-containing material remaining in the supernatant was determined.

D. Centrifugation and Filtration as an Indication of Vesicle Size

Sonicated dispersions of PC can be resolved by Sepharose 4B column chromatography into two vesicle populations. The one eluting at or near the void volume is precipitable by centrifugation at 160,000 × *g* for 3 hr, the other eluting at ~0.5 column volumes is not.[4] This has also been found for PA dispersions the pH of which has been raised to 10 to 11 and returned to 7.6[1] The results shown in Figure 7 are for mixed dispersions of PC and PA. The phospholipids were either dispersed in water and the pH then raised slowly (~ 1 min) to 7.6, or rapidly (~ 1 sec) to 10.5 to 11.0, or were dispersed in 10 m*M* sodium hydroxide; the pH after addition of sodium hydroxide was returned to 7.6. The percentage of small vesicles formed was taken as the amount of phosphate remaining in the supernatant after centrifugation. Slowly raising the pH to 7.6 results in a less than 50% yield of small vesicles. Rapidly raising the pH, or dispersion directly in sodium hydroxide, results in a ~95% yield of small vesicles, provided that the PA content is 30% or more. This is consistent with the data obtained from NMR.[3]

The filtration data plotted in Figure 8 are of two mixed dispersions of PA and PC, one at a weight ratio of 0.2 to 0.8 and the other of 0.8 to 0.2. The dried phospholipid mixture (1 mg) was dispersed in 1 mℓ of 10 m*M* sodium hydroxide. The vesicles formed from the mixture of 0.8 PA to 0.2 PC are smaller with about 50% of the phospholipid passing through filters with a mean pore size of 100 and 500 Å and 80 to 90% through filters of 2000 and 2200 Å. In contrast, about 50% of the phospholipid present in the 0.2 PA to 0.8 PC vesicles passes through a filter with a mean pore size of 6500 Å.

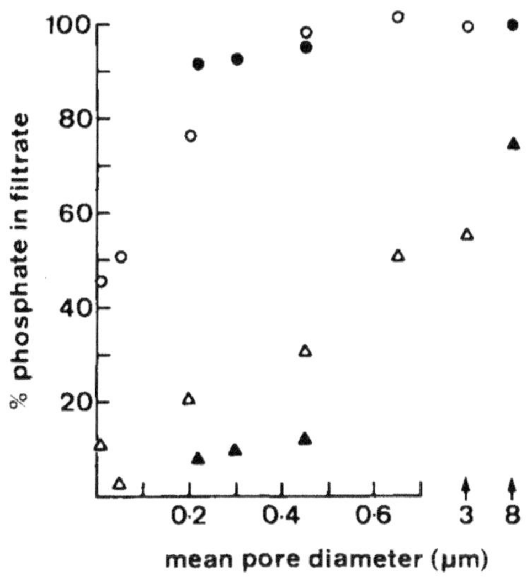

FIGURE 8. Filtration of PC-PA mixed dispersions through cellulose filters of various average pore size. 1 mℓ of 10 m*M* sodium hydroxide was added directly to the rotary evapoprated phospholipids (PA plus PC: ●, ○ 0.8 + 0.2 mg; ▲, △ 0.2 + 0.8 mg, respectively), and after readjusting the pH to 7.5 the dispersions were filtered through Millipore (●, ▲) or Sartorius (○, △) membranes. The phosphate-containing material in the filtrate was determined.

V. CONCLUSION

The title of our chapter suggests that unilamellar vesicles can in general be formed merely by adjusting the pH. However, the technique is almost certainly limited to charged phospholipids and mixtures of these with neutral phospholipids. So far only PA, with two potential negative charges, and mixtures of this with PC have been shown to produce SUV (< 600 Å). However, preliminary work indicates that the technique produces somewhat larger vesicles from acidic phospholipids that have a single negative charge. Although the molecular rearrangements that occur during vesiculation are not known, it seems probably that the size of the resulting vesicles is influenced by both the extent and the rate of change of the surface charge that is induced on increasing the pH.

ACKNOWLEDGMENTS

This work was supported by the Swiss National Science Foundation (Grant No. 3.570-0.79).

REFERENCES

1. **Hauser, H. and Gains, N.,** Spontaneous vesiculation of phospholipids: a simple and quick method of forming unilamellar vesicles, *Proc. Natl. Acad. Sci. U.S.A.*, 79, 1683, 1982.
2. **Gains, N. and Hauser, H.,** Characterization of small unilamellar vesicles produced in unsonicated phosphatidic acid and phosphatidylcholine-phosphatidic acid dispersions by pH adjustment, *Biochem. Biophys. Acta*, 731, 31, 1983.

3. **Hauser, H., Gains, N., and Müller, M.,** Vesiculation of unsonicated phospholipid dispersions containing phosphatidic acid by pH adjustment: physicochemical properties of the resulting unilamellar vesicles, *Biochemistry,* in press, 1983.
4. **Bahrenholz, Y., Gibbes, D., Litman, B. J., Goll, J., Thompson, T. E., and Carlson, F. D.,** A simple method for the preparation of homogeneous phospholipid vesicles, *Biochemistry,* 16, 2806, 1977.
5. **Huang, C.,** Studies on phosphatidylcholine vesicles. Formation and physical characteristics, *Biochemistry,* 9, 344, 1969.
6. **Pande, S. V.,** Liquid scintillation counting of aqueous samples using Triton-containing scintillants, *Anal. Biochem.,* 74, 25, 1976.
7. **Chen, P. S., Toribara, T. Y., and Warner, H.,** Microdetermination of phosphorus, *Anal. Chem.,* 28, 1756, 1956.
8. **Johnson, S. M.,** The effect of charge and cholesterol on the size and thickness of sonicated phospholipid vesicles, *Biochim. Biophys. Acta,* 307, 27,1973.
9. **Chapman, D., Williams, R. M., and Ladbrooke, B. D.,** Physical studies of phospholipids. VI. Thermothropic and lyotropic mesomorphism of some 1,2-diacyl-phosphatidylcholines (lecithins), *Chem. Phys. Lipids,* 1, 445, 1967.

Chapter 7

THE PREPARATION OF VARIABLY SIZED HOMOGENEOUS LIPOSOMES FOR LABORATORY, CLINICAL, AND INDUSTRIAL USE BY CONTROLLED DETERGENT DIALYSIS

Hans Georg Weder and Othmar Zumbuehl

TABLE OF CONTENTS

I. DETERGENT DIALYSIS: PRINCIPLE AND EQUIPMENT

A. Principle and Development of the Detergent Dialysis Method

According to Carey and Small,[1] phospholipids belong to the class of polar lipids and in particular the insoluble swelling amphiphiles. In water, these substances form stable mono layers at the water-air interface, whereas they form liquid crystals in the bulk phase. Liquid crystals are lipid bilayers with intervening water layers. These phospholipids form mixed micelles with soluble amphiphiles, called detergents. One group of these detergents forms liquid crystals in very high concentration and includes sodium and potassium salts of long-chain fatty acids and many of the ordinary anionic, cationic, and nonionic detergents. Another group does not form liquid crystals in water even at the highest concentrations possible and it includes bile salts among others. Figure 1 shows a phospholipid bile salt mixed micelle according to Mazer et al.[2]

It was first demonstrated by Kagawa and Racker[3] that the removal of the detergent from a mixed micellar solution ends up in the formation of liposomes. Subsequently, a series of methods for detergent removal were published.[4-8]

B. Production of a Mixed Micellar Solution

Table 1 gives the required conditions to be observed for the preparation of mixed micelles from various lipid compositions. This procedure begins with the formation of the lipid film containing the hydrophobic substances on a rotatory evaporator. For the preparation of egg yolk lecithin (EYL)/cholesterol mixed micelles using the detergent cholate and for some other lipid compositions, it is important to dissolve the detergent in the organic solvent. For the preparation of mixed micelles from synthetic lipids and some additives, for instance modified corticosteroids, it is necessary to prepare the lipid film about 10°C above the corresponding gel to liquid phase transition.

The required amount of detergent for the formation of a clear mixed micellar solution depends on the lipids to be solubilized and on their concentration. In our laboratory, sodium cholate and *n*-alkyl-β-D-glucopyranosides are used as detergents. Other detergents such as deoxycholate, sodium dodecylsulfate, Triton® X-100, trimethylammonium bromide, and pluronic proved to be useless because of inadequate solubilization capacity or too slow dialysis kinetics. After evaporation, the lipid film is dissolved or suspended in 10 m*M* phosphate buffer pH 7.1, containing 0.85% sodium chloride. The ionic strength amounts to 0.16. A different ionic strength leads to a different critical micellar concentration and therefore to a different solubilization capacity. If indicated, the detergent is dissolved in the buffer (see Table 1). During the evaporation of the organic solvent, the temperature should not be raised too much in order to prevent lipid oxidation. The buffer should be warmed up above the gel to liquid crystalline phase transition before the formation of the mixed micelles. Several mixed micellar solutions become immediately clear, especially those of *n*-alkyl-glucosides. High lipid to cholate molar ratios need more time. The stability of the mixed micellar solutions depends on the lipid composition. Stability for several days is normal, although mixed micelles composed of EYL, cholesterol (chol), and cholate show a precipitate after about 1 day.

C. The Dialysis Equipment Lipoprep®

The continuous detergent removal from equilibrated solutions of mixed lipid/detergent micelles, simple detergent micelles, and detergent monomers leads spontaneously to the formation of liposomes, provided that the intermicellar detergent concentration (IMC) touches a critical value. A heterogeneous population of liposomes is obtained, if the commonly used dialysis sac is applied. In our laboratory, we developed a flow-through dialyzer allowing a controlled and reproducible detergent removal at a defined temperature leading to a ho-

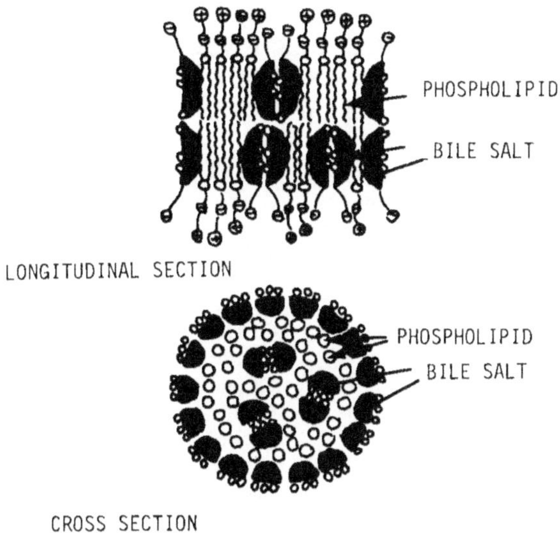

FIGURE 1. "Mixed disc" model of detergent (bile salt)-phospholipid mixed micelle according to Mazer et al. A lecithin bilayer disc is covered by bile salt molecules, which can be formed also in the interior of the bilayer. (From Mazer, N. A., Benedek, G. B., and Carey, M. C., *Biochemistry*, 19, 601, 1980. With permission.)

mogeneous population of liposomes with a wide range of vesicle diameter. A commercial apparatus, the Lipoprep®*), to form vesicles in this fashion has recently become available.

The principle is schematically shown in Figure 2. The features of the Lipoprep® equipment shown in Figure 3A are the high rate thermostatically controlled flow-through dialyzer, the microprocessor controlled temperature with digital selection operating in the range of 7 to 65°C, the chemically and physiologically inert materials of construction, and the easy operation and maintenance. Figure 3B shows the disassembled triple-cell dialyzer with the two flow-through cells and the middle cell with magnetic bar to prevent the formation of detergent concentration gradients during the dialysis process. The capacity of the middle cell is 8 mℓ with a lower limit of 4 mℓ. Two meshes protect the semipermeable membranes against mechanical destruction. The mixed micellar solution (preheated up to some degrees above transition temperature of the constituent lipids) is injected with a sampler. The dialysis fluid, e.g., buffer, always kept at ambient temperature, is first pumped through a heat regenerator and passes the flow-through cells at a defined temperature. The prepared liposomes leave the middle cell through an outlet tube practically without loss. The cleaning of the middle cell is carried out by flushing with water (several injections) and that of the flow-through cells by pumping water for about 10 min with the highest possible flow rate.

For clinical and industrial purposes we scaled up the above described laboratory technique. The detergent removal is performed with a counter-current dialyzer. Its special feature is a channel system with low cross-section. Five helically grooved discs for diffusate and dialysate with a total channel length of 36 m each are used to achieve a controllable high rate dialysis. Figure 4A shows the counter-current dialyzer unit of the pilot plant of which the ingenious sealing system without washers is worth mentioning. An air-thermostat and an electrical heat regenerator for diffusate and dialysate, both microprocessor controlled, allow the di-

* Manufacture: Diachema AG, CH-8135 Langnau a.A./Zurich, Switzerland. International sales: Dianorm-Geräte, P.O. Box 126, D-8000 München 65, Federal Republic of Germany.

Table 1
FORMATION OF MIXED MICELLES: REQUIRED MANNER OF DETERGENT ADDITION, TEMPERATURE, AND TIME OF MICELLE FORMATION FOR VARIOUS LIPID COMPOSITIONS

Lipid composition (molar ratio)	Lipid concentration (mg/mℓ)	Detergent	L/D[a]	Detergent[b] addition	TF[c] (°C)	TIF[d] (hr)	Ref.
EYL	13.0	Cholate	0.60	B	20	6	33
EYL	13.0	Cholate	0.25	B	20	2	33
EYL	13.0	Cholate	1.15	B	20	150	33
EYL/chol (7:3)	11.8	Cholate	0.51	A	20	i.	33
EYL/chol (8:2)	11.1	Cholate	0.60	A	20	i.	33
EYL/ganglioside(94.6)	14.6	Cholate	0.64	A	20	i.	33
EYL/PE (3:7)	12.4	Cholate	0.22	B	80	0.2	33
EYL/PI (8:2)	13.2	Cholate	0.60	B	20	3	33
EYL/PA (10:2)	15.3	Cholate	0.62	B	20	1	33
EYL/SA (10.2)	13.9	Cholate	0.62	B	20	6	33
EYL/cerebroside (10:1.5)	14.7	Cholate	0.68	A	20	i.	33
EYL/chol/PA (7:2:1) + sp2/HL antibody	8.2	Deoxy-cholate	0.70	B	20	i.	33
EYL/PA/HCP (7.3:1.8:0.85)	32.2	Cholate	0.60	C	20	i.	16
EYL/PA/PDP (8,63:0.95:0.41)	31.0	Cholate	0.60	B	20	i.	16
EYL/PA/BMV (10:1:1)	21.0	Cholate	0.60	B	20	i.	15
DMPC/PI (10:2)	12.0	Cholate	1.25	B	45	i.	15
DPPC/PA (10:1)	20.0	Cholate	0.20	B	50	i.	15
DPPC/DMPA (9:1)	15.0	Cholate	0.80	B	50	i.	15
DSPC/DCP (9:1)	15.3	Cholate	0.20	B	65	i.	15
DPPC/DMPA/BMV (10:1:1)	10.6	Cholate	1.00	A,B	50	i.	15
DPPC/DMPA/BMP (8:1:1)	11.0	Cholate	0.50	C	50	i.	15
DPPC/DMPA/PDP (8:1:1)	11.0	Cholate	0.80	C	50	i.	15
DPPC/DMPA/HCP (9:1:1.5)	17.0	Cholate	0.80	C	50	i.	16
EYL	10.0	[e]	0.2	B	20	i.	11
EYL	10.0	[f]	0.26/ 0.60	B	20	i.	11
EYL	10.0	[f]	0.40/ 0.35	B	20	i.	11
EYL	10.0	[f]	0.80/ 0.17	B	20	i.	11
EYL	10.0	[g]	0.13	B	20	i.	11
EYL	10.0	[h]	0.05	B	20	i.	11
Hydrogenated soya lecithin/ DMPA (10:1)	10.9	[e]	0.2	A	50	i.	9
DSPC/DCP (9.5:0.5)	10.3	[e]	0.2	B	60	i.	17
DSCP/Chol/DCP (8:1:1)	9.4	[g]	0.13	B	48	i.	17

[a] Molar lipid/detergent ratio.
[b] A, The detergent was dissolved together with the lipids in the organic solvent; B, the detergent was dissolved in the buffer prior to its addition to the lipid phase; C, the lipid mixture was suspended in buffer and the MLV were sonified during 1 hr at appropriate temperature. The detergent was added to the sonified suspension.
[c] Temperature of formation.
[d] Time of formation.
[e] *n*-Octylglucoside.
[f] *n*-Octyl/*n*-heptylglucoside.
[g] *n*-Heptylglucoside.

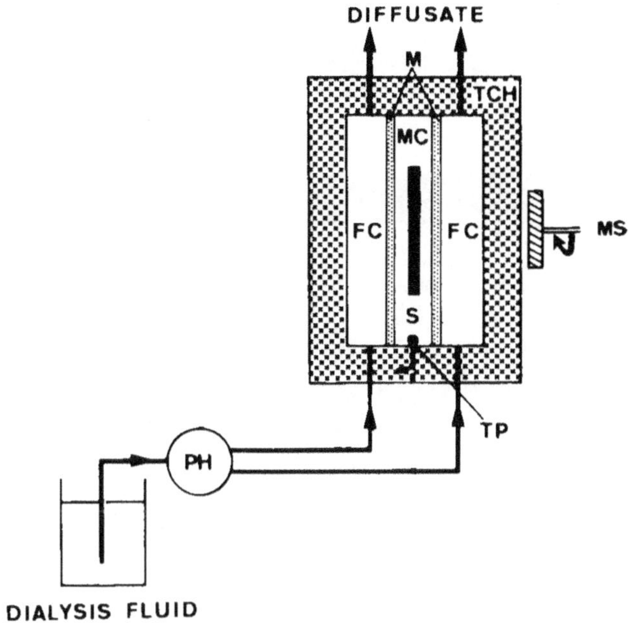

FIGURE 2. Instrumental set-up of the triple-cell dialyzer: Two flow-through cells (FC) and the middle cell (MC), stirring bar (S), micro temperature probe (TP), temperature controlled cell holder (TCH), magnetic stirrer (MS), and two pump heads (PH).

alyzer to operate at a defined temperature in the range of 20 to 65°C. High performance peristaltic pumps with digital operation are responsible for the transport of the mixed micellar solution (diffusate) and the dialysis fluid (dialysate). Figure 4B gives an impression of the semiautomatic pilot plant for the continuous preparation of liposomes which are in no way inferior to those prepared with the laboratory sized apparatus. The minimal production volume is 10 mℓ of liposome suspension. As the process is a continuous one, no upper limit exists. The production or output capacity amounts to 60 to 120 mℓ/hr, whereby the consumption of dialysis fluid per hour is twice as high. The commercial pilot plant, Lipoprep®-CLS, has now become available*).

Table 2 summarizes the specifications of the semipermeable membranes commonly used for liposome preparations. Experience has shown that these commercially available membranes are the most suitable for a fast and reproducible detergent removal. Practically no lipid loss occurs by adsorption to the membrane. Soaking and repeated washing (at least three times) in the buffer used is recommended. Routinely, the triple cell dialyzer and the counter-current dialyzer are assembled with wet membranes. In case of the counter-current dialyzer the membranes are multiusable over weeks.

Using the pilot plant first-order kinetics for both detergents are obtained, but for the Lipoprep® only *n*-octylglucoside shows a first-order process. In contrast, the first-order process of the cholate during about 80 min is followed by a process of higher order. It is evident from Figure 5 that the preparation of liposomes containing less than 2% residual detergent takes about 3.5 hr with the pilot plant whereas about 24 hr are necessary with the Lipoprep.® In both cases, extremely homogeneous liposome populations are produced. As a result of the faster dialysis kinetics of the pilot plant, the vesicle diameter is decreased

* Manufacture: Diachema AG, CH-8135 Langnau a.A./Zurich, Switzerland. International sales: Dianorm-Geräte, P.O. Box 126, D-8000 München 65, Federal Republic of Germany.

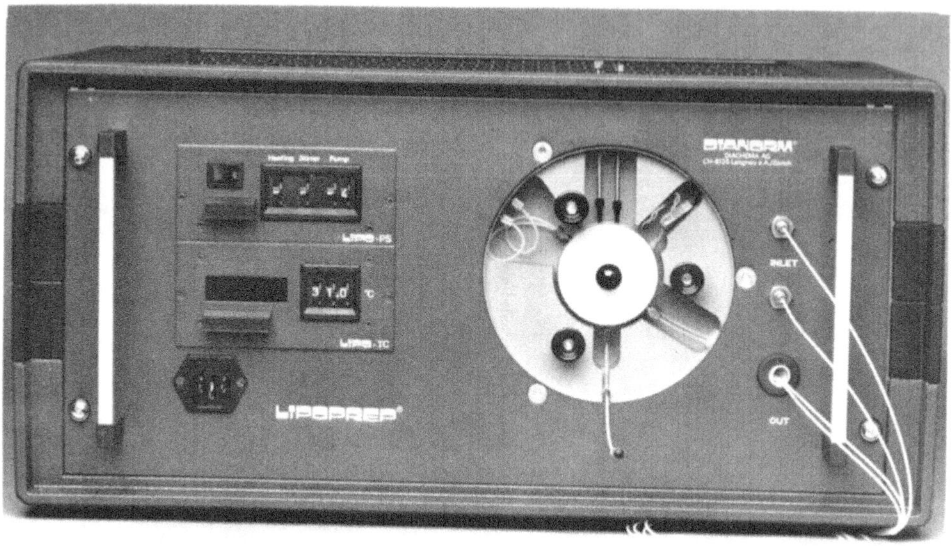

A

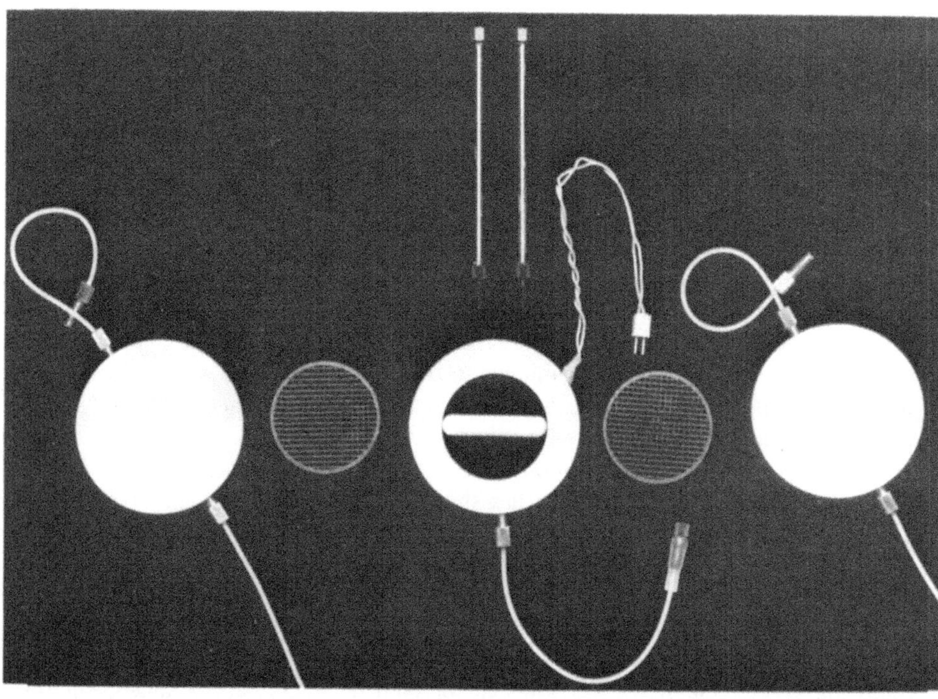

B

FIGURE 3. (A) Front view of the laboratory equipment Lipoprep.® (B) Disassembled triple-cell dialyzer.

compared to that of vesicles prepared by the Lipoprep.® This could also be shown with hydrogenated soya lecithin (pilot plant 135 ± 5 nm; Lipoprep®, 89 ± 3 nm diameter) produced at 47°C. The physicochemical properties of liposomes prepared with the pilot plant remain unchanged during the continuous production process. Furthermore the time of production is not limited. Depending on the lipid composition, the lipid loss ranges from 5 to 10% in the case of the Lipoprep® and less than 5% for the pilot plant.

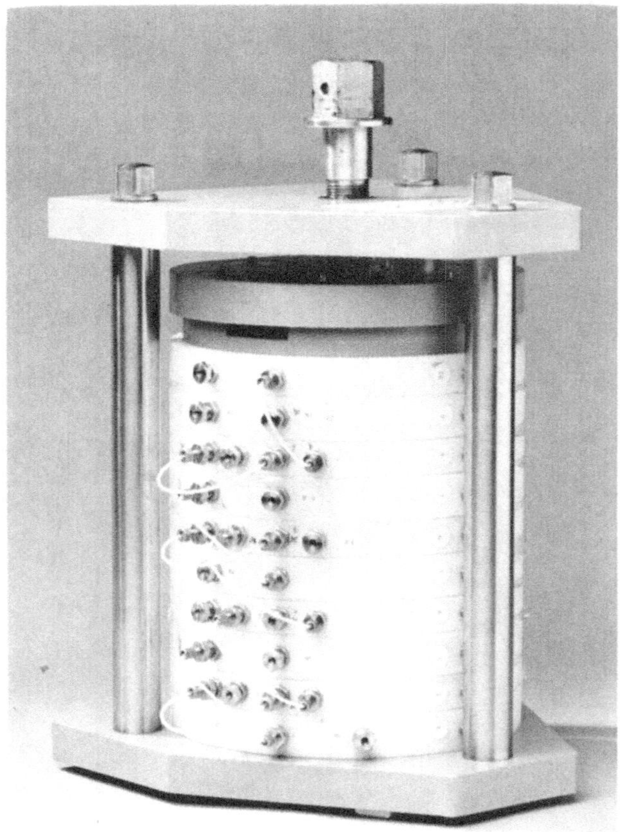

A

FIGURE 4. (A) Counter-current dialyzer of the liposome pilot plant (B) Front view of the liposome pilot plant.

D. Materials and Physicochemical Characterization of the Liposomes

Cholesterol (Chol), stearylamine (SA), dipalmitoyl-, distearoyl-, and dimyristoyl phosphatidylcholine (DPPC, DSPC, DMPC, respectively), and phosphatidic acid (PA) were obtained from Fluka, Buchs, Switzerland; ganglioside type III, cerebroside type II, dicetyl phosphate (DCP) and *n*-octyl-β-D-glucopyranoside from Sigma, phosphatidyl ethanolamine (PE) from egg yolk, phosphatidyl inositol (PI) from wheat germ and PA from Lipid products Ltd., Nutfield, U.K. Cortisol palmitate, prednisolon palmitate (PDP), betamethason valerate (BMV) betamethason palmitate (BMP) sodium cholate, sodium deoxycholate and insulin were from Merck, soya lecithin from Nattermann and 5(6)-carboxyfluorescein (CF) from Kodac. Iron-dextran (Imferron) was obtained from TAAB, U.K. and ferritin from Serva. [2,4-^{3}H]cholic acid (14 Ci/mmol), [1-^{14}C]dipalmitoyl phosphatidylcholine (100mCi/mmol), [glucose-U-^{14}C]-*n*-octyl-β-D-glucopyranoside (300 mCi/mmol) and [G-^{3}H]insulin (145.4 mCi/g) from New England Nuclear or Applied Science Laboratories. The gel materials Sephadex® Sepharose and DEAE-Sephacel® were supplied by Pharmacia and Ultrogel ACA 22 by LKB.

To produce *n*-hexyl- and *n*-heptyl-β-D-glucopyranoside, the method described by Keana and Roman[10] was used as slightly modified by Schwendener et al.[11] EYL was isolated according to the method of Singleton et al.[12] and the isolated product was checked for purity by thin layer chromatoagrophy (TLC). The oxidation degree was determined according to Klein.[13]

FIGURE 4B

A weakly opalescent solution of liposomes points to a small diameter of liposomes, a strongly opalescent or opaque solution to a large diameter. A yellow discoloration indicates oxidation of the lipid molecules. After the dialysis, the liposomes are centrifuged in order to remove large multilamellar liposomes or precipitated substances. The analytical ultracentrifuge is used to determine sedimentation and diffusion coefficients. From the results of the analytical ultracentrifuge, conclusions can also be drawn about the homogeneity of the liposomes. Gel chromatography on Sepharose 4B serves to determine the degree of homogeneity of the liposomes. Sephadex® columns are used for the separation of liposomes from unencapsulated materials. This procedure allows the estimation of encapsulated volume. Chromatography on DEAE-Sephacel® permits the separation of neutral from negatively charged liposomes. The portion of the encapsulated material is determined with atomic absorption spectroscopy in the case of iron-dextran and ferritin. In most cases, the quantitative determination of substances is performed with radioactively labeled materials. A Beckman LS 7800 liquid scintillation counter is used. The determination of the detergent remaining liposome associated after the dialysis and stability studies are based on radioacivity measurements.

Table 2
CELLULOSE MEMBRANES USED FOR DETERGENT REMOVAL

Membrane		Thickness (μm)	Diameter of disc (mm)	Permeability	Mol wt cut off
Diachema[a]	10.14	24.0	62	High	5,000
Diachema	10.15	17.5	62	Very high	10,000
Diachema	10.16	10.0	62	Extremely high	10,000
Diachema	10.17	10.0	210	Extremely high	10,000
Spectrapor[b]	6	32.2	Sheets of tubes	Low	2,000

[a] Dianorm-Geräte, P.O. Box 126, D-8000 München, FRG.
[b] Spectrum Medical Industries, Inc., 60916 Terminal Annex, Los Angeles, Calif. 90054.

Diffusion coefficients and particle diameters are determined among other methods by dynamic laser light scattering. The measurements are carried out with an argon ion laser (514.5 nm, Spectra Physics) scattering photometer at a scattering angle of 90° and a temperature of 23°C. The liposome concentration varies between 10 and 13 mg per milliliter. From the light-scattering data, evidence can be given about the homogeneity of a liposome preparation. The quality parameter Q for ideal scatterers, like latex spheres, amounts to 0.05 and runs up to 0.15 to 0.30 for sonicated liposomes. Several electron microscopy techniques are applied to liposomes or cells incubated with liposomes. Freeze-fracture gives information about homogeneity and size and freeze-substitution can tell us whether the liposomes are unilamellar, oligolamellar, or multilamellar and it also allows for the determination of the volume distribution. Scanning electron microscopy is used for the examination of cellular surfaces after incubation with liposomes. Light microscopy is not applicable to detergent dialysis liposomes because of their small diameter. However, it can be used to detect large multilamellar vesicles and precipitated materials.

II. THE DETERGENT SODIUM CHOLATE

Shankland[34] scrutinized the EYL/cholate mixed micelles and discovered that the mixed micellar radius decreased with decreasing lipid/detergent ratio. In our laboratories, lipid/cholate ratios between 0.6 and 1.15 are used. Clarification of the mixed micellar solution is assumed to be the end point of the mixed micelle formation. This formation time depends on the molarlipid/cholate ratio (L/C). Low ratios need less time than higher ratios. With a molar ratio above 1.15, a clear mixed micellar solution is not achieved even after a period of 15 days. The formation time can be shortened greatly if the cholate is dissolved in the organic solvent together with the EYL. If the lipid concentration is changed at constant molar EYL/cholate ratio, the mixed micellar diameter is changed too. This can be traced back to the constant cholate monomer and constant single micellar concentration in an aqueous solution. The ionic strength affects also the mixed micellar radius because of the different critical micelle concentrations at different ionic strengths. The formation of mixed micelles from EYL/cholate is performed at room temperature under nitrogen and kept away from light. The hydrodynamic radius from mixed EYL/cholate micelles is measured by dynamic laser light scattering at a molar ratio between 0.25 and 1.15. The radius increases from 1.0 to 7.2 nm with increasing molar EYL/cholate ratio (Figure 6). The measurements are carried out at a lipid concentration at 13 mg/mℓ. The mixed micellar solution must not be diluted for the measurement, otherwise, a new equilibrium is established between mixed micelles, simple micelles, and cholate monomers. This effects the mixed micellar size. A linear correlation (data not shown) exists between sedimentation coefficients and molar EYL/cholate ratios of mixed micelles. EYL/cholate mixed micelles are stable for several days if they are stored under nitrogen at 4°C and kept away from light.

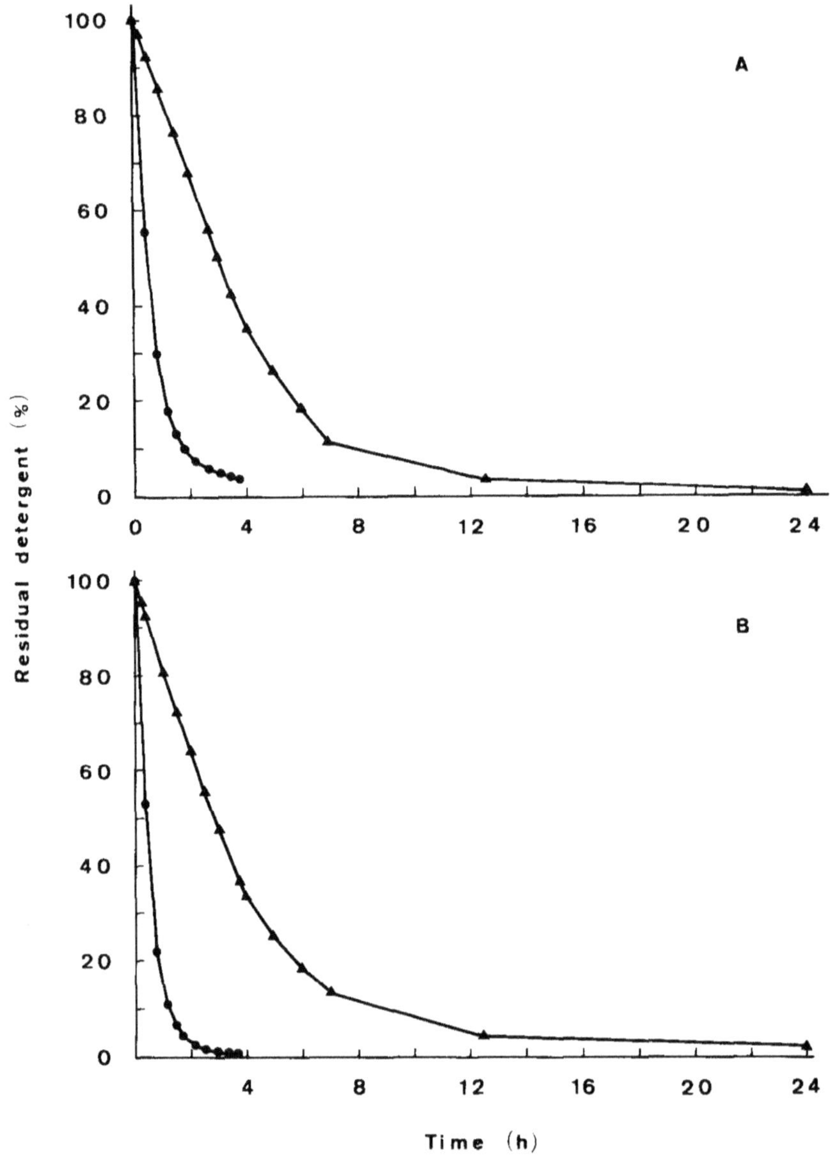

Detergent	Lipid/detergent molar ratio	Lipoprep® Liposome diameter (nm)	Lipoprep® $T_{1/2}$ (min)	Pilot Plant CLS Liposome diameter (nm)	Pilot Plant CLS $T_{1/2}$ (min)
(A) Cholate	0.6	55 ± 3	180	54 ± 2	28
(B) *n*-octylglucoside	0.2	180 ± 4	140	121 ± 4	23

FIGURE 5. Dialysis kinetics of the detergent removal during the preparation of unilamellar liposomes from mixed EYL/detergent micelles with (▲) the Lipoprep® (small sample volume) and (●) the pilot plant (counter-current dialyzer continuous production mode).[9] A lipid concentration of 5 mg/mℓ in 10 m*M* phosphate buffer, pH 7.1 was dialyzed at 22°C; cellulose membranes type Diachema 10.16 were used.

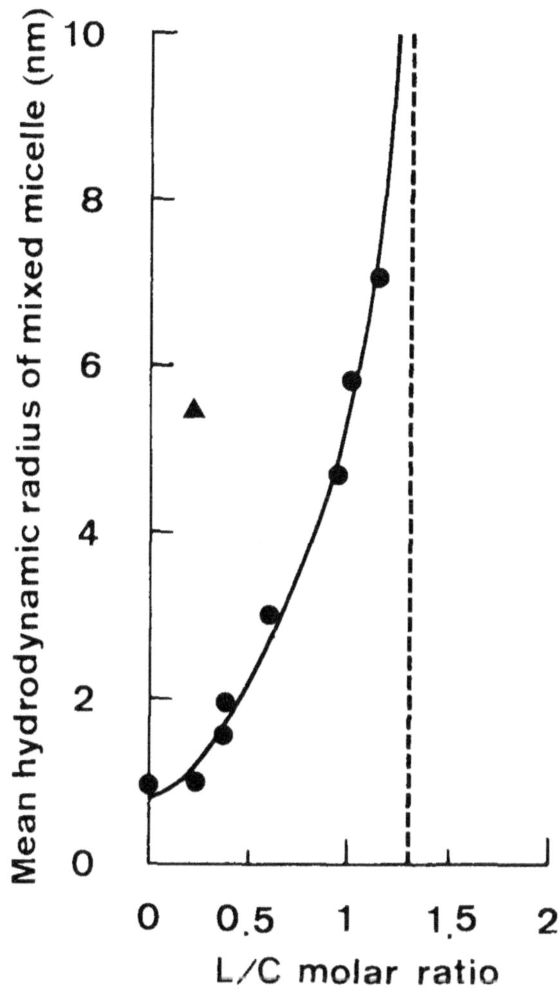

FIGURE 6. Mean hydrodynamic radii of egg phosphatidylcholine/cholate mixed micelles with different lipid/detergent molar measured by dynamic laser light scattering at constant lipid concentration (16.93 m*M*). ▲ of mixed micelles from EYL/*n*-octylglucoside (see Table 1).[9,42]

The dialysis of EYL/cholate mixed micelles is executed at room temperature during a 22- to 24-hr period using cellulose membranes of high permeability (molecular weight cut off 10,000). The intermediate stages during the formation of liposomes from mixed micelles are examined with the analytical ultracentrifuge (Figure 7). The second small peak after the removal of 30% cholate indicates the start of liposome formation. This formation seems to be terminated after the removal of 60% cholate. This fact has consequences with regard to the inclusion of water-soluble substances into liposomes and to the preparation of liposomes from lipids with high gel to liquid phase transitions. The cholate residue after the dialysis amounts to about 0.15 mg/mℓ, using a lipid concentration of 13 mg/mℓ and a L/C ratio of more than 0.6. The residue is larger, if smaller L/C ratios are applied. The cholate dialysis liposomes from EYL are eluted from Sepharose 4B in a single narrow peak. This homogeneity is emphasized by light-scattering measurements and electron microscopy pictures (Figure 8). The diameters of liposomes from various lipid compositions are listed in Table 3. The pertinent Q-values range between 0.03 and 0.12. The freeze-substitution picture in Figure

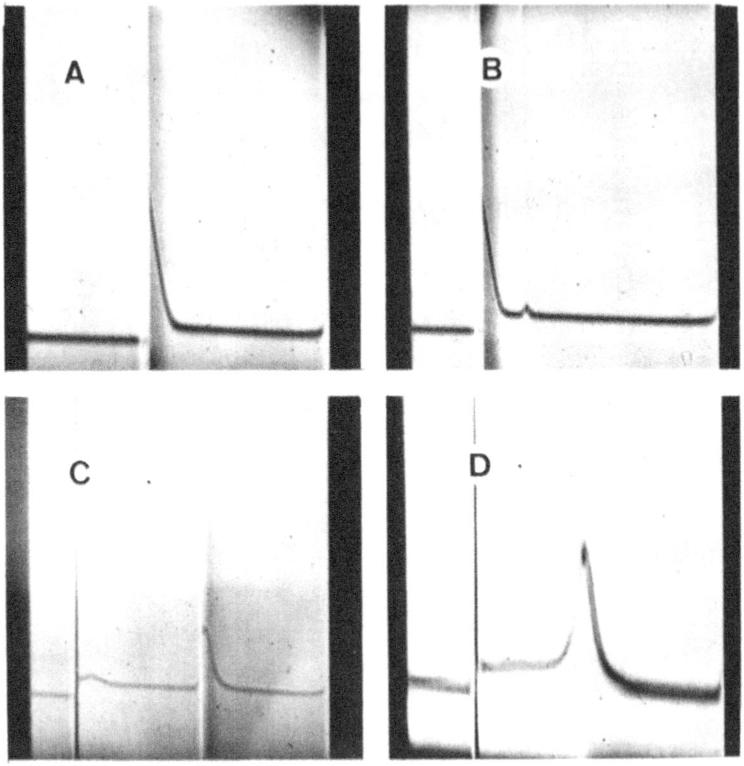

FIGURE 7. Sedimentation Schlieren patterns of the formation of homogeneous unilamellar liposomes during detergent removal at 20°C and 42,000 rev/min. (A) Mixed micelles after 10% cholate removal. (B) Mixed micelles and formed liposomes after 30% cholate removal. (C) Disappearing mixed micelles and growing liposomal fraction after 60% cholate removal. (D) Liposomes, greater than 99% cholate removal during 22 hr. (From Zumbühl, O., and Weder, H. G., *Biochim. Biphys. Acta*, 640, 252, 1981. With permission.)

9 proves the unilamellarity of these liposomes. The liposomes should be stored under nitrogen, at 4°C, and kept away from daylight.

If the dialysis rate of the cholate is ten times slower, the vesicle diameter is slightly increased by a factor of 1.4. A combined cholate removal method using dialysis and gel chromatography leads to the same liposome diameter as the simple dialysis method. After 2 hr of dialysis, more than 60% cholate is removed and therefore formation of liposomes is accomplished. Gel chromatography on Sepharose 4B can be employed in order to rapidly remove the residual detergent. Liposomes from EYL can also be prepared by an adequate dilution of the EYL/cholate mixed micellar solution and a subsequent removal of the detergent. This procedure leads to smaller liposomes than those obtained by dialysis without preceding dilution. Liposomes from EYL/PE, EYL/PA, and EYL/SA are made from organic solutions of these lipids. The following procedure is similar to that for the preparation of liposomes from EYL alone (see Table 3). The liposomes from EYL/PE are very susceptible to degradation by light.So far, we have not succeeded in preparing liposomes from DMPC, DPPC, DLPC (dilauroyl phosphatidycholine), or DSPC alone using the cholate dialysis technique. The dialysis of a mixed micelle from DLPC/cholate results in large multilamellar structures of homogeneous size (data not shown).The addition of 10 mol% PA, PE, PI, DCP (dicetylphosphate), or DSPC causes the formation of small unilamellar structures (see Table 3). Using synthetic phospholipids, the dialysis of the mixed micelles has to be carried

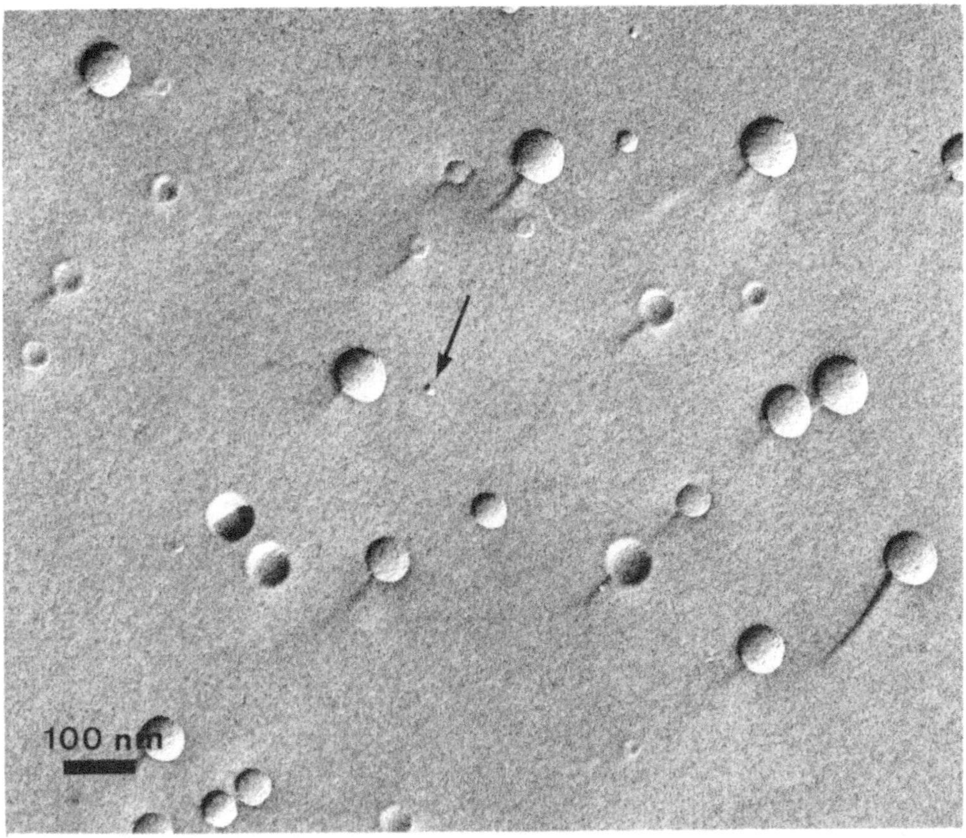

FIGURE 8. Freeze-fraction electron microscopy of egg phosphatidylcholine liposomes (13 mg/mℓ). The detergent removal from the mixed micellar solution was performed in the presence of 0.5% ferritin (arrow). (From Zumbühl, O. and Weder, H. G., *Biochim. Biophys. Acta*, 640, 252, 1981. With permission.)

out above the corresponding gel to liquid phase transitions. Because the formation of cholate dialysis liposomes is accomplished after the removal of 60% cholate, the dialysis can be performed at room temperature after about 2 hr of dialysis. After the complete removal of the detergent, these liposomes are preferably stored under nitrogen at room temperature.

III. THE DETERGENTS *n*-ALKYL-β-D-GLUCOPYRANOSIDES

The nonionic detergent *n*-octyl-β-D-glucopyranoside is easily separated from solubilized proteins because of this high critical micelle concentration (CMC).[21] This amounts to 0.025 mol/ℓ.[22] Therefore, this detergent is used for the reconstitution of biological membranes. The detergents *n*-hexyl-, *n*-heptyl-, and *n*-octyl-,β-D-glucoside are prepared by condensation of 2,3,4,6-tetra-*O*-acetyl-α-D-glucopyranosyl bromide (α-acetobromglucose) with the corresponding *n*-alkyl alcohol, followed by hydrolysis of the acetyl residues according to the slightly modified method described by Keana and Roman.[10,11] The purification of the products was extended by several washings of the crude glucosides with petrol ether in order to remove residual alcohol. The resulting detergents showed to be of high purity according to TLC, elementary analysis, melting points, and IR spectra.

The mixed micelles from EYL/*n*-alkylglucosides are prepared in the same manner as the mixed micelles from EYL/cholate. The detergent can be dissolved in the buffer prior to its addition to the lipid film or it can be added to the lipid suspension into the buffer. The

Table 3
PROPERTIES OF UNILAMELLAR LIPOSOMES OF VARIOUS LIPID MIXTURES PREPARED BY CONTROLLED DETERGENT REMOVAL. TOTAL LIPID CONCENTRATION VARIED BETWEEN 10 AND 13 MG/mℓ

Lipid composition (molar ratio	Detergent	Molar lipid/ detergent ratio	Liposome diameter (nm)	Ref.
EYL	Cholate	0.60	55 ± 3	42
EYL	Cholate	1.15	81 ± 4	42
EYL/chol (8:2)	Cholate	0.60	80 ± 4	42
EYL/chol (7:3)	Cholate	0.52	61 ± 4	42
EYL/PE (3:7)	Cholate	0.22	36 ± 2	42
EYL/PI (8:2)	Cholate	0.60	59 ± 2	42
EYL/PA (10:2)	Cholate	0.62	42 ± 2	43
EYL/SA (10:2)	Cholate	0.62	49 ± 2	43
EYL/cerebroside	Cholate	0.60	81 ± 4	42
EYL/ganglioside (10:0.6)	Cholate	0.64	63 ± 4	44
EYL/Chol ◇ PA (7:2:1) + sp2/HL antibody	Deoxycholate	0.74	75 ± 4	33
DMPC/PI (10:2)	Cholate	1.25	143 ± 7	15
DPPC/PA (10:1)	Cholate	0.20	101 ± 6	15
DPPC/DMPA (9:1)	Cholate	0.20	97 ± 4	15
DSPC/DCP (9:1)	Cholate	0.20	74 ± 4	15
DPPC/DCP (9:1)	Cholate	0.62	90 ± 4	15
EYL	*n*-Octyl-glucoside	0.20	176 ± 3	11
EYL	*n*-Octyl-glucoside + *n*-heptyl-glucoside	0.40 0.35	141 ± 8	11
EYL	*n*-Octyl-glucoside + *n*-heptyl-glucoside	0.80 0.17	117 ± 3	11
EYL	*n*-Heptyl-glucoside	0.13	79 ± 2	11
EYL	*n*-Hexyl-glucoside	0.05	59 ± 2	11
DSPC/DCP (10:1)	*n*-Octyl-glucoside	0.20	156 ± 10	17
DSPC/Chol/DCP (7:2:1)	*n*-Octyl-glucoside	0.20	79 ± 4	17
Hydrogenated soya lecithin /DMPA (9:1)	*n*-Octyl-glucoside	0.20	130 ± 5	9

preparation is performed at room temperature and a clear solution of mixed micelles is immediately obtained. These micelles are stable for several days. The solution of mixed micelles is dialyzed at room temperature. After 60 min of dialysis, the liposomes are already formed and the residual detergent content decreased to 50%. After additional 9 hr of dialysis, only 2% residual detergent is associated with the liposomes. Less than 10% of the initial amount of EYL is lost during the dialysis.

n-Octyl- and *n*-heptylglucosides in various ratios are combined together for the preparation of liposomes from EYL. The applied molar lipid/detergent ratios and the hydrodynamic radii of the corresponding liposomes are shown in Table 3. The lipid concentration amounts to 10 mg/mℓ. The Q values from the light-scattering measurements emphasize the homogeneity of the above described liposomes. The respective value for sonicated liposomes is 0.3. The same homogeneity for liposomes prepared by dialysis is evident from the freeze-substitution electron microscopy.[11] One can determine from pictures obtained the volume distribution of (three) liposome preparations with various sizes (Figure 10). By the freeze-substitution technique, the spherical liposomes are cut in any level. With the intention of overcoming this falsification of the diameter, the volume distribution is determined according to the method of Rose.[23] The size distribution of these liposomes is very narrow and the main

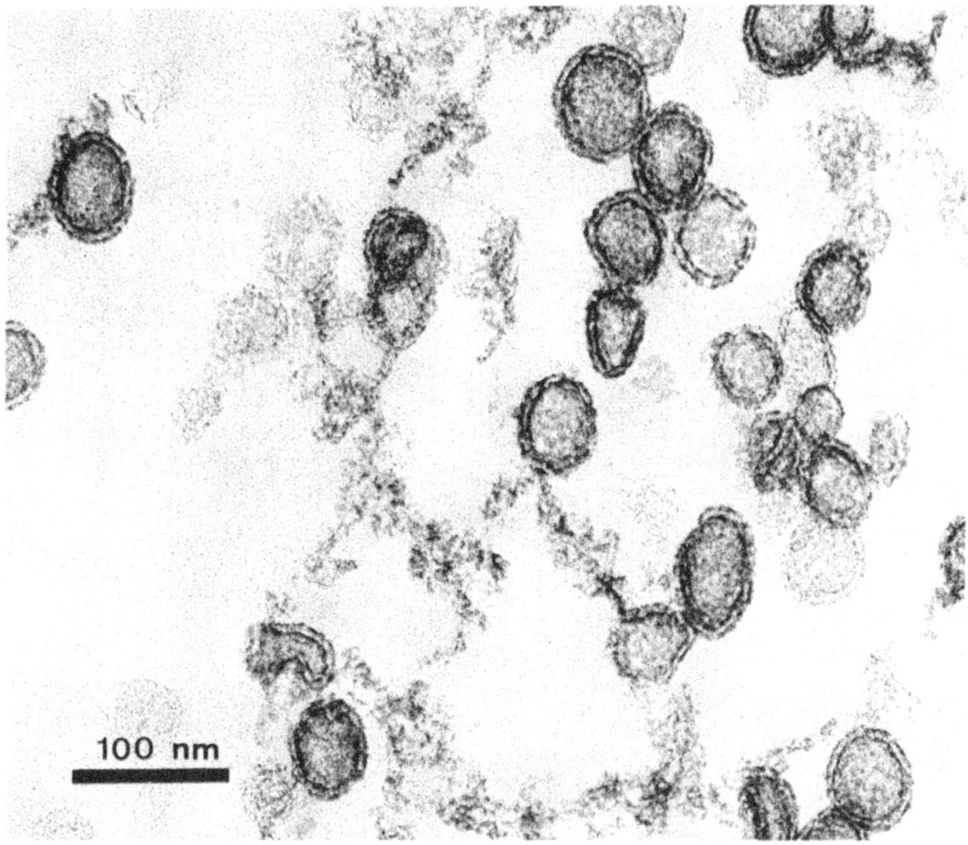

FIGURE 9. Freeze-substitution electron microscopy of unilamellar liposomes prepared by defined detergent removal from a mixed micellar solution of egg PC/cholate (molar ratio 0.60) in the presence of an iron-dextran complex (Imferron, 2.5 mg iron/mℓ).[42] The liposomes (13 mg/mℓ) were cryofixed in a propane-jet freezer.[18] Frozen samples were subjected to freeze-substitution at −90°C. The substituent consisted of methanol with 0.5% uranyl acetate, 1% osmium tetroxide, and 3% glutaraldehyde.[19] Dehydrated samples were embedded in araldite epon. These sections were stained with uranyl acetate and lead citrate.[20]

fraction is well in agreement with the diameter measured by dynamic laser light scattering. The liposomal diameter can be changed continuously between 80 and 180 nm by using accurately fixed mixtures of *n*-octylglucoside and *n*-heptylglucoside. A shorter alkyl chain of the glucoside leads to smaller liposomes. The preparation of liposomes with the *n*-hexylglucoside gives indeed vesicles with a diameter of 60 nm.

The preparation of liposomes from hydrogenated soya lecithin using the *n*-octylglucoside is feasible. The detergent is dialyzed at 47°C. With the intention of shortening the time for the preparation of liposomes using the *n*-alkylglucosides, the dialysis can be performed during 1.5 to 2 hr, followed by a separation of the preformed liposomes from the residual cholate on Sephadex® G-50. The residual detergent is in the range of 3 to 4%.

IV. INCORPORATION OF LIPOPHILIC COMPOUNDS INTO THE LIPID BILAYER

A. Naturally Occurring Membrane Constituents

EYL/cholesterol liposomes up to a molar ratio of 7:3 are produced using the detergent cholate. A clear mixed micellar solution is only obtained if the detergent is dissolved together

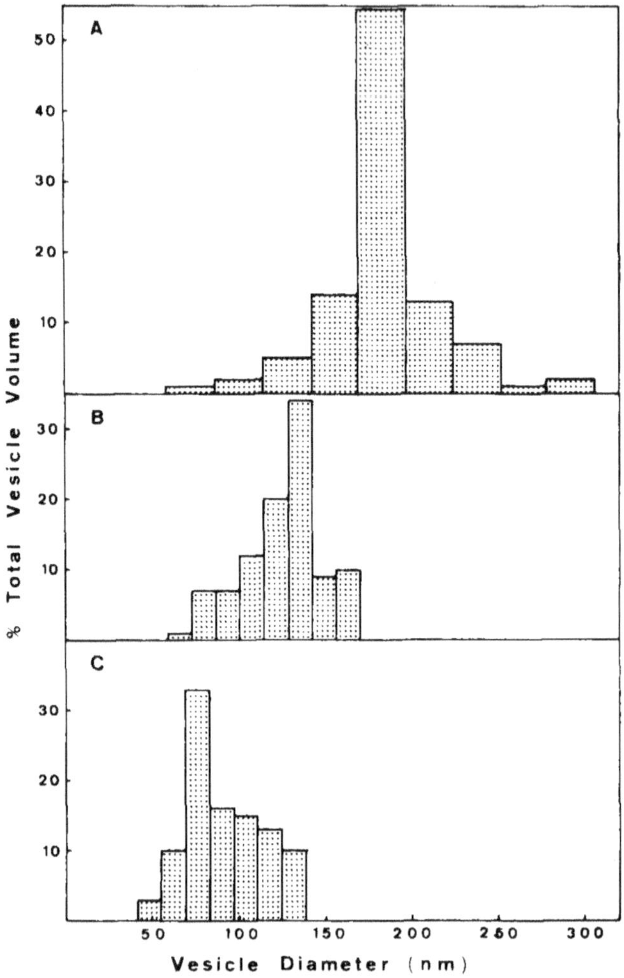

FIGURE 10. Volume distribution of detergent dialysis liposomes: (A) *n*-octylglucoside, (B) 50% *n*-octyl/*n*-heptylglucoside, (C) *n*-heptylglucoside. (From Schwendener, R. A., Asanger, M., and Weder, H.-G., *Biochem. Biophys. Res. Commun.*, 100, 1055, 1981. With permission.)

with the lipids in the organic solvent. A flocky precipitate is deposited from such solutions of mixed micelles after 1 to 2 days. Therefore, the dialysis has to be performed immediately after the preparation of the mixed micellar solution. The preparation of liposomes with more than 30% cholesterol was not attempted until now. Preliminary results show, that cholate is much more suitable for the preparation of liposomes from EYL/cholesterol than are *n*-alkylglucosides. This may be related to the fact, that cholate is used in the intestine for the solubilization of cholesterol and other lipids. Freeze-fracture electron microscopy showed the high homogeneity of the EYL/cholesterol liposomes (not shown). The quantitative incorporation of cholesterol into cholate dialysis liposomes from EYL is shown by the elution profile on Sepharose 4B (Figure 11). The molar lipid ratios and the absolute amount of lipid remain unchanged over a period of at least 20 days.

The incorporation of 15 mol % bovine brain cerebroside and 6 mol % bovine brain ganglioside is also performed with the detergent cholate. Gangliosides contain negative charges in their molecules belonging to the sialic acid residues. A 1:1 mixture of liposomes from EYL and liposomes from EYL/ganglioside is eluted from an ion-exchange chroma-

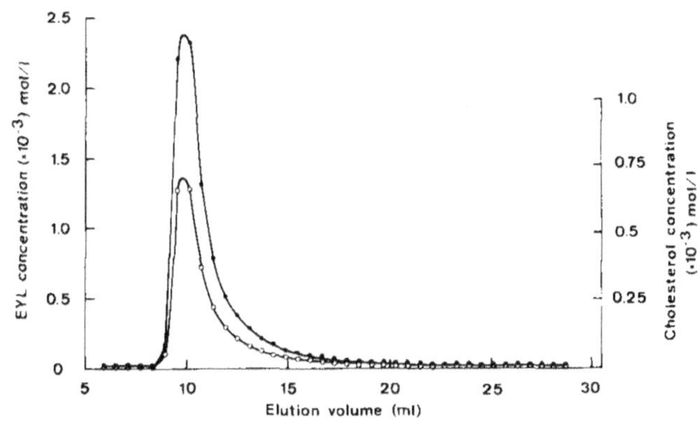

FIGURE 11. Elution profile of egg PC/cholesterol liposomes (molar ratio 8:2) on Sepharose 4B. (●), Egg PC, (○), cholesterol. (From Zumbühl, O. and Weder, H. G., *Biochim. Biophys. Acta*, 640, 252, 1981. With permission.)

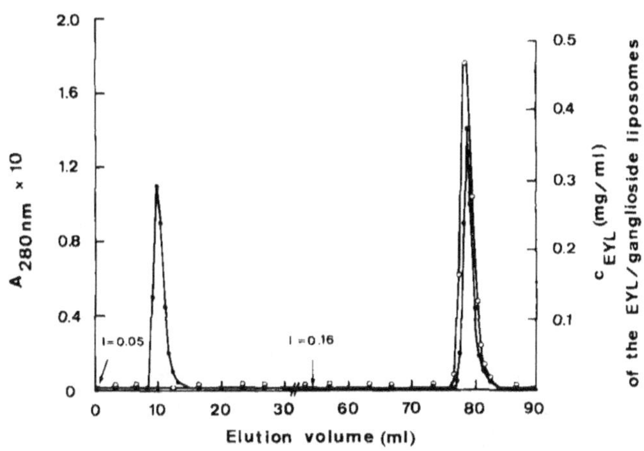

FIGURE 12. Separation of unlabeled EYL liposomes in Tris buffer from ^{14}C-labeled ganglioside/EYL liposomes (0.062 μmol/μmol) on DEAE Sephacel®. (●), UV absorption at 280 nm; (○), EYL concentration of the ganglioside/EYL liposomes.[44]

tography column, whereby neutral liposomes from EYL are eluted with an ionic strength of 0.05. But for the elution of liposomes made of EYL/ganglioside, an ionic strength of 0.16 is necessary (Figure 12). Thus, a negative charge must be on the surface of these liposomes which can only be attributed to the sialic acid residues of incorporated gangliosides. Freeze-fracture electron microscopy demonstrates the homogeneity of the liposomes from EYL/ganglioside (not shown). In Table 3 the diameters of liposomes from EYL/cholesterol, EYL/ganglioside, and EYL/cerebroside, determined by dynamic laser light scattering, are listed.

B. Modified Corticosteroids

Today, many corticosteroids are injected intraarticularly in the form of a crystal suspension. In order to overcome the disadvantages attached to this application form, the corticosteroids are incorporated into liposomes.[24] In our laboratories, the corticosteroids betamethasone, prednisolone, and hydrocortisone are used. In order to achieve a membrane integration of

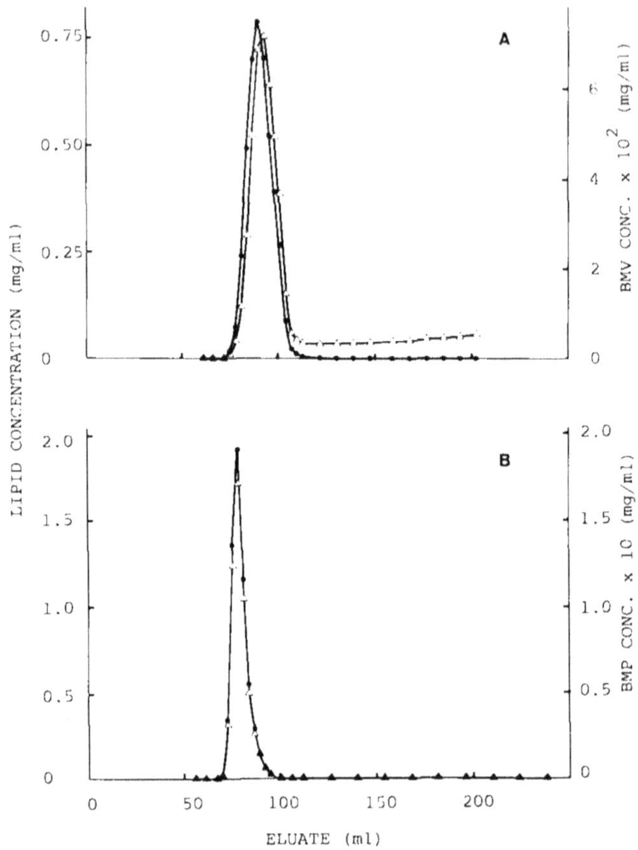

FIGURE 13. Elution profile of (A) BM-17-V (B) BM-21-P containing liposomes on Sephadex® G-25 coarse.[15] (A) DPPC/DMPA/BM-17-V (9:1:1), applied lipid concentration 32.9 mg/mℓ. (B) DPPC/DMPA/BM-21-P (8:1:1), applied lipid concentration 10.0 mg/mℓ.

these poorly water-soluble compounds, they were esterified with fatty acids of various chain lengths. Betamethason was substituted in the C_{17} position with valeric acid, in the C_{21} position with palmitic acid and prednisolone and hydrocortisone in the C_{21} position with palmitic acid. For incorporation and stability measurements, ^{3}H-labeled compounds were used. It is to be expected, however, that such a chemical modification of a drug will influence its pharmacological properties. It should, therefore, be investigated, whether such compounds are pharmacologically active and/or whether they are degraded to the initial substances under biological conditions. All esterified compounds show very low water solubility, which is responsible for the spontaneous formation of well-defined crystals.

The following lipid compositions were used: EYL/PA, hydrogenated soya lecithin/dimyristoyl phosphatidic acid (DMPA) and DPPC/DMPA. The amount of charged lipid to prevent self aggregation was generally around 10%. It was observed that there exists an upper limit for all corticosteroids tested for the amount incorporated into liposomes per weight lipid. This amount comes generally to about 1 mg esterified corticosteroid per 10 mg lipid. The micellization was always performed about 5 to 10°C above the corresponding gel to liquid phase transitions. For further details see Table 1. As already explained, the detergent removal was carried out above the gel to liquid phase transitions. The liposomal diameter ranged between 50 and 85 nm. The corresponding multilamellar liposomes were prepared according to Shaw et al.[25]

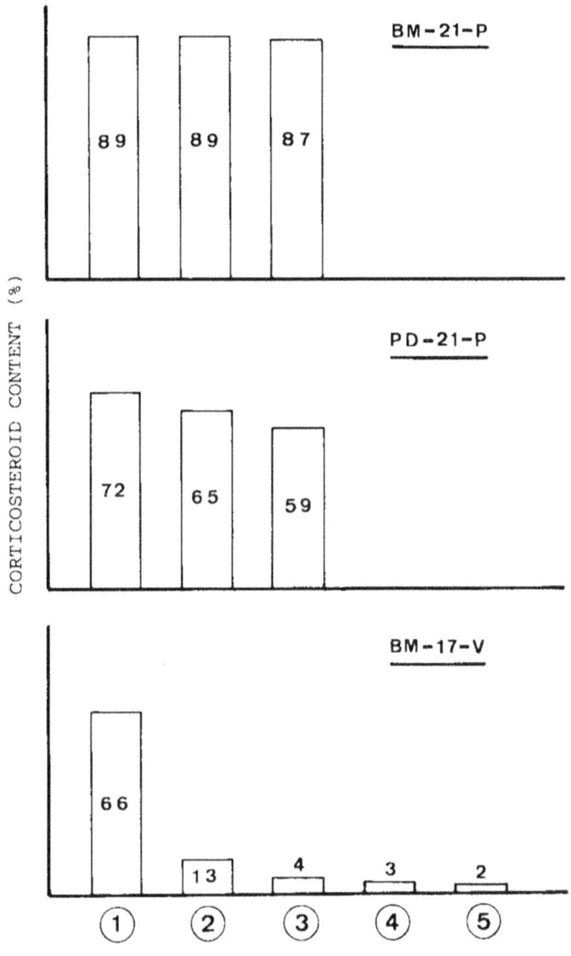

FIGURE 14. Corticosteroid content of liposomes from DPPC/DMPA/BM-21-P (8:1:1), DPPC/DMPA/PD-21-P (8:1:1), and DPPC/DMPA/BM-17-V (9:1:1) immediately after preparation (1) and after repeated gel filtrations (2 to 5).[15]

The lipids were dissolved in *tert*-butanol and lyophilized. After the addition of the buffer, rather homogeneous vesicles are formed in the range of 0.5 to 1 μm. The following stability tests are based on radioactivity measurements of the lipids and corticosteroids. Examination under light microscopy gives an impression of the degree of crystal formation. This crystal formation starts above the critical ratio of corticosteroid to lipids mentioned earlier.

Figure 13 shows the elution profile of liposomes from DPPC/DMPA containing: (1) BMV and (2) BMP (lipid to corticosteroid ratios see Table 1). An unstable incorporation of the BMV leads to a continuous leakage of this corticosteroid during the elution. In contrast, all the palmitic ester remains liposome associated. Palmitic acid seems to be an ideal membrane constituent for liposomes made of DPPC/DMPA, whereas the shorter valeric acid is not. This is also emphasized in Figure 14. Identical samples of detergent dialysis liposomes with incorporated BMV, MBP, and PDP, respectively, were several times applied to a Sephadex® G-50 coarse column and the released and still incorporated corticosteroids were determined. As expected, the BMV is released in large amounts, so that liposomes are lacking in this compound after the fifth elution. In contrast to this, BMP and PDP remain liposome associated after several elutions. 100% is the amount of the corticosteroid applied to the mixed

Table 4
CORTICOSTEROID RETENTION BY BM-17-V AND BM-21-P LIPOSOMES AFTER INCUBATION WITH MULTILAMELLAR LIPOSOMES MADE FROM EYL AT 20 AND 37°C[15,16]

Time of incubation (hr)	Retention (%) BM-17-V 20 / 37°[a]	BM-21-P 20°	BM-21-P 37°
1/4	30	96	94
2	30	92	86
4	32	91	83
6	30	92	84
18	30	90	84

[a] Incubations at 20 and 37°C gave identical values.

micellar solution. The loss prior to the first elution is due to dialysis of the corticosteroids during the vesicle preparation. Another attempt to determine incorporation stability of esterified corticosteroids into liposomes is illustrated in Table 4. Adequate portions of the liposomal samples were mixed with 6 mℓ of 10 m*M* phosphate buffer, pH 7.1, and incubated in the triple cell of the Lipoprep®. The released corticosteroids were measured both in the diffusate and in the dialysate and related to the initially incorporated corticosteroids (100%). The amounts of BMP released during the incubation correspond well with the related amounts released during the elution of the freshly prepared liposomes on Sephadex® G-25 coarse, indicating that the incubation over 4 days does not induce an additional release of BMP from liposomes. In the case of PDP/EYL/PA and PDP/DPPC/DMPA liposomes, the degree of binding (micrograms corticosteroid per milligram lipid) did not change after treatment of the corresponding liposomes during 20 min at 120°C and 1 atm.

A stability test more relevant for in vivo conditions was performed by measuring the time dependent transfer of corticosteroid from small unilamellar DPPC/DMPA liposomes (SUV) to multilamellar EYL liposomes (MLV) at room temperature and 37°C (Table 5). After the incubation, the SUV and the MLV fractions were separated by centrifugation and the radioactivity of the ^{3}H-labeled corticosteroid determined in the supernatant. Reference experiments showed that no sedimentation of liposomes made of DPPC occurred and that the MLV made of EYL were sedimented completely. The MBV is transfered from the SUV to MLV very fast and to a large extent. This could not be observed in the case of BMP. This transfer seems to be temperature dependent and these results correspond well with the above-discussed stability data.

C. Modified Antibodies

Interest in the attachment of antibodies to liposomes has grown because of the aim of a specific targeting of drugs to cells. The antibody can be attached by covalent coupling to liposomal PE[26,27] by coupling liposomal aldehyde to the activated amino groups of the antibody,[28,29] or by anchoring the antibody with a hydrophobic agent.[30,31]

In a report by Martin et al.,[32] the Fab fragments are covalently attached to liposomes via disulfide bonds. Harsch et al.[33] tried to incorporate unesterified monoclonal antibody sp2/HL into liposomes from EYL/DCP (7:1) in a lipid/protein ratio of 200:1 (w/w). They showed that the protein was not incorporated by this method (see Figure 15). The coupling method with carbodiimide proved to be useless.[14] The antibody could be attached to EYL/

Table 5
RETENTION OF ^{3}H-BM-17-V AND ^{3}H-BM-21-P BY UNILAMELLAR LIPOSOMES AFTER INCUBATION AT 37°C DURING 1 TO 4 DAYS[15,16]

Preparation			Age of preparation	Retention (%) (days) 1	2	3	4
EYL/PA	9:1 / 0.1%	BM-17-V	Freshly prepared	91	86	82	78
DPPC/DMPA	9:1 / 0.1%	BM-17-V	Freshly prepared	92	88	86	82
DPPC/DMPA	9:1 / 0.1%	BM-21-P	Freshly prepared	97	ND	ND	97
DPPC/DMPA	9:1 / 0.1%	BM-17-V	Incubated at 120°C[a]	68	62	59	57
DPPC/DMPA	9:1 / 0.1%	BM-21-P	Incubated at 120°C[a]	81	ND	ND	80

[a] 1 atm, 20 min; ND, not determined.

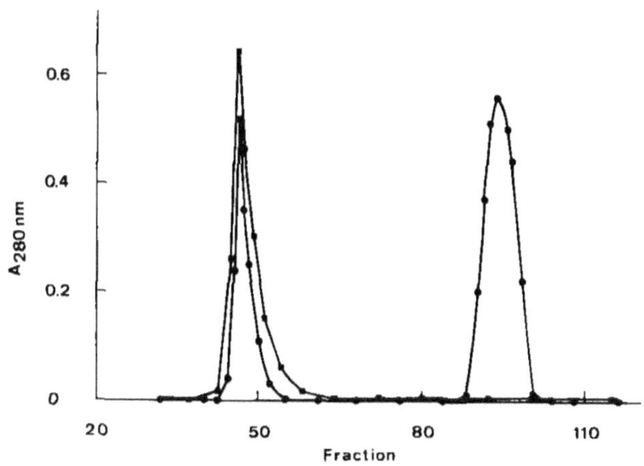

FIGURE 15. Chromatography pattern of sp2/HL coated liposomes. Detergent dialysis liposomes from EYL/Chol/PA (7:2:1, 6 mg/mℓ) were applied to an Ultrogel ACA 22 column. (■), with esterified monoclonal antibody; (●), with native antibody. (From Harsch, M., Walther, P., Hengartner, H., and Weder, H.-G., *Biochem. Biophys. Res. Commun.*, 103, 1069, 1981. With permission.)

PE liposomes, but the carbodiimide inactivated the protein. Therefore, attempts were made to attach the antibody to liposomes via a hydrophobic anchor.

The *N*-hydroxysuccinimide ester of palmitic acid was synthesized according to Lapidot et al.[35] Palmitic acid (20 mmol) was added to a solution of *N*-hydroxysuccinimide (20 mmol) in dry ethyl acetate (87 mℓ). A solution of dicyclohexylcarbodiimide (20 mmol) in dry ethyl acetate (6.7 mℓ) was then added and the reaction mixture left overnight at room temperature. Dicyclohexylurea formed was removed by filtration, and the filtrate was concentrated under reduced pressure to yield white crystals. Recrystallization from ethanol yielded the pure *N*-hydroxysuccinimide ester of palmitic acid. 2.5 mg of monoclonal IgG was added to 20 μg of palmitic acid ester in 1 mℓ 10 m*M* phosphate buffer pH 7.1 containing 0.9% NaCl and 2% deoxycholate. The reaction mixture was incubated in a rotating flask overnight at 37°C.

For the formation of unilamellar liposomes, an ethanol/chloroform solution (1:1; v/v) of 25.2 μmole EYL, 7.2 μmol chol, 3.6 μmol PA, and 40 μCi of (^{3}H-DPPC) was evaporated to dryness under reduced pressure at 30°C. The remaining lipid film was dispersed in 2 mℓ phosphate buffer under nitrogen and the detergent/monoclonal antibody solution (1 mℓ)

Table 6
ESTIMATED ENCAPSULATION EFFICIENCY OF SUV AND LUV LIPOSOMES FROM EYL PREPARED WITH THE DETERGENT DIALYSIS METHOD[11,44]

Liposome diameter (nm)	Lipid concentration (mg/mℓ)	Encapsulation efficiency (%)	Marker
65	50	12.4	Inulin
176	50	40.9	Inulin

added. The dispersion was gently stirred until clear at room temperature for about 5 min. The mixed micelles were dialyzed with the detergent dialyzer Lipoprep® against phosphate buffer for 22 hr at room temperature. After the incorporation of the antibody into the liposomes, the coated liposomes were separated from free protein on a Ultrogel ACA 22 column (Figure 15). The native antibody is not incorporated into the liposomes, whereas all the esterified antibody is eluted in the liposome fraction.

V. ENTRAPMENT OF WATER-SOLUBLE COMPOUNDS INTO THE AQUEOUS SPACE OF LIPOSOMES

Using the detergent dialysis method for the entrapment of water-soluble compounds into the aqueous space of liposomes, difficulties arise from the fact that the compounds to be entrapped dialyze depending on their three-dimensional size together with the detergent through the membrane. Dialysis kinetics depend also strongly on membrane properties such as molecular weight cut off and thickness. With the intention of overcoming this disadvantage, small water-soluble compounds are dissolved in about 200 mℓ of buffer in the same concentration as in the solution of the mixed micelle. This solution is used as dialysate during the first 2 hr of the dialysis, whereby a closed circuit of the dialysate is arranged. With the lipid/cholate system, the vesicle formation is accomplished after 2 hr and with the lipid/*n*-alkylglucoside after 1 hr. After these periods of time, the water-soluble compound is entrapped, and the compound is no longer required in the dialysate. By the preparation of liposomes from lipids with a gel to liquid phase transition above room temperature, the dialysis temperature can be lowered to room temperature or even below at the end of liposome formation. This lowering of temperature prevents the diffusion of water-soluble compounds out of the liposomes.

In the present experiments the markers inulin (mol wt 5000) and CF (mol wt 376) are used. In the case of CF, additional precautions have to be taken. Because CF is not soluble in acid solutions, it was dissolved by addition of sodium hydroxide, and the solution was adjusted with diluted sodium hydroxide to pH 7.14[15] After the entrapment of CF in the aqueous space of liposomes and the removal of untrapped materials, the osmotic pressure inside the liposomes exeeds that of the surrounding solution. Consequently, liposomes can be disrupted. To prevent this, the osmotic pressure in the dialysate has to be adjusted to the osmotic pressure in the aqueous compartment of the liposomes.

Table 6 shows the encapsulation efficiency of liposomes from EYL, prepared with the dialysis detergent method. The percent encapsulation efficiency is the portion of the encapsulated volume compared with the total volume. At a constant lipid concentration, the encapsulation efficiency is higher for larger liposomes. The encapsulation efficiency was simultaneously calculated from the corresponding sedimentation coefficients and diffusion constants. These values were applied to the Svedberg equation:

Table 7
HALF-LIFE OF THE CF RELEASE FROM DPPC/DCP AND DSPC/DCP LIPOSOMES AT VARIOUS STORAGE TEMPERATURES. THE RELEASE WAS OBSERVED OVER 130 DAYS.[15]

	(°C) CF release ($T_{1/2}$) (days)	
Storage temperature	**DPPC/DCP[a]**	**DSPC/DCP[a]**
4	165	5680
20	5.9	1095
37	0.4	4.1

[a] Total lipid 15 mg/mℓ (9:1).

$$M = \frac{R \cdot T \cdot S}{D(1 - \rho \cdot \bar{v})}$$

whereby M = particle weight, R = gas constant = $8.314 \cdot 10^7$ erg/grd·mole, T = temperature in K, S = sedimentation coefficient, D = diffusion coefficient, ρ = density of the water, $\bar{v}$ = partial specific volume, which was assumed to be 0.9848 mℓ/g according to Hauser and Irons.[36] From the estimated particle weight, the encapsulation efficiency was calculated assuming an outer hydration cost of 0.7 nm and a bilayer thickness of 3.7 nm according to Huang and Mason.[37] The difference between the determination by inulin entrapment and by calculation from sedimentation coefficient and diffusion constant comes only to 2%. Therefore, the vesicles are indeed unilamellar.

Table 7 shows the leakage of CF out of liposomes made of DSPC/DCP or DPPC/DCP and incubated at different temperatures. The following rule can be derived for both lipid compositions: The leakage of CF is reversibly proportional to the difference between the gel to liquid phase transition and the incubation temperature. This difference is larger in the case of DSPC than in the case of DPPC. Therefore, liposomes from DSPC/DCP are more effective in preventing CF leakage. This result corresponds well with the observations of Kanehisa and Tsong.[38] These authors observed that the leakage of small hydrophilic ions out of liposomes is highest at the gel to liquid phase transition and is zero above or below this boundary. This leads to the conclusion, that liposomes, loaded with water-soluble compounds, should be stored as much as possible below their gel to liquid phase transition temperature in order to prevent leakage.

VI. INTERACTIONS BETWEEN DETERGENT DIALYSIS LIPOSOMES AND CELLS

A. Cytotoxicity of Liposomes on *Trypanosoma brucei* and *Trypanosoma congolense* Bloodstream Forms in Vitro

For the following studies, the extracellular parasites *Trypanosoma brucei* and *Trypanosoma congolense* were used.[39] These parasites cause the african trypanosomiasis in game animals and cattle and are related to the parasites causing human sleeping sickness. The bilayer membrane of *T. brucei* and *T. congolense* bloodstream forms is covered by a membrane coat.[40] *T. brucei* STIB 366-A and *T. congolense* STIB 68-F-A were obtained from the Swiss Tropical Institute, Basel. *T. brucei* was cultivated in adult female ICR mice and *T. congolense* in adult female SIV rats. The blood was harvested from infected animals at the peak parasitemia period and mixed with 20% (v/v), 3.8% (w/v) sodium citrate. This

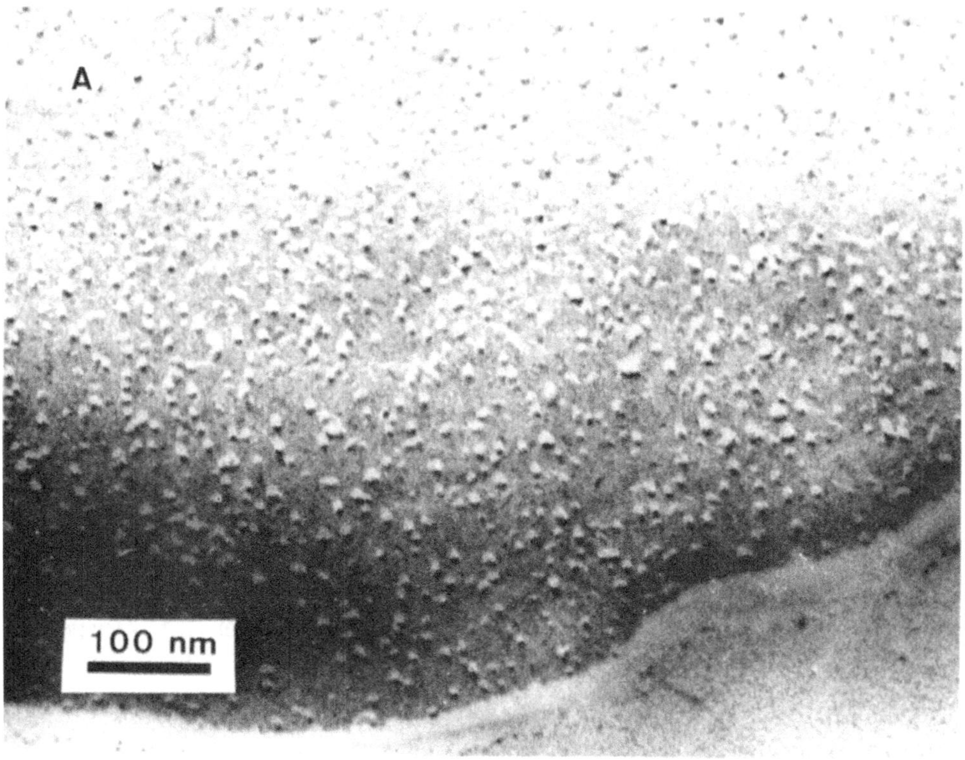

A

FIGURE 16. Electron micrograph of the freeze-fractured cell body membrane of *T. brucei* before and after incubation with liposomes from EYL.[39] (A) 90 μℓ parasite suspension (3.6 × 10^7 cells/mℓ) were incubated with 250 μℓ 10 m*M* phosphate buffer pH 7.1 and 34 μℓ fetal bovine serum for 45 min at 37°C. After the incubation, the parasites were washed three times with 10 m*M* phosphate buffer pH 7.1 and 10% (v/v) fetal bovine serum and centrifuged (1000 × *g*, 10 min). The pellet was then frozen in the propan-jet-freezer.[18] (B) 90 μℓ parasite suspension (see A) were incubated with 250 μℓ liposomes from EYL (13 mg/mℓ) in 10 m*M* phosphate buffer pH 7.1 and 34 μℓ fetal bovine serum for 45 min at 37°C. For further procedures see (A). The liposomes were prepared one day before the incubation.

suspension was mixed with 50% (v/v) PSG6:4[41] and centrifuged (50 × *g*, 15 min). The supernatant was then centrifuged (1000 × *g*, 10 min) and the parasite-containing pellet passed through a DEAE-cellulose column according to Lanham and Godfrey.[41] Isolated trypanosomes were kept on ice and used for the incubation within 1 hr. For these experiments, detergent dialysis liposomes of various lipid compositions were used as described in previous sections.

The freeze-fractured membrane shown in Figure 16A represents the membrane of viable trypanosomes with normal cell structure and motion as observed in the light microscope. Trypanosomes do not move and are totally lysed after interaction with liposomes. Corresponding membranes are represented by Figure 16B.

Liposomes from EYL induce aggregation of intramembranous particles of *T. brucei*. The number of particles per membrane area remains unchanged. Fusion between trypanosomes and liposomes could be responsible for intramembranous particle aggregation, whereas the areas free of intramembranous particles might represent liposomal lipid.

An increase of cytotoxicity is observed with increasing lipid concentration and age of liposomes made of EYL and EYL/ganglioside. During storage, the formation of lysolecithin could be promoted and therefore, the toxicity of the liposomes could be enhanced. Residual cholate and ganglioside micelles show no trypanocidal activity. Concerning the in vitro

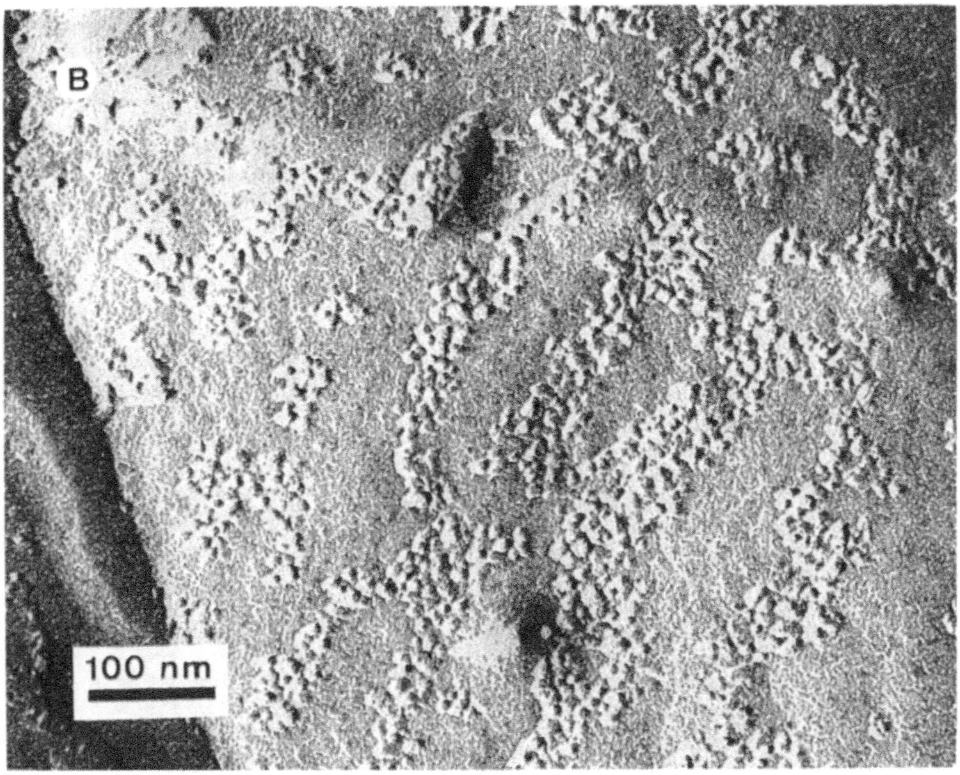

FIGURE 16B

incubations, the trypanosomes isolated from blood were more susceptible to destruction by these liposomes, compared to parasites in whole blood. No trypanocidal effect of liposomes could be shown in vivo when injected intravenously into the tail vein of mice infected with *T.brucei*, even at very high blood levels over 13 hr. Moreover, the cytotoxicity of liposomes from EYL could be reduced by the incorporation of 30 mol% cholesterol. These observations lead to the suggestion that the trypanocidal effect of the liposomes is based on the withdrawal of cholesterol from the parasite membrane. In vivo, initially cholesterol-free liposomes take up cholesterol from sources other than trypanosomes, which makes such liposomes ineffective against these parasites.

B. Targeting of Monoclonal Antibody-Coated Liposomes to Sheep Red Blood Cells

A monoclonal mouse anti-sheep red blood cell-specific antibody IgG_{2b} was esterified with palmitic acid which served as a hydrophobic anchor for successful incorporation into the liposomal membrane (see Section IV.C). The formation of antibody-coated liposomes by dialyzing the mixed antibody/lipid/detergent micelles against phosphate buffer was simplified by using the same detergent as for the antibody derivatization. No purification step of any intermediate product was necessary. Targeting of the resulting vesicles to sheep red blood cells occurred with high efficiency compared with control liposomes. The uptake was rapid and specific as demonstrated with sheep and horse red blood cells by the use of radioactively labeled liposomes and by scanning electron microscopy (see Figure 17).[33]

VII. FUTURE PROSPECTS

We now are on the point of establishing the technical feasibility that enables us to produce

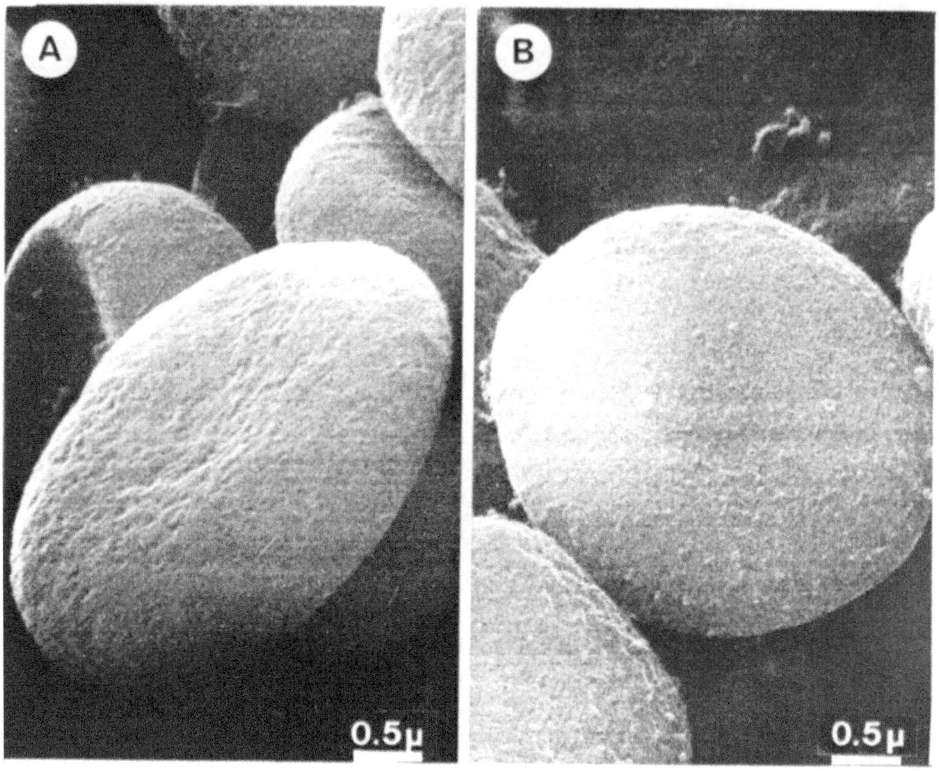

FIGURE 17. Scanning electron microscopy of sheep red blood cells in the absence (A) or presence of (B) noncoated and (C) coated liposomes. (A) Sheep red blood cells were washed three times, diluted to 3×10^8 cells/mℓ phosphate buffer, freeze dried, and examined in a Hitachi S-700 scanning electron microscope. Bar represents 0.5 μm. (B)/(C) Washed sheep red blood cells (3×10^8 cells/mℓ) were incubated with noncoated (B) or coated (C) liposomes (1 mg/mℓ), EYL/Chol/PA (7:2:1) at 37°C during 3 min. After incubation, cells were washed three times to remove unbound vesicles, followed by the preparation and examination of the cells. Bar represents 0.5 μm. (From Harsch, M., Walther, P., Hengartner, H., and Weder, H.-G., *Biochem. Biophys. Res. Commun.*, 103, 1069, 1981. With permission.)

liposome preparations of various homogeneous populations at quantities large enough for clinical and toxicological evaluation and for diagnostic tests. But in developing a liposome formulation, it must be assured that lipids and charged additives are available at a reasonable cost and with reproducible purity. Other methodological implications are the influence of incorporated compounds on the liposomal size, the long-term stability, and the elimination of contaminant infectious agents. Further efforts have to be made to prepare multilamellar liposomes of similar homogeneity to that achieved for unilamellar vesicles. It seems rather difficult to develop detergents that are more appropriate to the formation of liposomes. It must be borne in mind that the industrial manufacture of liposome formulations will require further optimization of the proposed production process.

REFERENCES

1. **Carey, M. C. and Small, D. M.,** The characteristics of mixed micellar solutions with particular reference to bile, *Am. J. Med.*, 49, 590, 1970.

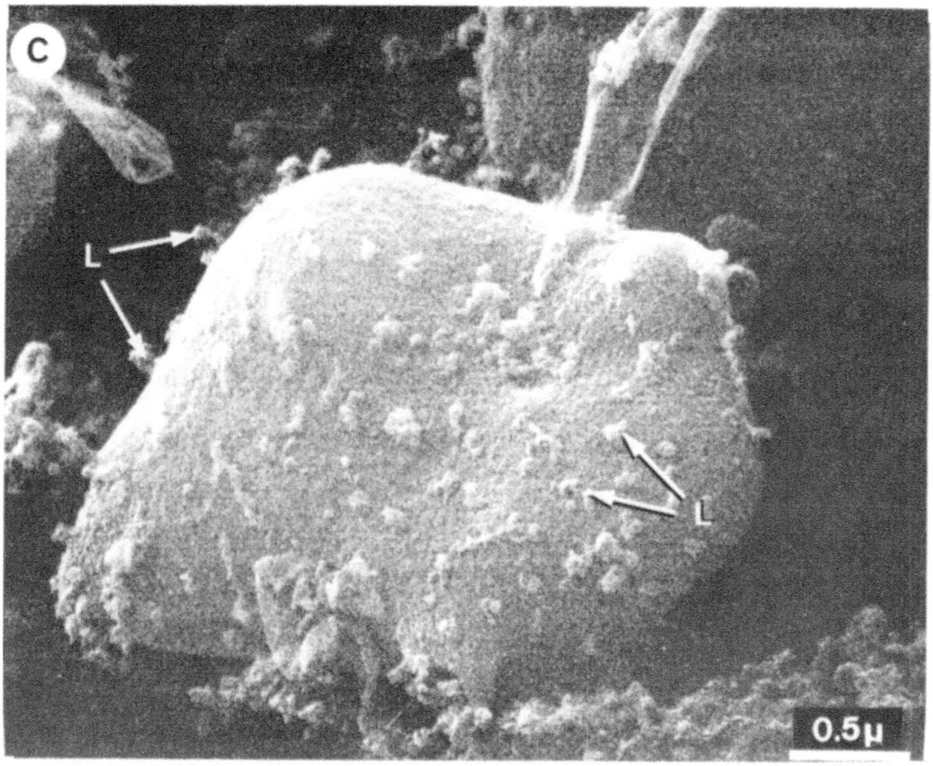

FIGURE 17C

2. **Mazer, N. A., Benedek, G. B., and Carey, M. C.,** Quasielastic light-scattering studies of aqueous biliary, lipid systems. Mixed micelle formation in bile salt-lecithin solutions, *Biochemistry*, 19, 601, 1980.
3. **Kagawa, Y. and Racker, E.,** Partial resolution of the enzymes catalyzing oxidative phosphorylation, *J. Biol. Chem.*, 246, 5477, 1971.
4. **Rhoden, V. and Goldin, S. M.,** Formation of unilamellar lipid vesicles of controllable dimensions by detergent dialysis, *Biochemitry*, 18, 4173, 1979.
5. **Brunner, J., Skrabal, P., and Hauser, H.,** Single bilayer vesicles prepared without sonication. Physico-chemical properties, *Biochim. Biophys. Acta.*, 455, 322, 1976.
6. **Enoch, H. G. and Strittmatter, P.,** Formation and properties of 100-Å-diameter, single-bilayer phospholipid vesicles, *Proc. Natl. Acad. Sci. U.S.A.* 76, 145, 1979.
7. **Gerritsen, W. J., Verkley, A. J., Zwaal, R. F. A., and Van Deenen, L. M.,** Freeze-fracture appearance and disposition of band 3 protein from the human erythrocyte membrane in lipid vesicles, *Eur. J. Biochem.*, 85, 255, 1978.
8. **Milsmann, M. H. W., Schwendener, R. A., and Weder, H.-G.,** The preparation of large single bilayer liposomes by a fast and controlled dialysis, *Biochim. Biophys. Acta*, 512, 147, 1978.
9. **Asanger, M.,** Dissertation, ETH-Zürich.
10. **Keana, J. F. W. and Roman, R. B.,** Improved synthesis of *n*-octyl-β-D-glucoside: a nonionic detergent of considerable potential in membrane biochemistry, *Membr. Biochem.*, 1, 323, 1978.
11. **Schwendener, R. A., Asanger, M., and Weder, H.-G.,** *N*-alkylglucosides as detergents for the preparation of highly homogeneous bilayer liposomes of variable sizes (60-240 nm) applying defined rates of detergent removal by dialysis, *Biochem. Biophys. Res. Commun.*, 100, 1055, 1981.
12. **Singleton, W. S., Gray, M. S., Brown, M. L., and White, J. L.,** Chromatographically homogeneous lecithin from egg phospholipids, *J. Am. Oil. Chem. Soc.*, 42, 53, 1965.
13. **Klein, R. A.,** The detection of oxidation in liposome preparations, *Biochim. Biophys. Acta*, 210, 486, 1970.
14. **Harsch, M.,** Liposomen als Arzneistoffträger für die orale Applikation und das immunologische Targeting, Dissertation ETH-Zürich, Nr. 7039, 1982.

15. **Bertschi, W.,** Dissertation, ETH-Zürich.
16. **Roos, K.,** Dissertation, ETH-Zürich.
17. **Schwendener, R.,** unpublished results.
18. **Müller, M., Meister, N., and Moor, H.,** Freezing in a propane jet and its application in freeze-fracturing, *Mikroskopie,* 36, 129, 1980.
19. **Zalokar, M.,** A simple freeze-substitution method for electron microscopy, *J. Ultrastruct. Res.,* 15, 469, 1966.
20. **Reynolds, E. S.,** The use of lead citrate at high pH as an electron-opaque stain in electron microscopy, *J. Cell Biol.,* 17, 208, 1963.
21. **Baron, C. and Thompson, T. E.,** Solubilization of bacterial membrane proteins using alkyl glucosides and dioctanoyl phosphatidylcholine, *Biochim. Biophys. Acta,* 382, 276, 1975.
22. **Shinoda, K., Yamanaka, T., and Kinoshita, K.,** Surface chemical properties in aqueous solutions of nonionic surfactants: octyl glycol ether, α-octyl glyceryl ether and octyl glucoside, *J. Phys. Chem.,* 63, 648, 1959.
23. **Rose, P. E.,** Improved tables for the evaluation of sphere size distributions including the effect of section thickness, *J. Microsc.,* 118, 135, 1980.
24. **Dingle, J. T., Gordon, J. L., Hazleman, B. L., Knight, C. G., Page Thomas, D. P., Phillips, N. C., Shaw, I. H., Fildes, F. J. T., Oliver, J. E., Jones, G., Turner, E. H., and Lowe, J. S.,** Novel treatment for joint inflammation, *Nature (London),* 271, 372, 1978.
25. **Shaw, J. H., Knight, C. G., and Dingle, J. T.,** Liposomal retention of a modified anti-inflammatory steroid, *Biochem. J.,* 158, 473, 1976.
26. **Leserman, L. D., Barbet, J., Kourilsky, F., and Weinstein, J. N.,** Targeting to cells of fluorescent liposomes covalently coupled with monoclonal antibody or protein A, *Nature (London),* 288, 602, 1980.
27. **Heath, T. D., Macher, B. A., and Papahadjopoulos, D.,** Covalent attachment of immunoglobulins to liposomes via glycosphingolipids, *Biochim. Biophys. Acta,* 640, 66, 1981.
28. **Heath, T. D., Fraley, R. T., and Papahadjopoulos, D.,** Antibody targeting of liposomes: cell specificity obtained by conjugation of F $(ab')_2$ to vesicle surface, *Science,* 210, 539, 1980.
29. **Dunnick, J. K., McDougall, I. R., Aragon, S., Goris, M. L., and Kriss, J. P.,** Vesicle interactions with polyamino acids and antibody: *in vitro* and *in vivo* studies, *J. Nucl. Med.,* 16, 483, 1975.
30. **Huang, A., Huang, L., and Kennel, S. J.,** Monoclonal antibody covalently coupled with fatty acid, *J. Biol. Chem.,* 255, 8015, 1980.
31. **Torchilin, V. P., Omel'yanenko, V. G., Klibanov, A. L., Mikhailov, A. I., Gol'danskii, V. I., and Smirnov, V. N.,** Incorporation of hydrophilic protein modified with hydrophobic agent into liposome membrane, *Biochim. Biophys. Acta,* 602, 511, 1980.
32. **Martin, F. J., Wayne, L. H., and Papahadjopoulos, D.,** Immunospecific targeting of liposomes to cells: a novel and efficient method for covalent attachment of Fab' fragments via disulfide bonds, *Biochemistry,* 20, 4229, 1981.
33. **Harsch, M., Walther, P., Hengartner, H., and Weder, H.-G.,** Targeting of monoclonal antibody-coated liposomes to sheep red blood cells, *Biochem. Biophys. Res. Commun.,* 103, 1069, 1981.
34. **Shankland, W.,** The equilibrium and structure of lecithin-cholate mixed micelles, *Chem. Phys. Lipids,* 4, 109, 1970.
35. **Lapidot, Y., Rappoport, S., and Wolman, Y.,** Use of esters of N-hydroxysuccinimide in the synthesis of *N*-acylamino acids, *J. Lipid Res.,* 8, 142, 1967.
36. **Hauser, H. and Irons, L.,** The effect of ultrasonication on the chemical and physical properties of aqueous egg yolk lecithin dispersions, *Hoppe-Seyler's Z. Physiol. Chem.,* 353, 1579, 1972.
37. **Huang, C. and Mason, J. T.,** Geometric packing constraints in egg phosphatidylcholine vesicles, *Proc. Natl. Acad. Sci. U.S.A.,* 75, 308, 1978.
38. **Kanehisa, M. I. and Tsong, T. Y.,** Cluster model of lipid phase transitions with application to passive permeation of molecules and structure relaxations in lipid bilayers, *J. Am. Chem. Soc.,* 100, 424, 1978.
39. **Zumbühl, O. and Weder, H.-G.,** Cytotoxic effect of unilamellar detergent dialysis liposomes on *Trypanosoma brucei* and *Trypanosoma congolense* bloodstream forms *in vitro, Biochem. Biophys. Res. Commun.,* 1982, in press.
40. **Vickerman, K.,** Polymorphisme and mitochondrial activity in sleeping sickness trypanosomes, *Nature (London),* 208, 762, 1965.
41. **Lanham, S. M. and Godfrey, D. G.,** Isolation of salivarian trypanosomes from man and other mammals using DEAE-cellulose, *Exp. Parasitol.,* 28, 521, 1970.
42. **Zumbühl, O. and Weder, H. G.,** Liposomes of controllable size in the range of 40 to 180 nm by defined dialysis of lipid/detergent mixed micelles, *Biochim. Biochys. Acta,* 640, 252, 1981.

43. **Milsmann, M.,** Entwicklung und Problematik der Anwendung eines liposomalen Arzneistoffträgers am Beispieldes Mutterkornalkaloids Dihydroergokryptin, Dissertation ETH-Zürich, Nr. 6563, 1980.
44. **Zumbühl, O.,** Cholatdialyseliposomen: Herstellung und trypanozide Wirkung, Dissertation ETH-Zürich, Nr. 6969, 1981.

Chapter 8

REMOVAL OF DETERGENT AND SOLVENT TRACES FROM LIPOSOMES

T. M. Allen

TABLE OF CONTENTS

I. INTRODUCTION

Many of the techniques used in the formation of liposomes or liposome-protein recombinants involve initial solubilization of the lipid or lipid-protein mixture in detergents or organic solvents. The detergent or solvent is subsequently removed by any one of a number of techniques with resulting formation of phospholipid bilayers, which depending on the experimental design, may contain proteins or chemicals associated with either their lipid bilayer or their aqueous interior.

The presence of traces of detergents or solvents in liposomes, following their preparation by the techniques discussed below, may have important consequences on experimental outcome. Several authors have reported that detergents can increase the permeability of liposomes to entrapped solutes at concentrations substantially lower than their critical micelle concentration (CMC).[1-7] Others have reported increases in electrical conductances of bilayer lipid membranes at low (0.05 m*M* Triton® X-100) detergent concentrations.[8-11] Rice et al.[12] have shown that residual cholate (>2 wt %) will have significant disordering effects on lipids as determined by ^{2}H and ^{31}PNMR studies of protein-lipid complexes. Gerritsen et al.[13] have reported that levels of Triton® X-100 as low as 0.01% caused extensive aggregation of intramembrane particles of human erythrocytes. Kramer et al.[14] report that transmembrane movement of phosphatidylcholine (PC) was a relatively fast process in liposomes prepared from cholate in contrast to the half-time of translocation of PC of several days for sonicated liposomes containing no cholate. They report this effect at residual cholate levels of 1 molecule per 275 molecules of PC. It is apparent from the above studies that residual detergent or solvent present in liposomes at concentrations several-fold below their CMC can have significant effects on a number of membrane properties. The sensitivity of liposomes to perturbation by detergent is influenced by their composition, size, cholesterol content, and phase transition.[2,5,6]

Care must be taken to reduce residual levels of detergents or solvents to low levels in order to ensure both accuracy and uniformity of experimental results. For most experiments a level of residual detergent or solvent of one molecule per 100 molecules (1 mol% of phospholipid should be adequate to ensure minimal interference by detergent, but in experiments designed to measure membrane conductance, transmembrane movement of phospholipids, or other such sensitive techniques, it is desirable to achieve even lower residual detergent levels. The techniques discussed below are, for the most part, capable of reducing levels of residual detergent to less than 1 molecule per 100 molecules of phospholipid.

Detergents used in the formation of liposomes fall into two main categories, the ionic detergents such as the anionic bile salts cholate and deoxycholate (DOC), and the nonionic detergents, many of them polyoxyethyleneglycol derivatives, such as Triton®-X-100, Lubrol WX, Tween® 20, and the more recent nonionic detergents, *n*-octyl-β-D-glucoside which appears to share some of the properties of both groups.

The most important considerations in the removal of detergents from liposomes are the CMC and the aggregation number. The CMC is a measure of the concentration of amphiphile above which it associates to form micelles or stable colloidal aggregates, and it determines the concentration of free monomer in solution. The aggregation number determines micelle size and micellar weight. A detailed discussion of detergent properties with tables of CMC and aggregation number can be found in the review by Helenius and Simons[15] and the properties of some commonly used detergents are summarized in Table 1. With the exception of octylglucoside, nonionic detergents have low CMCs and high aggregation numbers making them difficult to remove during liposome formation. Ionic detergents and octylglucoside on the other hand, have high CMCs and low aggregation numbers and can be removed with relative ease from detergent-phospholipid solutions with the resulting liposomes containing low levels of residual detergent.[17-22] The ionic detergents and octylglucosides are therefore the detergents of choice in the preparation of liposomes from detergent solution.

Table 1
SOME PROPERTIES OF DETERGENTS COMMONLY USED IN LIPOSOME PREPARATION[a]

Detergent	*Mol wt*	Aggregation *number*	Micellar *mol wt*	CMC (m*M*)	Approx. half-times for dialysis (hr)
Sodium cholate	431	2—4	900—1800	13—15	3[c,d]
Sodium DOC	415	4—10	1700—4200	4—6	10[d]
Triton® X-100	642	140	90,000	0.240	50[d]
Octyglucoside	292	—	—	25[b]	2[e]
Decylglucoside	320	—	—	2.2[b]	—
Dodecylglucoside	348	—	—	1.9[b]	—

[a] Data taken from Reference 15 except where indicated.
[b] Data taken from Reference 16.
[c] Estimated from Reference 45.
[d] Estimated from Reference 20.
[e] Estimated from Reference 23.

However, one of the important applications of liposome formation from detergent solution involves protein reconstitution, and experience has shown that many membrane proteins will not retain their structural integrity when extracted by ionic detergents as compared to nonionic detergents. Octylglucoside, with its high CMC, appears to be useful for some membrane proteins[22-25] but other membrane proteins may require detergents with longer alkyl chains (lower CMCs).[22] In order to reconstitute into liposomes proteins which have been extracted in nonionic detergents such as Triton® X-100, the nonionic detergent, in most instances, must be replaced by an ionic detergent. A discussion of the techniques for removing (or replacing) detergents from protein is beyond the scope of this chapter but several excellent reviews discuss this topic.[26-30]

Examples of liposome preparations from ionic detergent solutions include the detergent dialysis method of Racker et al.[31] as applied by a number of investigators,[32-35] the hollow fiber dialysis of Rhoden and Goldin,[19] and the dialysis cell method of Milsmann et al.[18] Liposomes have also been prepared from ionic detergent solutions by a gel filtration method,[17,20,36,37] and from nonionic detergent by gel filtration[20] or following passage of solution over, or dialysis against a polymeric adsorbent such as Biobeads SM-2 or Amberlite® XAD-2.[20,35,38,39] In the dilution method, phospholipid-detergent micelles are diluted into 30- to 60-fold excess of detergent-free buffer.[40] Liposomes have been formed from octylglucoside by dialysis or by gel filtration.[21-24]

Solvents used in liposome preparation include those solvents such as hexane, ethanol, chloroform, methanol, and benzene in which most phospholipid preparations are stored and which are removed immediately prior to lipid use, and solvents such as ethanol, diethyl isopropylether, chloroform, and methanol, which are used in the actual preparation of liposomes. One of the important considerations in the removal of solvents during liposome preparation is the boiling point of the solvent. The solvent should be capable of being vaporized under low vacuum at temperatures where protein denaturation or sample breakdown will not occur and at temperatures below the vaporization point of the buffer system employed in the preparation. Most of the solvents currently in use for liposome preparations have boiling points between 35 and 65°C at atmospheric pressure and at 40 torr (53.2 Pa) will vaporize between −30 to +5°C, well below the temperature at which protein denaturation occurs or at which water (buffer) vaporizes (34°C at 40 torr). Ethanol is an exception and cannot easily be removed from aqueous solutions by rotary evaporation as codistillation of ethanol and water occur. In this case its ability to be removed from the sample by dialysis against an aqueous buffer is important.

Examples of liposome preparation using organic solvents include the ethanol-injection method of Batzri and Korn,[41,42] the ether infusion method of Deamer and Bangham,[43] and the reverse-phase-evaporation (REV) method of Szoka and Papahadjopoulos.[44]

II. METHODOLOGY

The initial step in most phospholipid preparations is the removal of organic solvent from lipid to form a dried lipid film coating the interior of the reaction vessel. This procedure is usually accomplished by rotary evaporation of the lipid preparation followed by a 1- to 18-hr period of exposure of the dried film to a high vacuum ($< 10^{-2}$ torr or 1.33 PA) such as that generated in a freeze-drying apparatus, in order to remove residual traces of solvent. This period of exposure to high vacuum is important, as most house vacuum systems or vacuums generated by water pressure are not high enough to remove all residual solvent, particularly when high concentrations of lipid are involved (solvent pockets may be left behind in the dried lipid). Many readers are no doubt familiar with the problems arising out of solvent traces in the dried lipid film. Following hydration, these preparations are difficult to sonicate to clarity, taking up to several-fold the time of a well-dried lipid preparation. In some instances it is not possible to sonicate samples containing residual organic solvents to clarity.

Detergent levels in liposome preparations are most easily measured by incorporation of radiolabeled detergents into the sample prior to liposome formation. Radiolabeled detergents are available for all of the most commonly used detergents (cholate, DOC, Triton® X-100, octylglucoside) and for many of the organic solvents. We have not yet found a commercial source for radiolabeled ether. These radiolabeled compounds are frequently used without further purification but trace amounts of impurities are present in the preparations and if it is critical to have an accurate estimate of residual detergent, the radiolabeled source should be purified. One way of doing this is to add the label to a concentrated solution of cold detergent (cholate or DOC), recrystallize the detergent, and use the labeled, recrystallized detergent in subsequent experiments. High-molecular-weight contaminants can be removed by dialyzing the radiolabeled detergent exhaustively against a concentrated solution of cold detergent and lyophilizing the solution external to the dialysis bag for use in subsequent experiments.

Many of the nonionic detergents of the Brij and Lubrol series are not available as radiolabeled compounds and therefore, for these substances it will be difficult to measure residual levels of detergent. A rapid, crude measure of residual Triton® X levels can be obtained from $O.D._{280}$ readings, as Triton®X absorbs strongly at this wave length.

A. Dialysis

1. Ionic Detergents

In a typical liposome preparation made by removing ionic detergents by dialysis, from 2 to 6 mol of detergent (cholate or DOC) are required to solubilize 1 mol of phospholipid although some investigators have formed liposomes successfully with ratios as low as 0.5.[48] The resulting solution is clear or very slightly turbid. This solution is placed in a bag made from dialysis tubing which has been presoaked and rinsed in distilled water in order to remove glycerine and other impurities. Weiner[51] boils the dialysis tubing in 10 m*M* EDTA twice, followed by boiling twice in distilled water and stores the tubing in 70% ethanol, 1 m*M* EDTA. This procedure is useful when removal of trace metal and other impurities from the dialysis tubing is critical, e.g., in enzyme reconstitution. A convenient technique for dialysis is to knot one end of the dialysis tubing, introduce the sample, and clamp the other end with a clamp similar to those available from Spectra/Por. This allows ease of sampling during the dialysis procedure. The dialysis bag is placed in at least 100 volumes of detergent-free buffer at 4°C and continuously stirred. The detergent-free buffer should be changed at frequent intervals. The solutions in the bag will start to become turbid after a few hours of dialysis indicating the formation of lipid bilayers. The half-time for detergent removal depends on the molecular weight cut off of the dialysis tubing used and the wall thickness of the tubing. The older type of cellulose dialysis tubing are classified only according to wall thickness. For dialysis tubing with a given molecular cut off, for example 12,000 to 14,000 mol wt, dialysis will be more rapid the thinner the walls of the tubing. The higher the molecular weight cut off the more rapidly will dialysis of detergent take place. Kagawa et al.[45] have reported a half-time for cholate removal of 3 hr with cellulose tubing of 0.04 mm wall thickness. Similar results have been reported by Allen et al.[20] (Figure 1A). DOC, which has a higher aggregation number, and therefore a higher molecular weight has a longer half-time for removal (10 hr) (Figure 1A). The kinetics of removal differ for the two detergents (Figure 1A). The explanation may lie in the initial removal of cholate as both micelles and monomers during dialysis. The low micellar molecular weight of cholate suggests this possibility.

Even though liposome formation takes place from cholate or DOC solutions after less than 1 day dialysis, significant amounts of detergent will remain associated with the phospholipid membranes. Using radiolabeled cholate and DOC Allen et al.[20] report that 4 mol cholate per 100 mol phospholipid (4 mol %) remain liposome associated after 50-hr dialysis. The corresponding figure for DOC is 30 mol %. It requires 150 hr of dialysis to get levels of both detergents down to 1 mol/100 mol phospholipid (1 mol %). Many authors report remaining detergent only in terms of percent of original detergent. Expressing the data in terms of mole detergent per mol phospholipid (or protein) is more precise and allows comparisons between different detergents to be easily made. The last few molecules of detergent are removed only very slowly from phospholipid bilayers by dialysis (Figure 1A) and it may prove to be too time consuming to dialyze samples for up to 1 week or longer. Other methods like gel filtration (see below) could be used in combination with dialysis to further reduce detergent levels.

Liposomes prepared by sample dialysis show considerable size heterogeneity. Rhoden and Goldin[19] have used rapid (hollow fiber) dialysis to achieve residual cholate levels of 8 mol % after 16-hr dialysis. In their experiments the mean diameter of the vesicles could be varied from 34 to 128 nm depending on the conditions of pH and mole fraction cholesterol. Milsmann et al.,[18] using a stirred three-compartment dialysis cell, have achieved cholate levels of less than 2 mol % in 24-hr dialysis with resulting homogeneous unilamellar liposomes of a mean diameter of 60 nm.

2. *Nonionic Detergents*

a. Triton® X-100 and Similar Detergents

Because of their low CMC and high aggregation numbers, Triton® X-100 and similar detergents will dialyze extremely slowly. Robinson and Tanford[46] report that Triton® X-100 solutions above their CMC (> 0.02%) could not be equilibrated by the techniques of equilibrium dialysis in any reasonable time period, even in the absence of a protein which binds Triton® X-100. Figure 1B demonstrates the extremely slow equilibration of Triton® X-100. Dialysis times for Triton® X-100 and similar detergents can be reduced by dialyzing against buffer containing suspended Biobeads SM-2 or Amberlite® XAD-2.[38,39] Allen et al.[20] report levels of 16 mol % Triton® X-100 (7% of the original 0.2% detergent) associated with liposomes after 4 days dialysis against Biobeads SM-2, a considerable reduction in the 2 weeks or more of dialysis it would require to achieve similar levels of detergent in the

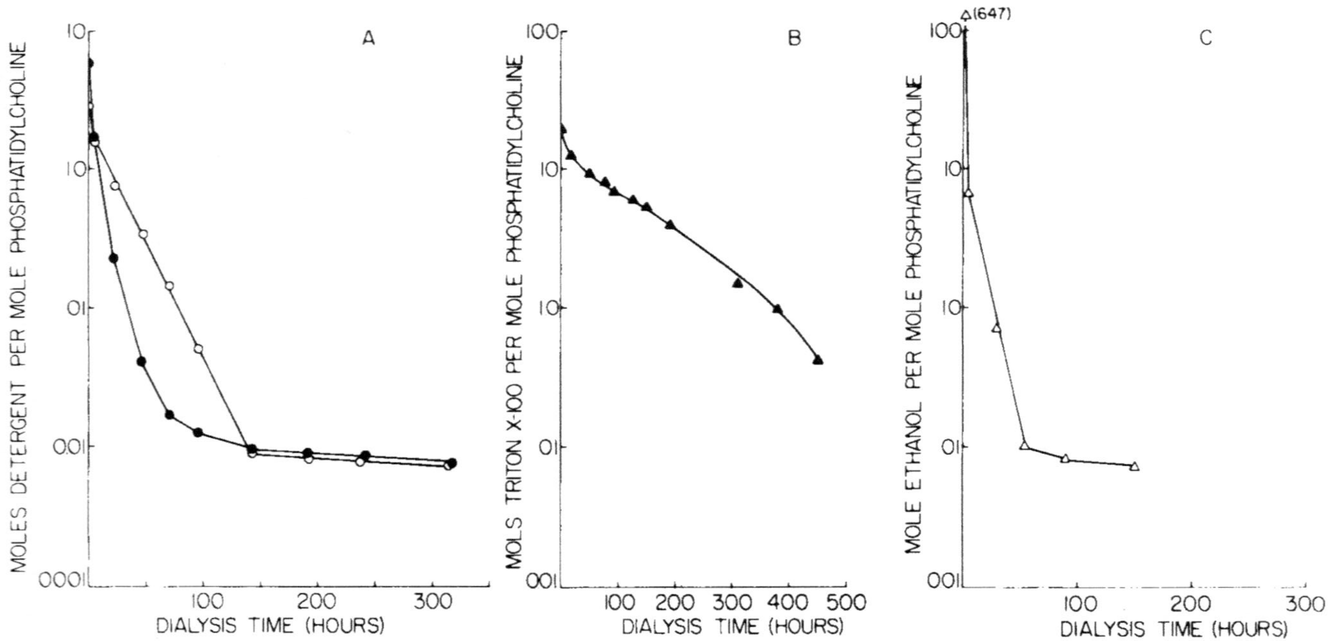

FIGURE 1. Time course for removal of detergents or ethanol from 5 mg of egg phosphatidylcholine in 2 mℓ of detergent solution or 5 mg of phosphatidylcholine in 250 *M* ethanol squirted into 4.75 mℓ 10 m*M* Tris, pH 8.2. Samples were dialyzed against several changes of 1 ℓ of 10 m*M* Tris buffer, pH 8.2 at 4°C. (A) Cholate (●), DOC (○); (B) Triton® X-100 (▲); (C) Ethanol (△). (From Allen, T. M., et al., *Biochim. Biophys. Acta*, 601, 328, 1980. With permission.

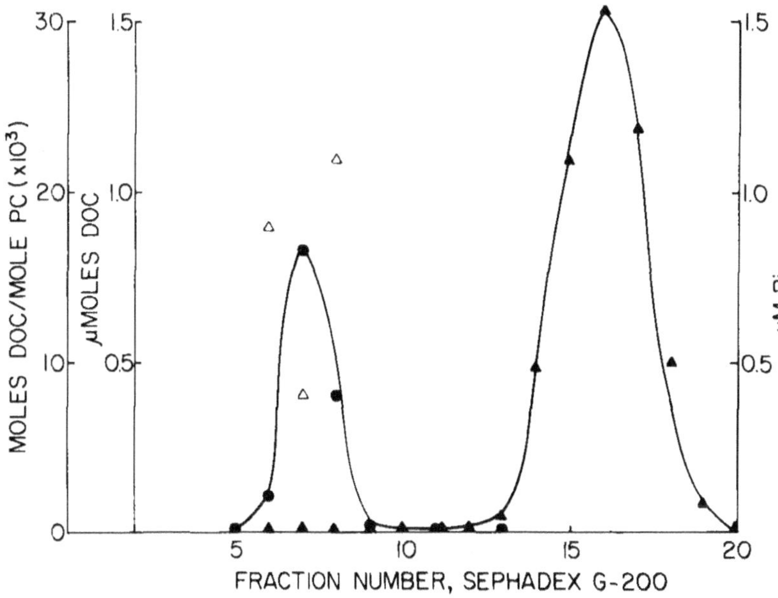

FIGURE 2. Gel filtration on 1 cm × 40 cm Sephadex® G-200 of 200 μℓ of 20 mM Hepes buffer containing 1.3 μM egg PC (●) and 4.8 μM sodium DOC (▲) labeled with ^{14}C-deoxycholate. 2 mℓ fractions were collected. DOC:phospholipid molar ratio (△).

absence of the beads. Regenerating the beads by washing with 4 volumes of methanol, rinsing them in distilled water and resuspending them in detergent-free buffer will speed up dialysis time when high levels of detergent (1 to 2% Triton® X-100) are being dialyzed. It may not be possible to achieve low levels of these detergents in the presence of hydrophobic membrane proteins (such as fusion protein from Sendai virus) which are capable of tightly binding the detergent. In these situations detergent exchange with an ionic detergent such as cholate may be effective.[30,47]

b. N-Octyl-β-D-Glucopyranoside (Octylglucoside)

Octylglucoside occupies a unique position among the nonionic detergents in that it has a high CMC (Table 1) and is easily dialyzed. Although less common, *n*-hexyl and *n*-heptyl-β-D-glucoside may also be used in liposome preparation.[21] A 3:1 ratio of octylglucoside to lipid is sufficient to obtain a clear solution, but for optimal liposome formation, molar ratios of 10:1 to 15:1 detergent:lipid are required.[22] Homogeneous populations of liposomes varying from 60 to 240 nm result from the dialysis of *n*-alkylglucosides.[21] Helenius et al.[23] report that 95% of octylglucoside was dialyzed away from a sample of egg PC/Semliki forest virus in 10 hr, and less than 0.06% of the octylglucoside was present in the dialysis bag after 36-hr dialysis. Based on these figures the half-time for removal of octylglucoside appears to be approximately 2 hr, slightly less than cholate and much less than DOC. The dialysis half-time for shorter chain glycosides will also be short, but longer chain glycosides like *n*-decyl or *n*-dodecyl, with much lower CMCs (Table 1) will have dialysis half-lives of several hours.

3. Ethanol

In the ethanol-injection method of preparing liposomes, lipids are dissolved in a small quantity of ethanol and squirted through a 22-gauge needle into rapidly stirred buffer to give a maximum of 7.5% ethanol. A homogeneous liposome population of 25 nm diameter is formed.[41] Ethanol is rapidly removed from the sample by dialysis with approximately the

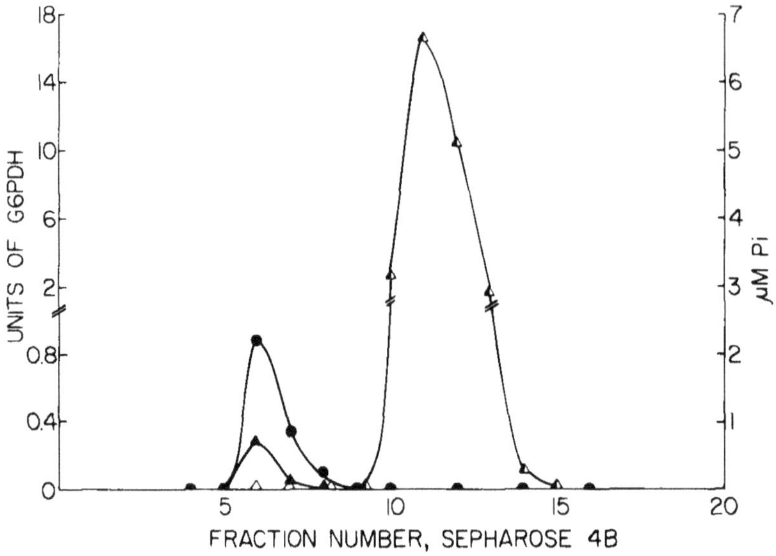

FIGURE 3. Trapping of glucose-6-phosphate dehydrogenase (G6PDH) from Leuconostoc mesenteroides in large unilamellar liposomes formed by gel filtration. 0.5 mℓ of 1.3% DOC was added to 20 mg PC and 0.5 mℓ of 100 units/mℓ G6PDH and the mixture run over a 0.7 cm × 10 cm Sephadex® G-50 column. 1.5 mℓ from the void volume was run over a 1 cm × 40 cm Sepharose 4B column. 2 mℓ fractions were collected and analyzed for phosphate (●) and G6PDH in the presence (▲) and absence (△) of Triton® X-100. The reaction mixture contained 0.05 m*M* NAD and 1.5 m*M* glucose-6-PO_4. The vesicle-trapped internal volume averaged 5 ℓ/mol phospholipid indicating an average vesicle diameter of approximately 150 nm.

same time course as for cholate removal (Figure 1C). However, there is a small pool of ethanol (10 mol%) which cannot be removed from the bilayer vesicles even after extensive (> 300 hr) dialysis.[20]

B. Gel Filtration

1. Ionic Detergents

In this technique, liposomes are formed by gel filtration of phospholipid-detergent solutions (prepared as in the dialysis method) on gels such as Sephadex® G-50, G-200, Sephacryl® S-200, Biogel® P-30, or other similar gels having a molecular exclusion limit of roughly 5000 to 500,000 daltons. Cholate or DOC-phospholipid solutions (1 to 5 mℓ depending upon column size) are loaded onto a column which has been presaturated with phospholipid (by passing a few moles of liposomes over it and washing with the appropriate buffer). The column is then eluted at a flow rate of 8 to 30 mℓ/hr with detergent-free buffer. Liposomes appear in the void volume (V_0) while detergent is eluted at larger volumes (Figure 2). Residual cholate levels in the liposomes formed by gel filtration were reported by Brunner et al.[17] to be 2 mol % after Sephadex® G-50 chromatography and by Allen et al.[20] to be 8 and 5 mol % after a single passage over Sephadex® G-50 or G-200 columns, respectively. A second passage of the liposomes over Sephadex® G-50, G-200, Sepharose 4-B, or a 12-hr dialysis reduced detergent levels to below 1 mol %.[17,20] Residual DOC levels were 1 mol % in liposomes after a single passage over Sephadex® G-200 or G-50 and were further reduced by a second passage over either of the aforementioned columns or Sepharose 4B.[20] The amount of residual detergent was independent of phospholipid headgroups or the presence of an intrinsic membrane protein (erythrocyte glycophorin).[20] Molar ratios of detergent to phospholipid from 2 to 4 are optimum for this technique. It is possible, in some circumstances to overload the column, for example at high detergent concentrations, and when this happens substantial amounts of detergent co-elute with the phospholipid.

Brunner et al.[17] report unilamellar liposomes of 30-nm diameter formed by this technique. We have found the liposomes to be 100 to 150 nm in diameter, trapping 3 to 4 ℓ/ mol phospholipid in their aqueous interior. Similar results have been reported by Enoch and Strittmatter.[48] Liposomes of similar size have been obtained independent of whether Sephadex® G-25, G-50, G-100, or G-200 are used.[52]

We have observed that liposome formation takes place in the first 1 to 2 cm of a Sephadex® G-50 column at detergent to phospholipid molar ratios of around 4:1 and have put this observation to practical use to incorporate proteins or chemicals into the aqueous interior. To incorporate a large molecule such as glucose-6-PO_4-dehydrogenase we dissolve the enzyme in the detergent-phospholipid mixture and pass the solution over Sephadex® G-50. Both free enzyme and liposome-entrapped enzyme elute in the void volume. Liposome-entrapped enzyme can be separated from free enzyme by chromatography over Sepharose 4B (Figure 3). Small molecules such as NAD, calcein, cytosine arabinoside, etc. can also be trapped by this technique with the following modification. The phospholipid-detergent-chemical solution is loaded onto a column which has been preloaded with an equal volume of detergent-free chemical (e.g., preload with 1 mℓ of calcein, then add 1 mℓ of phospholipid:detergent:calcein to a 1 cm × 40 cm Sephadex® G-50 column). The phospholipid-detergent solution moves faster through the column than the free chemical and will over take the preloaded free chemical 1 to 2 cm down the column where liposome formation and entrapment of chemical take place. In this instance, liposomes containing trapped chemical appear in the void volume, while free chemical and detergent elute at larger volumes. Maximizing the gel surface area increases the trapping efficiency. We have used 2.5 cm × 15 cm Sephadex® G-50 columns to trap cytosine arabinoside.[49]

Membrane proteins have also been efficiently incorporated into phospholipid bilayers by a gel filtration technique.[20,36,37] This technique is a more rapid means of reconstitution than dialysis, but may not always result in an active preparation. Kagawa et al.[45] report that the enzymes catalyzing oxidative phosphorylation were inactive following rapid removal of cholate by Sephadex® chromatography, but active following the slower removal of detergent by dialysis. Warren et al.[35] report that during reconstitution of a Ca^{++} pump, the lipid-protein complex formed by rapid detergent removal by sucrose-gradient centrifugation was unable to accumulate Ca^{++}, but reconstitution by the slower method of detergent dialysis resulted in Ca^{++} accumulation at a rate comparable to native sarcoplasmic reticulum.

2. *Nonionic Detergents*

a. Triton® X-100

Triton® X-100, at detergent to phospholipid molar ratios of 2 to 4 (0.2% original concentration) can be removed from phospholipid with resulting liposome formation following gel filtration on Sephadex® G-50 or G-200. The resulting liposomes contain from 10 to 20 mol % of detergent[20] after a single passage over Sephadex® G-50 or G-200, approximately ten-fold the level of residual detergent found with the anionic bile salts. Higher levels of Triton® X-100, such as those found in many membrane protein preparations could not be effectively removed by gel filtration, and Triton® X-100 micelles co-eluted with protein and lipid in the void volume.

b. N-Alkylglucosides

Mimms et al.,[22] using similar techniques to those described above for ionic detergents, have formed unilamellar liposomes of 230 nm average diameter containing an average internal volume of 7.5 ℓ/ mol phospholipid following removal of octylglucoside by Sephadex® G-25 or G-50 chromatography. After a single passage over Sephadex® G-25, residual detergent levels were 0.5 mol % and these levels were halved by a subsequent passage over Sepharose 4B-CL. Cytochrome C could be trapped in the aqueous interior of the liposomes by this technique and the membrane protein glycophorin could be reconstituted into liposomes. Octylglucoside appears, therefore, to be very similar to the anionic bile salts in its behavior upon gel filtration.

3. Organic Solvents

Ethanol could be removed from ethanol-injection liposomes by gel filtration. After a single passage over a Sephadex® G-200 column only 0.1% of ethanol remained associated with liposomes. However, since the molar ratio of ethanol to phospholipid in ethanol-injection liposomes is initially high (700 to 3000), this represents a substantial amount of ethanol (290 mol %).[20] Much lower levels (7 mol %) have been achieved by dialysis (see above). It is possible that a second passage over Sephadex® would lower the levels closer to that achieved by extensive dialysis.

Gel filtration has also been used to remove residual ether from ether-injection liposomes. Following gel filtration on Sephadex® G-50, Deamer[43] has reported that, using ^{14}C-diethylether, no significant counts over background could be detected. However this method was only capable of detecting 2 m*M* ether (2 μmol/mℓ) and in this method lipid concentrations are in the order of 1 μmol lipid per milliliter, so significant residual ether could be present and remain undetected.

In the reverse-phase-evaporation (REV) method of liposome preparation from ether solution the REV solution was found to contain a few parts per million of ether by NMR. The REVs, before further treatment, also showed unexpected toxicity with cells in culture, which could be due to residual ether, or to peroxides present in the ether.[53] In Papahadjopoulos' laboratory they wash the ether in advance with the buffer to be used in liposome preparation and store the ether in bottles over water containing bisulfite. When they add an equal volume of buffer to the liposome preparation after the initial evaporation procedure and evaporate for a further 20 min they find no traces of ether detectable by NMR. Gel filtration of the liposomes over Sephadex® will also reduce the trace amounts of ether present in REVs but no estimate of residual ether is available.[53] REVs prepared in our laboratory[52] and in that of Mayhew[54] have been found to be a few percent more leaky to entrapped aqueous solutes than small unilamellar liposomes prepared in the absence of detergent. This effect could conceivably be due to traces of ether in the REVs.

C. Polymeric Adsorbents

Polymeric adsorbents such as Biobeads SM-2 and Amberlite® XAD-2 can be used to remove detergents such as cholate, DOC, and Triton® X-100 from liposomes. Addition of the absorbent to the buffer used in dialysis procedures will speed up the removal of detergent. Alternatively, the resin can be packed into a chromatography column and the detergent-phospholipid solution passed over the column. Liposomes will appear in the void volume of the column and detergent will bind to the column. The column can be regenerated by washing in 4 column volumes of methanol and rinsing in distilled water. A Biobeads SM-2 column (0.5 cm × 15 cm) was sufficient to remove 1 mℓ of 0.2% Triton® X-100 from 1 mg PC. Residual Triton® X-100 was 16 mol %[20] similar to levels found after Sephadex® G-50 chromatography. Cheetham[39] gives the maximum binding capacities for Triton® X-100 for a number of polymeric adsorbents.

D. Detergent-Dilution

This technique is generally applied to the reconstitution of proteins into phospholipid bilayers.[25,40] In this technique a mixture of protein, phospholipid, and detergent (usually cholate) is diluted into 30 to 60 volumes of detergent-free buffer. The resulting liposomes can be concentrated by ultrafiltration (e.g., Amicon) or centrifugation and freed of residual buffer by dialysis or gel filtration.

III. DISCUSSION

Each of the techniques for detergent removal and liposome formation outlined have advantages and disadvantages depending on the desired application. A discussion of these follows.

The dialysis method can be applied to very large as well as small sample volumes. Given sufficient time, all detergents can be removed to negligible levels by this method with little or no change in the initial sample volume. A number of proteins have been successfully reconstituted by the dialysis method.[31-35] On the negative side, dialysis is generally time consuming and cumbersome, particularly if several samples are involved. Proteins are exposed to detergents over long time periods which could cause denaturation, but this is unlikely with the anionic bile salts and Triton® X-100.[50] On the other hand, the slow removal of detergent during reconstitution of some systems was necessary for active preparations.[35,45] The dialysis method is not a good method for entrapping small molecules in the aqueous interior of liposomes. Trapping efficiencies are extremely low and the recovery of nonentrapped solute (e.g., when an expensive solute is involved) is tedious. Trapping of large molecules, which do not cross the dialysis membrane, in the interior of liposomes formed by the dialysis method can be achieved by addition of the molecule, in the desired concentration, to the mixture in the dialysis bag. Trapping efficiencies are again low. The dialysis method achieves most use in the reconstitution of proteins.

The detergent-dilution method shares many of the disadvantages of the dialysis method. Its principle advantage lies in the rapid and reproducible way in which some membrane proteins can be reconstituted into liposomes. This is particularly useful when many samples must be processed. However, in this method liposomes are present in very large volumes of buffer containing significant quantities of residual detergent, necessitating additional procedures to concentrate the sample and remove detergent.

Gel filtration is a versatile method enabling formation of homogeneous populations of unilamellar liposomes of a variety of sizes, depending on the conditions employed. Most detergents and solvents can rapidly be removed and the resulting liposomes contain low levels of residual detergents. Small or large molecules can be trapped in the aqueous interior and many proteins can be reconstituted by this method. The main disadvantages of gel filtration in liposome preparation are the following. Unlike dialysis, it is possible to process only relatively small volumes (a few milliliters) at one time. Gel filtration results in at least a twofold dilution of the sample each time it passes over a column. Phospholipid and protein may be absorbed to column material, but the amounts of absorbed material can be greatly diminished by presaturating the column with phospholipid and bovine serum albumin. Liposomes formed by gel filtration will entrap only a small percentage (1 to 8%) of solutes in their aqueous interior making this an inefficient means of trapping drugs or enzymes, as compared to the REV method. Gel filtration is useful for trapping substances when the chemical or protein to be entrapped is susceptible to breakdown or inactivation in solvents such as ether. For example, we have been able to trap active glucose-6-PO_4-dehydrogenase in liposomes using gel filtration of DOC:phospholipid:enzyme mixtures, but have found the enzyme to be inactivated if exposed to ether in the REV method or the ether-injection method.

The REV method provides by far the most efficient means of entrapping aqueous solutes in large unilamellar liposomes, with up to 60% capture achieved. Problems may arise with this method with some proteins which are susceptible to denaturation in an ether (or other solvent) environment. Traces of ether remain in liposomes prepared by this method, but to date there is no definitive data on the levels of residual ether, or the possible effects of this on membrane properties. Because of the increasing popularity of the REV method, additional data is needed in this regard.

The use of polymeric adsorbants such as Biobeads SM-2 and Amberlite® XAD-2 appear to offer some advantage in the removal of large concentrations of nonionic detergents such as Triton® X-100 from phospholipids and proteins. However, problems with the concomitant absorption of phospholipid and proteins to these materials exist if they are being used in column form. When present in the detergent-free buffer in a dialysis experiment they will greatly increase the rate at which detergents can be removed from solution.

The ethanol-injection method does not appear to offer any advantages over any of the above methods, except perhaps in case of the reconstitution or trapping of a protein which is denatured in other organic solvents. A small pool of ethanol appears to remain membrane associated and is very difficult to remove.

The techniques for detergent removal and liposome formation have by and large been discussed separately in this chapter, but obviously much can be gained in time and convenience by using combinations of methods, for example, detergent dialysis to form liposomes, followed by gel filtration to rapidly remove residual detergent.

ACKNOWLEDGMENT

I would like to thank Dr. Joel Weiner of the Department of Biochemistry, University of Alberta for helpful comments and Ms. Rosalind Millar for typing the manuscript.

REFERENCES

1. **Weissmann, G., Sessa, G., and Weissmann, S.,** The action of steroids and Triton X-100 upon phospholipid cholesterol structures, *Biochem. Pharm.*, 15, 1537, 1966.
2. **Inoue, K. and Kitagawa, T.,** Effect of lipid composition on sensitivity of lipid membranes to Triton X-100, *Biochim. Biophys. Acta*, 426, 1, 1976.
3. **Kinsky, S. C., Haxby, J. A., Zopf, D. A., Alving, C. R., and Kinsky, C. B.,** Complement-dependent damage to liposomes prepared from pure lipids and Forssman Hapten, *Biochemistry*, 8, 4149, 1969.
4. **Hertz, R. and Barenholz, Y.,** The relation between the composition of liposomes and their interaction with Triton X-100, *J. Coll. Interface Sci.*, 60, 188, 1977.
5. **Richards, M. H., and Gardner, C. R.,** Effects of bile salts on the structural integrity of liposomes, *Biochim. Biophys. Acta*, 543, 508, 1978.
6. **Castellino, F.J. and Violand, B. N.,** ^{31}P-nuclear magnetic resonance and ^{31}P[^{1}H]nuclear overhauser effect analysis of mixed egg phosphatidylcholine-sodium taurocholate vesicles and micelles, *Arch. Biochem. Biophys.*, 193, 543, 1979.
7. **Alonso, A., Villena, A., and Goni, F. M.,** Lysis and resembly of sonicated lecithin vesicles in the presence of Triton X-100, *FEBS Lett.*, 123, 200, 1981.
8. **Seufert, W. D.,** Induced permeability changes in reconstituted cell membrane structure, *Nature (London)*, 207, 174, 1965.
9. **Van Zutphen, H., Merola, A. J., Brierley, G. P., and Cornwall, D. G.,** The interaction of nonionic detergents with lipid bilayer membrane, *Arch. Biophys. Biochem.*, 152, 755, 1972.
10. **Schlieper, P. and de Robertis, E.,** Triton X-100 as a channel-forming substance in artificial lipid bilayer membranes, *Arch. Biochem. Biophys.*, 184, 204, 1977.
11. **Bangham, J. A. and Lea, E. J. A.,** The action of detergents with bilayer lipid membranes, *Biochim. Biophys. Acta*, 511, 388, 1978.
12. **Rice, D. M., Hsung, J. C., King, T. E., and Oldfield, E.,** Protein-lipid interactions. High-field deuterium and phosphorus nuclear magnetic resonance spectroscopic investigation of the cycochrome oxidase - phospholipid interaction and the effects of cholate, *Biochemistry*, 18, 5885, 1979.
13. **Gerritsen, W. J., Verkleij, A. J., and Van Deenen, L. L. M.,** The lateral distribution of intramembrane particles in the erythrocyte membrane and recombinant vesicles, *Biochem. Biophys. Acta*, 555, 26, 1979.

14. **Kramer, R. M., Hasselbach, H. J., and Semenza, G.,** Rapid transmembrane movement of phosphatidylcholine in small unilamellar lipid vesicles formed by detergent removal, *Biochim. Biophys. Acta,* 643, 233, 1981.
15. **Helenius, A. and Simons, K.,** Solubilization of membranes by detergents, *Biochim. Biophys. Acta,* 415, 29, 1975.
16. **Shinoda, K., Yamaguchi, T., and Hori, R.,** The surface tension and the critical micelle concentration in aqueous solution of β-alkylglucosides and their mixtures, *Bull. Chem. Soc. Jpn.,* 34, 237, 1961.
17. **Brunner, J., Skrabal, P., and Hauser, H.,** Single bilayer vesicles prepared without sonication. Physico-chemical properties, *Biochim. Biophys. Acta,* 435, 322, 1976.
18. **Milsmann, M. H. W., Schwendener, R. A., and Weder, H. G.,** The preparation of large single bilayer liposomes by a fast and controlled dialysis, *Biochim. Biophys. Acta,* 512, 147, 1978.
19. **Rhoden, V. and Goldin, S.M.,** Formation of unilamellar lipid vesicles of controllable dimensions by detergent dialysis, *Biochemistry,* 18, 4173, 1979.
20. **Allen, T. M., Romans, A. Y., Kercret, H., and Segrest, J. P.,** Detergent removal during membrane reconstitution, *Biochim. Biophys. Acta,* 601, 328, 1980.
21. **Schwendener, R. A., Asanger, M., and Weder, H. G.,** *n*-Alkylglucosides as detergents for the preparation of highly homogeneous bilayer liposomes of variable sizes applying defined rates of detergent removal by dialysis, *Biochem. Biophys. Res. Commun.,* 100, 1055, 1981.
22. **Mimms, L. T., Zampighi, G., Nozaki, Y., Tanford, C., and Reynolds, J. A.,** Phospholipid vesicle formation and transmembrane protein incorporation using octylglucosides, *Biochemistry,* 20, 833, 1981.
23. **Helenius, A., Fries, E., and Kartenbeck, J.,** Reconstitution of Semliki forest virus membrane, *J. Cell Biol.,* 75, 866, 1977.
24. **Engelhard, V.H., Guild, B. C., Helenius, A., Terhorst, C., and Strominger, J. L.,** Reconstitution of purified detergent-soluble HLA-A and HLA-B antigens into phospholipid vesicles, *Proc. Natl. Acad. Sci. U.S.A.,* 75, 3230, 1978.
25. **Racker, E., Violand, B., O'Neal, S., Alfonzo, M., and Telford, J.,** Reconstitution, a way of biochemical research; some new approaches to membrane-bound enzymes, *Arch. Biochem. Biophys.,* 198, 470, 1979.
26. **Kagawa, Y.,** Reconstitution of oxidative phosphorylation, *Biochim. Biophys. Acta,* 265, 297, 1972.
27. **Razin, S.,** Reconstitution of biological membranes, *Biochim. Biophys. Acta,* 265, 241, 1972.
28. **Gulik-Kryzwicki, T.,** Structural studies of the associations between biological membrane components, *Biochim. Biophys. Acta,* 415, 1, 1975.
29. **Tanford, C. and Reynolds, J. A,** Characterization of membrane protein in detergent solutions, *Biochim. Biophys. Acta,* 437, 133, 1976.
30. **Furth, A. J.,** Removing unbound detergent from hydrophobic proteins, *Anal. Biochem.,* 109, 207, 1980.
31. **Kagawa, Y. and Racker, E.,** Partial resolution of enzymes catalyzing oxidative phosphorylation. XXV. Reconstitution of vesicles catalyzing^{32} P_i-adenosine triphosphate exchange, *J. Biol. Chem.,* 246, 5477, 1971.
32. **Slack, J. R., Anderton, B. H., and Day, W. A.,** A new method for making phospholipid vesicles and the partial reconstitution of ($Na^+ \cdot K^+$)-activated ATPase, *Biochim. Biophys. Acta,* 323, 547, 1973.
33. **Racker, E. and Kandrach, A.,** Partial resolution of the enzymes catalyzing oxidative phosphorylation. XXXIX. Reconstitution of the third segment of oxidative phosphorylation, *J. Biol. Chem.,* 248, 5841, 1973.
34. **Michaelson, D. M. and Raftery, M. A.,** Purified acetylcholine receptor: its reconstitution to a chemically excitable membrane, *Proc. Natl. Acad. Sci. U.S.A.,* 71, 4768, 1974.
35. **Warren, G. B., Toon, P. A., Birdsall, N. T. M., Lee, A. G., and Metcalfe, J. C.,** Reconstitution of a calcium pump using defined membrane components, *Proc. Natl. Acad. Sci. U.S.A.,* 71, 622, 1974.
36. **Wacker, H., Muller, F., and Semenza, G.,** Incorporation of hydrophobic aminopeptidase from hog kidney into egg lecithin liposomes: number and orientation of aminopeptidase molecules in the lecithin vesicles, *FEBS Lett.,* 68, 145, 1976.
37. **Hall, E. R. and Brodbeck, V.,** Human erythrocytes membrane acetylcholine esterase. Incorporation into the lipid bilayer structure of liposomes, *Eur. J. Biochem.,* 89, 159, 1978.
38. **Holloway, P. W.,** A simple procedure for removal of Triton X-100 from protein samples, *Anal. Biochem.,* 53, 304, 1973.
39. **Cheetam, P. S. J.,** Removal of Triton X-100 from aqueous solution using Amberlite XAD-2, *Anal. Biochem.,* 92, 447, 1979.
40. **Racker, E., Chien, T. F., and Kandrach, A.,** A cholate-dilution procedure for reconstitution of the Ca^{++} pump, $^{32}P_i$-ATP exchange, and oxidative phosphorylation, *FEBS Lett.,* 57, 14, 1975.
41. **Batzri, S. and Korn, E. D.,** Single bilayer liposomes prepared without sonication, *Biochim. Biophys. Acta,* 298,1015, 1973.
42. **Kremer, J. M. H., V.d., Esker, M. J. W., Pathmamanoharam, C., and Wiersema, P. H.,** Vesicles of variable diameter prepared by a modified injection method, *Biochemistry,* 16, 3932, 1977.

43. **Deamer, D. W.,** Preparation and properties of ether-injection liposomes, *Ann. N. Y. Acad. Sci.*, 308, 250,1978.
44. **Szoka, F., Jr. and Papahadjopoulos, D.,** Procedure for preparation of liposomes with large internal aqueous space and high capture by reverse-phase evaporation, *Proc. Natl. Acad. Sci. U.S.A.*, 75, 4194, 1978.
45. **Kagawa, Y., Kandrach, A., and Racker, E.,** Partial resolution of the enzymes catalyzing oxidative phosphorylation. XXVI. Specificity of phospholipids required for energy transfer reactions, *J. Biol. Chem.*, 248, 676, 1973.
46. **Robinson, N. C. and Tanford, C.,** The binding of deoxycholate, Triton X-100, sodium dodecylsulfate and phosphatidylcholine vesicles to cytochrome b_5, *Biochemistry*, 14, 369, 1975.
47. **Hsu, M. C., Scheid, A., and Choppin, P. W.,** Reconstitution of membranes with individual paramyxovirus glycoproteins and phospholipid in cholate solution, *Virology*, 95, 476, 1979.
48. **Enoch, H. G. and Strittmatter, P.,** Formation and properties of 1000Å-diameter, single bilayer phospholipid vesicles, *Proc. Natl. Acad. Sci. U.S.A.*, 76, 145, 1979.
49. **Allen, T. M., McAllister, L., Mausolf, S., and Gyorffy, E.,** Liposome-cell interactions: a study of the interactions of liposomes containing entrapped anti-cancer drugs with the EMT_6, S49 and AE_1 (transport-deficient) cell lines, *Biochim. Biophys. Acta*, 643, 346, 1981.
50. **Makino, S., Reynolds, J. A., and Tanford, C.,** The binding of deoxycholate and Triton X-100 to proteins, *J. Biol. Chem.*, 248, 4926, 1973.
51. **Weiner, J.,** personal communication.
52. **Allen, T.,** unpublished results.
53. **Papahadjopoulos, D.,** personal communication.
54. **Mayhew, E.,** personal communication.

Chapter 9

POLYMERIZED LIPOSOMES AND VESICLES

D. S. Johnston and Dennis Chapman

TABLE OF CONTENTS

I. INTRODUCTION

Surfactant vesicles have been extensively investigated because of their usefulness as models for biological membranes and their potential application as carriers for drug delivery. However, preparations of small sonicated vesicles tend to aggregate and fuse. Three approaches have been developed to prevent fusion and prolong vesicle lifetime: (1) incorporation of a second surfactant molecule which by itself does not form vesicles, i.e., cholesterol,[1] (2) modification of the surface of vesicles with charged groups, i.e., β amino galactose,[2] and (3) covalently linking the individual lipid molecules which form the membrane. The purpose of this review is to briefly summarize progress in this last area.

II. METHODS EMPLOYED TO COVALENTLY LINK SURFACTANT MOLECULES

The methods used to link membrane molecules are those of polymer science. Functional groups which are known to polymerize have been incorporated into a wide range of surfactant molecules. To date, six different functional groups have been employed. They are shown below with the means used to initiate polymerization and the structure of the covalent linkage formed. In most cases, electron microscopy has been used to show that the polymerizable surfactant in question forms vesicles when dispersed in water and that the vesicular structure is retained after polymerization. Ideally, polymeric surfactant assemblies should combine the beneficial properties of stable polymers with the fluidity of vesicles.

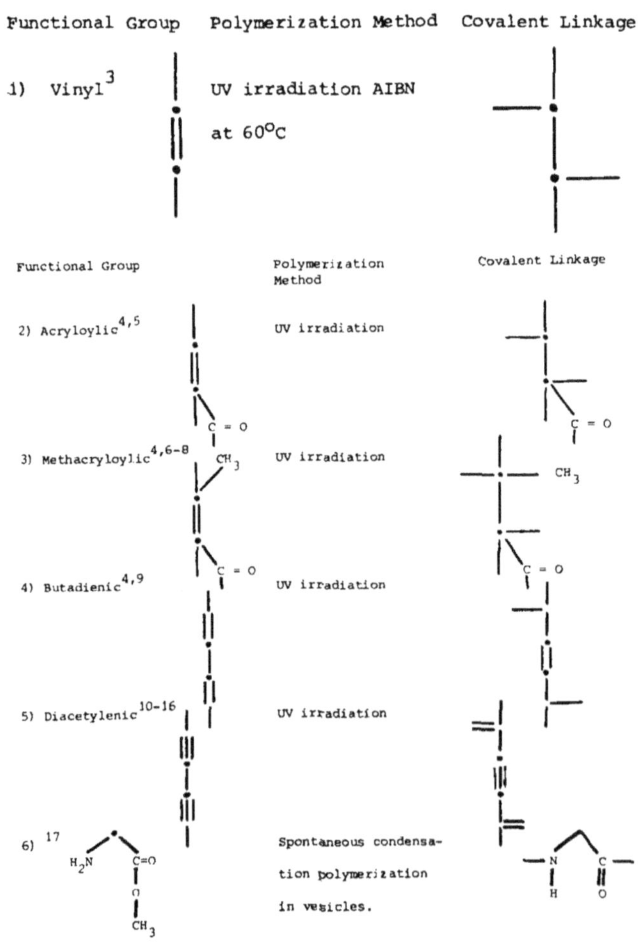

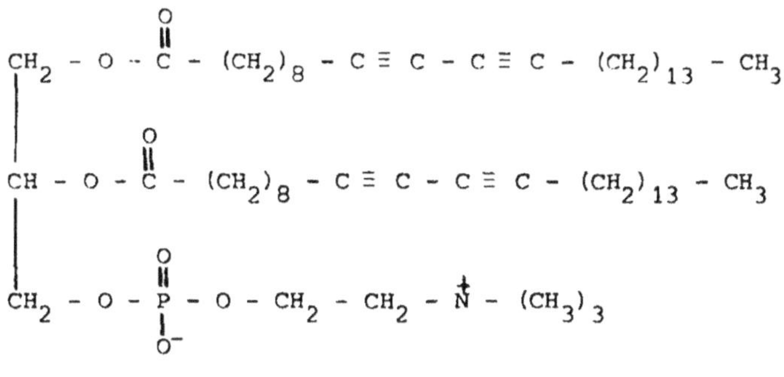

A

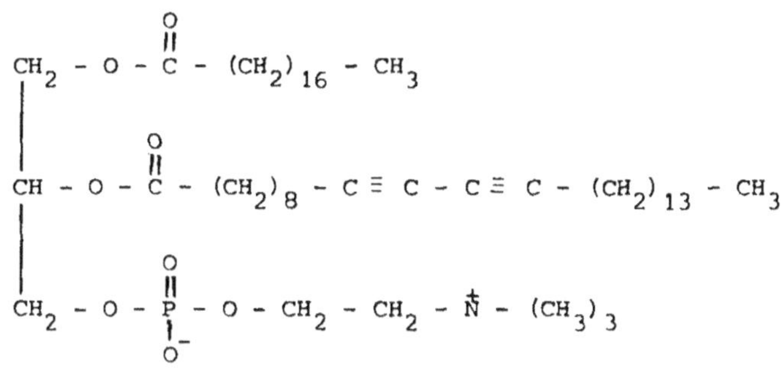

B

FIGURE 1. Structures of polymerizable phosphatidylcholines. (A) C_{27} same-chain PC. (B) C_{27} mixed-chain PC.

The reactivity of the polymerizable group is influenced by the state of the bilayer. Lipids containing groups 1 to 4 and 6 are polymerized at a temperature Tc above that at which the fluid-gel transition of the lipid bilayer occurs. Lipids containing the diacetylenic linkage only polymerize when the lipid bilayers are cooled below this temperature.

III. SURFACTANT TYPES

Probably the most studied polymerizable surfactants are those based on compounds containing diacetylene groups. The polydiacetylene linkage is composed of conjugated carbon-carbon multiple bonds and is strongly colored. Phospholipids of this type containing diacetylene groups in one and also in both acyl chains (see Figure 1) have been synthesized and polymerized. Johnston et al.[10] showed that vesicle size can influence the polymerizability of these lipids. A schematic diagram of the polymer is shown in Figure 2. Phosphatidylcholines (PCs) containing the diacetylene group polymerize when dispersed as large multi- or unilamellar vesicles but not as the small unilamellar type. When the surfactant has a chiral polar group such as L-glycerophosphorylcholine the polymer has circular dichroism bands in the visible region of the spectrum[12] (see Figure 3). From the electronic absorption spectra it is possible to measure the degree of conversion of monomer to polymer and derive information about the physical state of the bilayer.

Polymerizable groups have been placed in all parts of surfactant molecules, from the polar head group to the extremity of the hydrocarbon chain. It is possible therefore to selectively link any region of a lipid bilayer. Tundo et al.[3] demonstrated this in an elegant way. They made vesicles from a surfactant with a vinyl carbon-carbon double bond as part of the polar

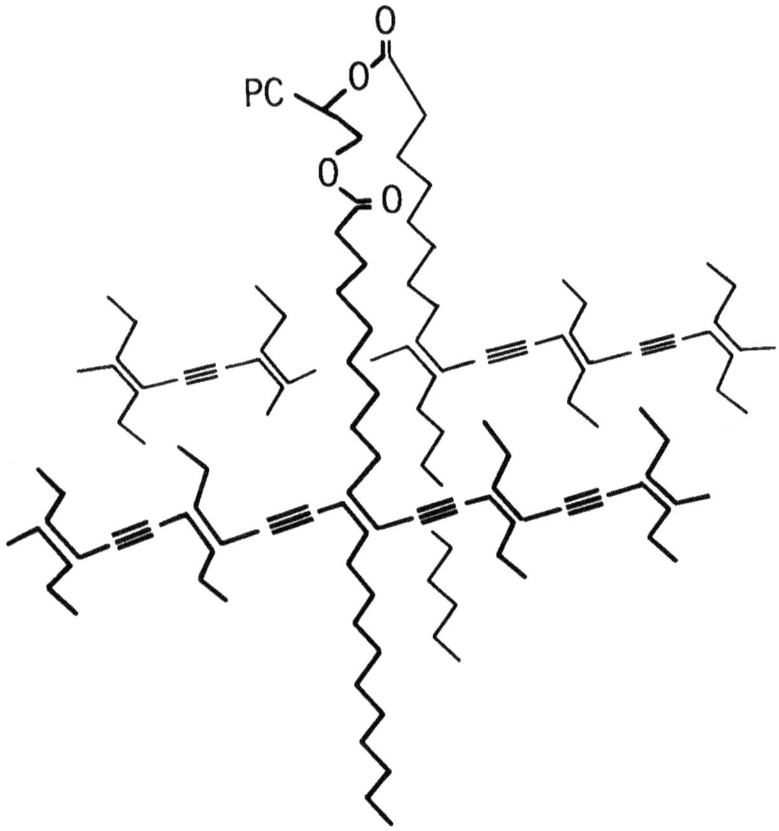

FIGURE 2. Schematic diagram of cross-bridges formed between linear polymer through the glycerol of a same-chain PC. The vertical chain segments represent part of the acyl chain of neighboring molecules. No attempt has been made to show precise bond angles in the head-group region.

head group. Treatment of the lipid dispersion with AIBN at 60°C resulted in a loss of only 60% of the double bonds, while under UV irradiation all double bonds reacted. AIBN, because it cannot cross the bilayer, can only initiate the polymerization of reactive groups which lie on the exterior of vesicles. In this way it is possible to selectively link one side of the bilayer.

Polymerizable surfactants have been synthesized with polar head groups of biological and nonbiological origin. The natural polar group usually employed is phosphorylcholine, but glycolipids[16] and phosphatidic acids[13] have also been made. Nonbiological surfactants are usually based on ammonium or phosphate ions.

IV. PROPERTIES OF POLYMERIZED SURFACTANT VESICLES

Covalently linking the constituent molecules of a vesicle markedly increases its stability. Rather than fusing within hours as unpolymerized preparations do, polymeric vesicles have shelf lives of weeks or even months. In addition: (1) they have a greater resistance to osmotic shock,[13] (2) it is possible to obtain scanning electron micrographs of them, something not possible with ordinary vesicles,[13] (3) the bilayer structure remains intact in 50% ethanol,[6,13] and (4) the leakage rate of entrapped substances may be greatly reduced.[3,6,7,9] The influence of polymerization on all these properties depends on the nature of the polymerizable group

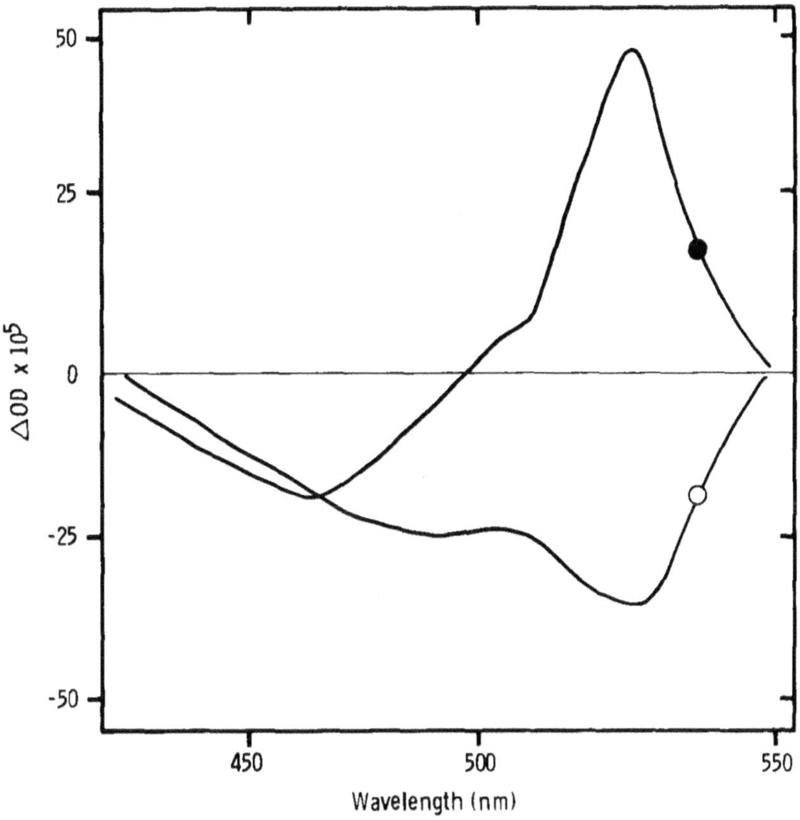

FIGURE 3. Circular dichroic spectra of polymerized C_{27} mixed-chain PC aqueous dispersions. Dispersions were prepared and polymerized at 4°C. Circular dichroic spectra were recorded at 4°C (●) and 25°C (○). G factor (Δ OD)/(OD) at 525 nm = 10^{-3}.

and its position in the surfactant molecule. For instance, Tundo et al.[3] have shown that in reducing OH^- permeability linking at the head group is less effective than linking in the hydrophobic core. Johnston et al.[10] found that the thermotropic phase transition, which occurs with a series of diacetylenic PCs (see Table 1) is removed by the polymerization process. Lipid molecules joined by all the other linkages exhibit phase transitions. This difference probably stems from the great rigidity of the polydiacetylene chains.

Polymerized vesicles with biological head groups retain their activity. Polymeric liposomes with a glucose head group were precipitated by the lectin, concanavalin A.[16] Membrane proteins reconstituted into polymeric liposomes may also remain active.[15]

V. FUTURE PROSPECTS

One of the anticipated benefits of covalently linking the molecules making up vesicles, the inhibition of fusion, has now been realized. Such vesicles should therefore be better suited to drug delivery in vivo. It is hoped that polymerization may also help overcome two other limitations of synthetic vesicles used for this purpose, their destruction by membrane-lysing agents in the body and their permeability.

Membranes provide a unique environment for chemical reactions and the courses a chemical reaction follows in homogeneous solution and when the reactants are associated with a membrane may be very different. For example while esters of long-chain α-amino acids spontaneously polymerize when dispersed as vesicles in water, in bulk only 2,5-diketopi-

Table 1
GEL-LIQUID CRYSTALLINE PHASE TRANSITION TEMPERATURES OF AQUEOUS DISPERSIONS OF DIACETYLENE-CONTAINING PHOSPHATIDYLCHOLINES

Lipid class	Chain length of diacetylene-containing acyl chain	Phase transition temp (°C)
Same chain PC	C_{23}	38
	C_{27}	60
Mixed-chain PC	C_{23}[a]	23
	C_{23}[b]	20
	C_{27}[b]	33

[a] Synthesized from stearoyl-lysophosphatidylcholine.
[b] Synthesized from egg yolk lysophosphatidylcholine.

deridine is formed.[17] The use of membranes to direct the course of chemical reactions requires that the membranes be stable. In this respect polymeric membranes will have a distinct advantage over their nonpolymeric counterpart.

The influence membrane fluidity has on protein activity has long been a source of conjecture. The ability to covalently link membrane lipids makes it possible to constrain their freedom of motion isothermally. Two groups have found that this operation may have different effects on different proteins. In one case activity was markedly enhanced,[15] in the other activity was lost completely.[18]

Chemical dissymmetry between the inner and outer surface of cell membranes has vital biochemical functions. Synthetic vesicles with similar characteristics may have novel applications. Polymerization could provide the stability necessary for the creation of long-term dissymmetry of this type.

REFERENCES

1. **Larrabee, H. L.,** Time-dependent changes in the size distribution of distearoyl phosphatidylcholine vesicles, *Biochemistry,* 18, 3321, 1979.
2. **Wu, P.-S., Tin, G. W., Baldeschwieler, J. D., Shen, T. Y., and Ponpipom, M. M.,** Effect of surface modification on aggregation of phospholipid vesicles, *Proc. Natl. Acad. Sci. U.S.A.,* 78, 6211, 1981.
3. **Tundo, P., Kippenberger, D. J., Klahn, P. L., Prieto, N. E., Jao, T.-C., and Fendler, J. H.,** Functionally polymerized surfactant vesicles, synthesis and characterisation, *J. Am. Chem. Soc.,* 104, 456, 1982.
4. **Akimoto, A., Dorn, K., Gros, L., Ringsdorf, H., and Schupp, H.,** Polymer model membranes, *Angew. Chem. Int. Ed. Engl.,* 20, 90, 1981.
5. **Kunitake, T., Kakashima, N., Takarabe, K., Nagai, M., Tsuge, A., and Yanagi, H.,** Vesicles of polymeric bilayer and monolayer membranes, *J. Am. Chem. Soc.,* 103, 5945, 1981.
6. **Regen, S. L., Czech, B., and Singh, A.,** Polymerised vesicles, *J. Am. Chem. Soc.,* 102, 6638, 1980.
7. **Regen, S. L., Singh, A., Oehme, G., and Singh, M.,** Polymerized phosphatidylcholine vesicles, synthesis andcharacterisation, *J. Am. Chem. Soc.,* 104, 791, 1982.
8. **Regen, S. L., Singh, A., Oehme, G., and Singh, M.,** Polymerized phosphatidyl-choline vesicles, stabilised and controllable time-release carriers, *Biochem. Biophys. Res. Commun.,* 101, 131, 1981.
9. **Gros, L., Ringsdorf, H., and Schupp, H.,** Polymeric antitumor agents on a molecular and on a cellular level, *Agnew. Chem. Int. (Ed. Engl.),* 20, 305, 1981.

10. **Johnston, D. S., Sanghera, S., Pons, M., and Chapman, D.,** Phospholipid polymers — synthesis and spectral characteristics, *Biochim. Biophys. Acta,* 602, 57, 1980.
11. **Pons, M., Johnston, D. S., and Chapman, D.,** A study of the spectra of diacetylenic phospholipid polymers in solvents and dispersions, *J. Polym. Sci.,* 20, 513, 1982.
12. **Pons, M., Johnston, D. S., and Chapman, D.,** The optical activity and circular dichroic spectra of diacetylenic phospholipid polymers, *Biochim. Biophys. Acta,* 693, 461, 1982.
13. **Hub, H.-H., Hupfer, B., Koch, H., and Ringsdorf, H.,** Polymerisable phospholipid analogues — new stable biomembrane and cell models, *Angew. Chem. Int. (Ed. Engl.),* 19, 938, 1980.
14. **Lopez, E., O'Brien, D. F., and Whitesides, T. H.,** Structural effects on the photopolymerisation of bilayer membranes, *J. Am. Chem. Soc.,* 104. 305, 1982.
15. **Wagner, N., Dose, K., Koch, H., and Ringsdorf, H.,** Incorporation of ATP synthetase into long-term stable liposomes of a polymerisable synthetic sulpholipid, *FEBS Lett.,* 132, 313, 1981.
16. **Bader, H., Ringsdorf, H., and Skura, J.,** Liposome from polymerisable glycolipids, *Angew. Chem. Int. (Ed. Engl.),* 20, 91, 1981.
17. **Folda, T., Gros, L., and Ringsdorf, H.,** Formation of oriented polypeptides and polyamides in monolayers and liposomes, *Makromol. Chem. Rapid Commun.,* 3, 167, 1982.
18. **Johnston, D. S., Restall, C. J., and Chapman, D.,** unpublished data, 1982.

Chapter 10

THE PREPARATION OF STERILE DRUG-CONTAINING LIPOSOMES

Jürgen Freise

TABLE OF CONTENTS

I. WHAT DOES STERILE MEAN?

Extensive work in animals[1-5] and first clinical trials[6-11] have significantly advanced the prospect of using liposomes in human medicine. With increasing therapeutic application sterile batches of liposomes will have to be prepared on a routine basis.

The demand that injection preparations be sterile has been made in every pharmacopoeia for many years. The term "sterility" can be interpreted in different ways, however (see Table 1). These examples give an indication of the problems associated with such sterility tests which are only applied to a certain, specified number of randomly chosen samples. The validity of such tests moreover depends on a variety of other factors such as the culture medium used, the incubation temperature, and the observation period.

There is a great deal of difference between the requirement of the GALL 1965 (must be sterile) and that of the European Pharmaceutical Register ("to comply with the tests for sterility"); for the tester this difference is also of legal significance. "To be sterile" applies with no preconditions whereas with the requirement "To comply with the tests for sterility" certain qualifications are made. The quantity tested for example is fixed. This amount (approximately 1 g) should contain no microorganisms capable of reproducing themselves even after repeated testing. The product would still fulfill the requirements of the test if one microorganism capable of reproduction is found in a 100-g sample whereas according to GALL 1965 it should be rejected as unsterile.

In Europe the coexistence of national and European regulations on definitions and test conditions in pharmacy still causes confusion. The European Pharmaceutical Register (Ph.Eur.) is currently being compiled under the auspices of the European Parliament in Strassburg and this should slowly take the place of each national pharmacopoeia. Information published in the Ph.Eur. was used as the basis for the following reflections on sterility.

How many containers or how much of each batch to be tested for sterility depends on the number produced and varies between 2 to 10%. As liposomes have only been applied systemically in individual patients as yet, 10% of the batch must be tested for sterility before injection. When stipulating the number of samples to be tested, the Ph.Eur. does not take into account the kind of procedure used to reduce the number of microorganisms. In the U.S. the sample number tested depends on the reliability of the antimicrobial treatment.

The results of sterility testing depend to a large extent on the properties of the nutrient medium. The Ph.Eur. does not specify any particular culture medium but, although not prescribed by law, it has been found advantageous in most European countries to use all three of following culture media (see Table 2).

For each injection preparation the volume which has to be tested for bacterial contamination depends on the volume contained in each ampule. The entire ampule contents must be tested with volumes up to 1 mℓ, half the contents with volumes up to 10 mℓ, 5 mℓ with volumes up to 50 mℓ, and 10 mℓ with volumes over 50 mℓ. If substances with antimicrobial properties such as liposomes containing penicillin or similar drugs are to be tested for sterility then inactivating substances must be added. The test schedule is complicated and will not be described here as details are to be found in the Ph.Eur. Table 3 gives an outline of the possible situations which result in the release of the product, repetition of the sterility test, or product rejection.

In addition to these general tests for the absence of microorganisms, a special test for *Mycobacterium tuberculosis* is required in several European countries but is not always necessary provided the raw materials have been heated at 60°C for over 30 min during the aseptic production procedure. If this has not taken place then the test for *M. tuberculosis* involves the use of culture media as well as experimental animals. The culture media used are not specified in the Ph.Eur.; mostly there are not even national regulations. The TB-Boullion to Dubos is suitable as are the Kirchner substrate, the TB-medium according to

Table 1
THE INTERPRETATION OF "STERILITY" BY VARIOUS PHARMACOPOEIA

Pharmacopoeia	Term	Interpretation
Ph. Eur.	Sterility	To comply with the tests for sterility.
Usp 1970	Sterility	It meets the requirements under sterility tests.
Brit. 1968	Sterility	To comply with the tests for sterility. The substances or preparation passes the test if a growth of microorganisms does not occur in any tube before the end of the seven days.
Dab 7	Sterilität	ist Anzunehmen, falls im Test . . . kein Wachstum nachzuweisen ist
Gall. 1965	Sterile	Les preparations injectables doivent être steriles - le produit examiné est considéré comme stérile, si . . . . .

Table 2
CULTURE MEDIA

Components and methods production	Thioglycolat-bouillon	Sabouraud-medium	Soybean-casein-peptone medium
Caseinpeptone	15.00 g	5.00 g	17.00 g
Soybeanpeptone			3.00 g
Lab beef extract		5.00 g	
Yeast extract	5.00 g		
Dextrose	5.50 g	20.00 g	2.50 g
NaCl	250 g		5.00 g
KH_2PO_4			2.50 g
L-cystine-HCl	0.50 g		
Na-Thioglycolat	0.50 g		
Resazurine-Na	0.001 g		
Agar	0.75 g		
Dist. water	1000 mℓ	1000 mℓ	1000 mℓ
Sterilization	20′ 121°	20′ 121°	20′ 121°
pH after sterilization	7.1	5.7	7.3
Preincubation (sterility control)	3 days 37°	3 days 22°	3 days 32°
Storage at (until use)	20-30° (protected from light	4°	4°
Growth control (inoculate with)	*Staph. aureus* *Clostridium sporogenes*	*Candida albicans*	*Staph. aureus* *Candida alb.*
Difco (Nordwald Hamburg)	0 256	0 642	
Merck Darmstadt	8191	7315	5459
Oxoid Wesel/Rhein	CM 173	CM 147	CM 129

Lowenstein-Jensen, or the egg medium according to Hohn. Animal trials are performed with 5 tuberculin-negative guinea pigs whereby at least 2 animals must survive 60 days. No tubercular changes should be found in any of the surviving animals at autopsy.

II. PYROGEN TESTING

Pyrogens are defined as substances which cause fever in humans and certain experimental animals within a short time after very small amounts have been administered parenterally. According to their origin, a distinction is generally made between pyrogens from bacteria, those produced by viruses, and endogenous pyrogens. Bacterial pyrogens are for the most

Table 3
ANALYSIS OF STERILITY

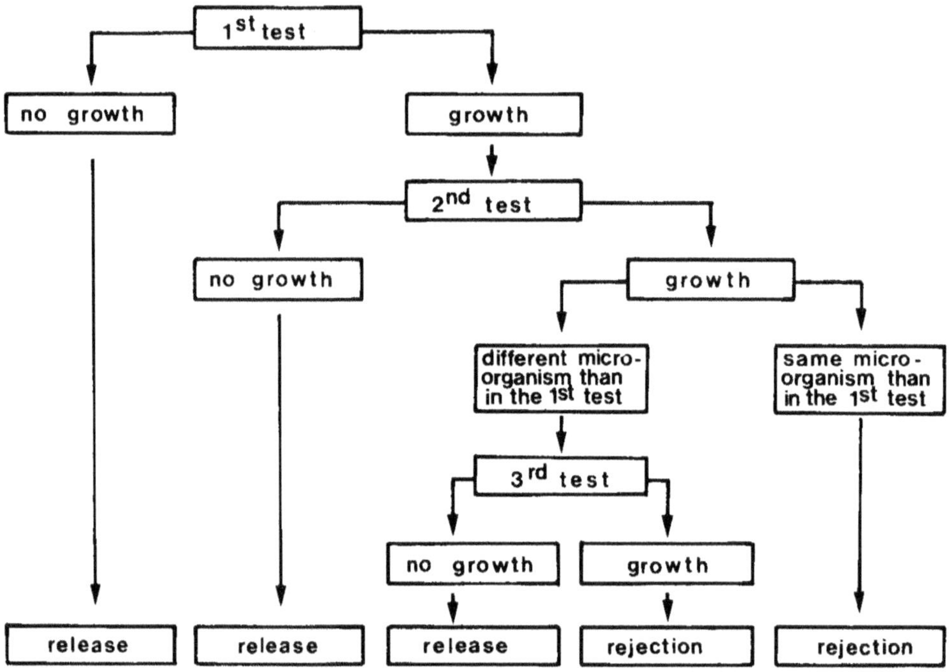

part produced by Gram-negative bacteria such as *Pseudomonas*, *Salmonella*, and *Escherichia coli* and 90% are fixed in the cell wall. Pyrogens are lipopolysaccharides and very heat resistent. They are only destroyed by long periods at temperatures over 200°C.

Pyrogens are widespread in raw materials and often found in demineralized water as well where they can only be removed at great expense with physical and chemical methods. Thus in the production of drugs which are to be applied systemically only fresh distilled water should be used.

Rabbits weighing 1.5 to 4 kg are used for pyrogen testing and they must be maintained according to certain specifications. After the i.v. injection of the test substance the rectal temperature should not increase by more than 2.65°C over a 3-hr period.

III. STERILITY AND ABSENCE OF PYROGENS IN LIPOSOME PREPARATIONS

A thorough survey of the literature on liposomes including a search with DIMDI using the key words "liposomes and sterility" or "liposomes and application in humans" showed that with a few exceptions[10,13,14] no information has been provided about sterility or even the absence of pyrogens in the batches of liposomes.

Where mention of sterility was made it was limited to the remark that the liposomes were produced under sterile conditions or that bacteria were removed before injection with Millipore filters. In the sterilization procedure with ionizing radiation recommended by Gregoriadis,[13] no experimental results are given and moreover pyrogens cannot be removed in this way. The problem of sterility has not played a role up till now as in most countries liposomes cannot be used in human medicine until the problem of toxicity has been adequately clarified. There have even been reports that liposomes per se, in nearly all cases positively charged, can be very toxic under certain experimental conditions.[15-21] There is no longer

Table 4
INCLUSION OF ^{3}H-MTX BEFORE AND AFTER FILTRATION THROUGH 0.22 μm MILLIPORE FILTERS. PART OF THE LIPOSOME SUSPENSION (MLV, NEGATIVELY CHARGED) IS NOT FILTERABLE (0.3-18.6%) DUE TO THE LARGER LIPOSOMES (0.1-0.7 μm). AFTER FILTRATION THE ^{3}H-MTX IS STILL IN ENCAPSULATED FORM.

Liposome batch	Activity in CPM/μℓ Before – Filtration	– After	Difference (%)	Rate of inclusion Before – Filtration	– After
1	8071	7843	2.8	100	99
2	3702	3014	18.6	100	100
3	3513	3247	7.6	100	100
4	22068	22010	0.3	100	100
5	15584	13184	15.4	100	100
6	2229	2160	3.1	100	100
7	2609	2373	10.5	100	99

doubt however that for certain diseases (e.g., leishmaniasis, enzyme defects, heavy metal poisoning, etc.) the therapeutic effectiveness of the pharmacons used can be increased significantly (up to a factor of 1 to 1500) if they have been included in liposomes.[8,22-34] The use of liposomes in human medicine not only demands that the preparation be nontoxic but also sterile and pyrogen free. As described in detail earlier, this must be established with specific control procedures which for the most part are prescribed by law.

Sterility adequate for animal experiments can be achieved for small and medium-sized liposomes up to approximately 0.4 μm by passage through a Millipore filter. The pore size must not be more than 0.22 μm as coarser filters no longer retain all bacteria. Our experiments have shown that liposomes are elastic enough to retain the enclosed substances when compressed to half their size and do not develop leaks (Table 4).

Liposomes which have been prepared without any particular precautions for sterility are contaminated without exception, in our laboratory with *Pseudomonas aerugenosa, Proteus,* Gram-negative bacilli, yeasts, and diphtheroid bacteria. After Millipore filtration these microorganisms are absent. The loss of liposomes and their contents fluctuated between 0.3 and 18.6% for the same liposome preparation. This variable loss can be attributed to the pestle pressure which is applied manually and thus never the same. Liposomes larger than 0.5 μm cannot be sterilized by filtration as their integrity is destroyed and their contents set free when forced through 0.22-μm pores. Other methods for sterilizing liposomes are hardly possible as all chemical or physical procedures would lead to the destruction of the integrity of their double membrane structure. Thus if liposomes are to have their use in human medicine then precautions for sterility must be taken during the entire production process, that is, (1) the use of raw materials which are sterile and pyrogen free, (2) preparation in sterile positive pressure rooms so that no unfiltered air can enter through doors and ventilation systems: working areas equipped with Laminar flow, if possible with vertical air flow, (3) the use of sterile containers and preparation techniques, and (4) when small numbers are produced in each batch (under 100) 10% must be subjected to the sterility testing described previously.

With all the various production procedures (ultrasonic treatment, ethanol or ether injection method, reverse phase evaporation, or controlled dialysis) drug-containing liposomes can easily be prepared under sterile conditions. Gel chromatography, the most common method of separating drug-containing liposomes from the excess free drug, is of limited use however.

The gel is easily contaminated with bacteria and provides an excellent medium for microorganisms. Contamination is easier to avoid if dialysis, centrifugation, or ultracentrifugation is the method of separation. The precautionary addition of antimicrobial substances such as penicillin to injection solutions is not permitted in human medicine as bacterial strains resistant to the particular antibiotic could easily develop.

In summary, the preparation of sterile liposomes is possible. Germ-free preparations which are adequate in most cases for animal experiments and experiments with cell suspensions can be obtained by filtering liposome suspensions through Millipore filters with a pore size of 0.22 μm. The application of liposomes in human medicine however requires that the raw materials are both germ free and pyrogen free and that the entire preparation takes place under sterile conditions. Before injection each batch must be subjected to strict sterility and pyrogen testing which satisfies the requirements of the Ph.Eur.

REFERENCES

1. **Gregoriadis, G.,** The carrier potential of liposomes in biology and medicine, *N. Engl. J. Med.,* 295, 704 and 765, 1976.
2. **Finkelstein, M. C. and Weissman, N. G.,** The introduction of enzymes into cells by means of liposomes, *J. Lipid Res.,* 19, 289, 1978.
3. **Ryman, B. E. and Tyrrell, D. A.,** Liposomes-bags of potential, *Essays Biochem.,* 16, 49, 1980.
4. **Tom, B. H. and Six, H. R., Eds.,** *Liposomes and Immunobiology,* Elsevier/North-Holland, New York, 1980.
5. **Gregoriadis, G. and Allison, A. C., Eds.,** *Liposomes in Biological Systems,* John Wiley & Sons, New York, 1980.
6. **Ryman, B. E., Jewkes, R. F., Jegasingh, K., Osborne, M. P., Patel, H. M., Richardson, V. J., Tattersall, M. H. N., and Tyrell, D. A.,** Potential applications of liposomes to therapy, *Ann. N. Y. Acad. Sci.,* 308, 281, 1978.
7. **Gregoriadis, G., Swain, C. P., Wills, E. J., and Tavill, A. S.,** Drug–carrier potential of liposomes in cancer chemotherapy, *Lancet,* 1, 1313, 1974.
8. **Belchetz, P. E., Braidman, I. P., Crawley, J. C. W., and Gregoriadis, G.,** Treatment of Gaucher's disease with liposome-entrapped glucocerebroside: β-glucosidase, *Lancet,* 2, 116, 1977.
9. **Gregoriadis, G., Neerunjun, E. D., Meade, T. W., Goolamali, S. K., Weereratne, H., and Bull, G.,** Experiences after long-term treatment of a type I Gaucher disease patient with liposome-entrapped glucocerebroside: β-glucosidase, in *Enzyme Therapy in Genetic Disease,* Desnick, R. L., Ed., Alan R. Liss, New York, 1980, 383.
10. **Tyrrell, D. A., Ryman, B. E., Keeton, B. R., and Dubowitz, V.,** Use of liposomes in treating type II glucogenosis, *Br. Med. J.,* 88, 1976.
11. **Ivey, H., Kattwinkel, J., and Roth, S.,** Nebulisation of sonicated phospholipids for treatment of respiratory distress syndrome of infancy, in *Liposomes and Immunobiology,* Tom, B. H., and Six, H. R., Eds., Elsevier/North–Holland, New York, 1980, 301.
12. **Seyfarth, H.,** in *Europäisches Arzneibuch,* Vol. 1 and 2, Böhme, H. and Karte, K., Eds., Wissenschaftliche Verlagsgesellschaft mbH, Stuttgart, 1978, 179.
13. **Gregoriadis, G.,** Medical applications of liposome-entrapped enzymes, *Meth. Enzymol.,* 44, 698, 1976.
14. **Freise, J., Magerstedt, P., and Schmidt, F. W.,** Die in vivo und in vitro Stabilität des Einschlusses von Methotrexat in negativ geladenen Liposomen nach Sterilfiltration, *Z. Naturforsch.,* 34, 114, 1979.
15. **Chawla, R. S., Kellaway, I. W., Hunneyball, I. M., and Stevens, J.,** The effect of liposomal charge on drugs toxicity and efflux, *J. Pharm. Pharmacol.,* 31, 86, 1979.
16. **Dunnick, J. K., Kallman, R. F., and Kriss, J. P.,** Lipid vesicle interaction with EMT-6 tumor cells and effect on subsequent cell growth, *Biochem. Biophys. Res. Commun.,* 73, 619, 1976.
17. **Panzner, E. A. and Jansons, V. K.,** Control of in-vitro cytotoxicity of positively charged liposomes, *J. Cancer Res. Clin. Oncol.,* 95, 29, 1979.
18. **Adams, D. H., Joyce, G., Richardson, V. J., Ryman, B. E., and Wisniewski, H. M.,** Liposome toxicity in the mouse central nervous sytem, *J. Neurol. Sci.,* 31, 173, 1977.
19. **de Barsy, T., Devos, P., and van Hoof, F.,** A morphologic and biochemical study of the fate of antibody-bearing liposomes, *Lab. Invest.,* 34, 273, 1976.

20. **Steger, L. D. and Desnick, R. J.,** Enzyme therapy. VI. Comparative in-vivo fates and effects on lysosomal integrity of enzyme entrapped in negatively and positively charged liposomes, *Biochim. Biophys. Acta,* 464, 530, 1977.
21. **Tyrrell, D. A., Heath, T. D., Colley, C. M., and Ryman, B. E.,** New aspects of liposomes, *Biochim. Biophys. Acta,* 457, 259, 1976.
22. **Alving, C. R., Steck, E. A., Chapman, W. L., Waits, V. B., Hendricks, L. D., Swartz, G. M. and Hanson, W. L.,** Therapy of leishmaniasis: superior efficacies of liposome-encapsulated drugs, *Proc. Natl. Acad. Sci. U.S.A.,* 75, 2959, 1978.
23. **Alving, C. R., Steck, E. A., Chapman, W. L., Waits, V. B., Hendricks, L. D., Swartz, G. M., and Hanson, W. L.,** Liposomes in leishmaniasis: therapeutic effects of antimonial drugs 8-aminoquinolines and tetracycline,*30Life Sci.,* 26, 2231, 1980.
24. **Alving, C. R., Steck, E. A., Hanson, W. L., Loizeaux, P. S., Chapman, W. L., and Waits, V. B.,** Improved therapy of experimental leishmaniasis by use of a liposome-encapsulated antimonial drug, *Life Sci.,* 22, 1021, 1978.
25. **Black, C. D. V., Watson, G. J., and Ward, R. J.,** The use of pentostam liposomes in the chemotherapy of experimental leishmaniasis, *Trans. Soc. Trop. Med. Hyg.,* 71, 550, 1977.
26. **New, R. R. and Chance, M. L.,** Treatment of experimental cutaneous leishmaniasis by liposome-entrapped pentostam, *Acta Trop. (Basel),* 37, 253, 1980.
27. **New, R. R., Chance, M. L., Thomas, S. C., and Peters, W.,** Antileishmanial activity of antimonials entrapped in liposomes, *Nature (London),* 272, 55, 1978.
28. **Rahman, Y. E. and Rosenthal, M. W.,** A new approach to the therapy of metal poisoning: liposome encapsulation of chelating agents, *Radiat. Res.,* 55, 516, 1973.
29. **Rosenthal, M. W., Rahman, Y. E., Moretti, E. S., and Cerny, E. A.,** Removal of polymeric plutonium by DTPA directed into cells by liposome encapsulation, *Radiat. Res.,* 63, 262, 1975.
30. **Rahman, Y. E., Rosenthal, M. W., Cerney, E. A., and Moretti, E. S.,** Preparation and prolonged tissue retention of liposome-encapsulated chelating agents, *J. Lab. Clin. Med.,* 83, 640, 1974.
31. **Jonah, M. M., Cerny, E. A., and Rahman, Y. E.,** Tissue distribution of EDTA encapsulated within liposomes of varying surface properties, *Biochim. Biophys. Acta,* 401, 336, 1975.
32. **Rahman, Y. E. and Wright, B. J.,** Liposomes containing chelating agents, *J. Cell Biol.,* 65, 112, 1975.
33. **Rahman, Y. E.,** Liposomes and chelating agents, in *Liposomes in Biological Systems,* Gregoriadis, G. & Allison, A. C., Eds., John Wiley and Son, New York, 1980, 265.
34. **Braidman, I. and Gregoriadis, G.,** Preparation of glucocerebroside-β-glucosidase for entrapment in liposomes and treatment of patients with adult Gaucher's disease, *Biochem. Soc. Trans.,* 4, 259, 1976.

Chapter 11

LIPID PEROXIDATION IN LIPOSOMES

A. W. T. Konings

TABLE OF CONTENTS

I. INTRODUCTION

Although liposomes have been used as model membranes in biophysical studies for already a long period of time, it is in the last 12 years that these vesicles have become extremely popular for several applications in biology and medicine (for recent reviews see References 1 to 5).

Most of the phospholipid liposomal preparations used in the past contained unsaturated acyl chains as part of the molecular structure. These acyl chains are vulnerable to oxidative degradation, a process called lipid peroxidation. These reactions may occur during the preparation, storage, or actual use of liposomes. The accompanying change in membrane structure may then be of influence on the applications of the liposomes or on experiments under investigation. In the author's laboratory liposomes and micelles have been successfully used as model systems in the study of radiation-induced lipid peroxidation. In living cells polyunsaturated fatty acyl chains in membrane phospholipids are among the critical biological target molecules easily attacked by products of ionizing radiation. Radiation-induced lipid peroxidation as well as autoxidation of unsaturated fatty acids largely occurs via free radical mechanisms. The main purpose of this chapter is to provide background for the process of lipid peroxidation in liposomes and to describe experiences with these reactions during the preparation (sonication), incubation, and radiation of liposomes.

II. LIPID PEROXIDATION, A FREE-RADICAL PROCESS

A. Free-Radical Reactions in Aqueous Systems

A free radical is a species containing a single unpaired electron. It may be charged (positive, negative) or neutral and may be produced by homolytic dissociation of a covalent bond:

$$L{:}H \rightarrow L\cdot + H\cdot \qquad (1)$$

In this example the molecule LH is dissociated in two free radicals L· + H·. The free-radical character is conventionally illustrated by the bold point (·) to indicate an odd number of electrons. The energy required to initiate homolytic fission of covalent bonds can be provided in different ways. Electromagnetic radiation, heat, and redox reactions coupled with a source of free radicals are the most well-known examples.

Generally, free radicals react very energetically and as a consequence have only transient life times. However, a number of free radicals are stable, for instance, oxygen (stable diradical). Free radicals can easily initiate relatively nonspecific hydrogen abstraction and chemical addition reactions. An important feature of free radicals is their ability to enter into chain reactions. Only a few initiation reactions may be necessary to decompose a large number of molecules of the starting material. This will be shown further on in the section on propagation of the lipid peroxidation process. The free radical reactions of interest to the topic of this chapter occur in aqueous solutions where the hydroxyl radical (OH·) and the atomic hydrogen radical (H·) are the most important reactive species. The hydroxyl radical has a powerful oxidative capacity while the hydrogen atom is a strong reductant.

B. Lipid Peroxidation

1. Oxidation of Unsaturated Fatty Acids

The oxidative deterioration of lipids involves primarily autoxidative reactions accompanied by various secondary reactions having an oxidative and a nonoxidative character. Oxidation of unsaturated fatty acids has been well reported [6–8] and, unless mediated by other oxidants or enzyme systemts, proceeds through a free-radical chain mechanism. The early contributions in lipid oxidation were summarized in a symposium, "Lipids and their Oxidation", held at Oregon State University in 1961.[9]

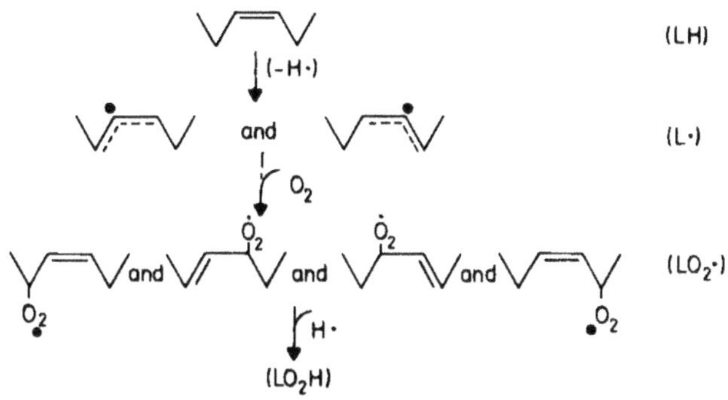

FIGURE 1. Scheme of the mechanism of oleate peroxidation.

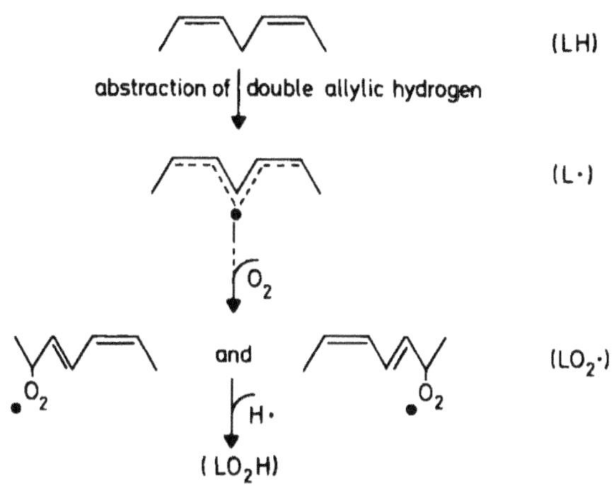

FIGURE 2. Scheme of the mechanism of linoleate peroxidation.

In this chapter the abbreviated form for designating fatty acids gives the number of carbons followed by the number of double bonds:

16:0 *n*-hexadecanoic acid	palmitic acid
18:0 *n*-octadecanoic acid	stearic acid
18:1 *cis*-9-octadecenoic acid	oleic acid
18:2 *cis,cis*-9-12-octadecadienoic acid	linoleic acid
18:3 all-*cis*-6,9,12-octadecatrieonic acid	linolenic acid
20:4 all-*cis*-5,8,11,14-eicosatetraenoic acid	arachidonic acid
22:6 all-*cis*-4,7,10, 13,16,19-docosahexanoic acid	clupanadonic acid

The susceptibility of unsaturated fatty acid molecules to the initiation of lipid peroxidation varies according to the lability of their allylic hydrogens. It has been recognized for some time that the divinyl methane structure (doubly allylic hydrogen) as present in most polyunsaturated fatty acids is particularly susceptible to hydrogen abstraction by free radicals. The chemical process of lipid peroxidation is defined as the reaction of an oxidant initiator with an unsaturated fat (LH) to form a lipid free-radical intermediate (L·), a peroxyl free radical (LO_2·), and hydroperoxides (LO_2H).

A simplified mechanism for the formation of hydroperoxides (LO_2H) from oleate is given in Figure 1. The most accepted reactions in the autoxidation of 18:1 (LH) involve hydrogen abstraction on carbon-8 and carbon-11 with the formation of two allylic radicals (L·). These radicals react with oxygen to lipid peroxyl free-radical intermediates (LO_2·) to form a mixture of allylic hydroperoxides (LO_2H) after addition of atomic hydrogen (H·).

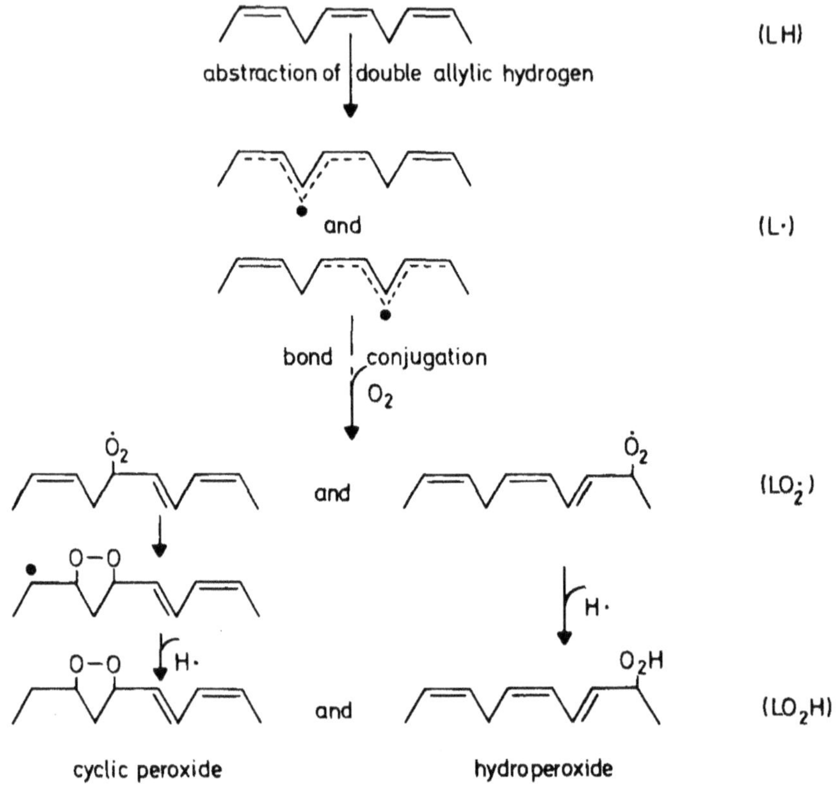

FIGURE 3. Scheme of the mechanism of linolenate peroxidation.

Figure 2 illustrates the mechanism of linoleate (18:2) autoxidation. Hydrogen abstraction on the reactive doubly allylic carbon-11 of the fatty acid (LH) produces a pentadienyl radical (L·). This short living free radical reacts rapidly with oxygen to produce the peroxyl radical (LO_2·) and after H· addition, lipid conjugated diene hydroperoxides (LO_2H) are formed.

When fatty acids with more than two double bonds are subjected to lipid peroxidizing conditions, more diverse reaction products are formed as compared to 18:1 and 18:2. In Figures 3 and 4 a selection of proposed[10-12] mechanisms for the oxidation of polyunsaturated fatty acids (PUFA) containing three or more double bonds is outlined.

In Figure 3 autoxidation of linolenate, 18:3, is shown. Hydrogen abstraction of the two active methylenes on carbon-11 and carbon-14 of the fatty acid (LH) produces two pentadienyl radicals (L·). These free radicals react with oxygen to produce a mixture of conjugated diene-triene lipid peroxide radical intermediates (LO_2·), yielding cyclic peroxides and hydroperoxides. The cyclic peroxides are either formed in a free radical cyclization process as shown or after cyclization of hydroperoxides.

The scheme for the autoxidation of arachidonic acid (20:4) is shown in Figure 4. Here three active methylenes can be distinguished which give cause to a larger number of reaction pathways as in the case with 18:3. The oxygen attack at the end carbons will lead to normal hydroperoxides while the reactions on the other locations may yield cyclic peroxides. The mechanism for the autoxidation of clupanadonic acid (22:6) is essentially as for arachidonic acid (not shown). Most cyclic- and hydroperoxides are relatively easy to decompose. A very complicated set of reaction paths has been recognized[13] for hydroperoxide decomposition. The breakdown products include: aldehydes, aldehyde esters, oxo acids, hydrocarbons, alcohols, ketones, hydroxy and keto acids, lactones, and dimeric compounds.

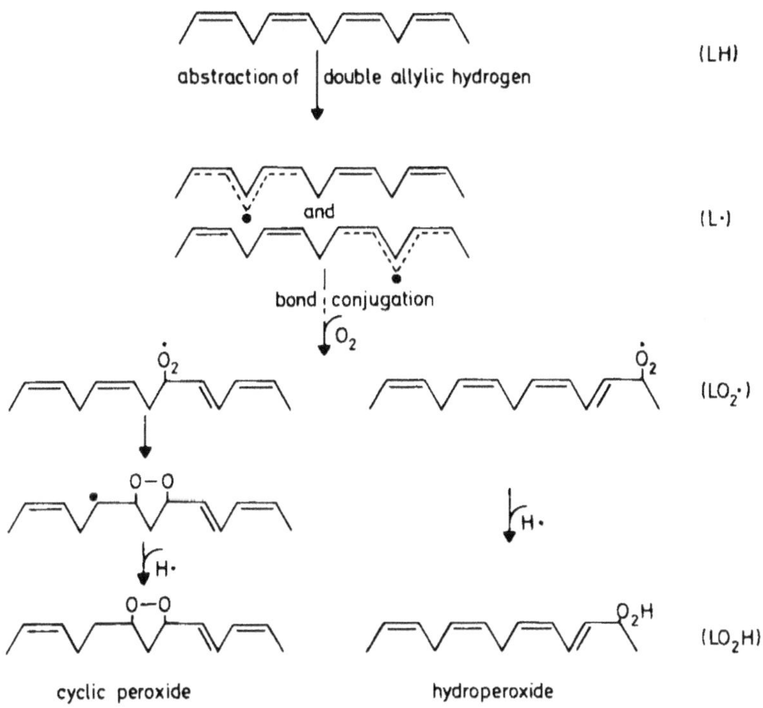

FIGURE 4. Scheme of the mechanism of arachidonate peroxidation.

2. Reaction Mechanisms

Direct reaction of oxygen with unsaturated fatty acids is thermodynamically improbable.[9,13] The ground state of the unsaturated fatty acids is of singlet multiplicity. The reaction with oxygen is spin forbidden because ground state oxygen is of triplet multiplicity. The activation of oxygen by electronic excitation forms singlet oxygen ($'O_2\cdot$), which reacts directly with the double bond of the fatty acid. These nonradical reactions are important when the lipid is exposed to light in the presence of a photosensitizer. When oxygen is univalently reduced, superoxide anion ($O_2^-\cdot$) is formed which is not very reactive toward the unsaturated fatty acids.[14-16] In the presence of transition metal ions such as Fe^{2+}, and when also complexed, sufficient oxidizing strength may be present to react with the unsaturated lipid.

The kinetic equations for lipid peroxidation can be divided into: (1) initiation reactions resulting in a lipid alkyl radical (L·); (2) propagation reactions in which the lipid alkyl radical reacts with oxygen to form a lipid peroxyl radical ($LO_2\cdot$). When $LO_2\cdot$ is in the proximity of another unsaturated fatty acid (LH) the peroxyl radical ($LO_2\cdot$) will abstract an allylic hydrogen from that fatty acid thus yielding a lipid hydroperoxide (LO_2H) and a new alkyl radical (L·), which may react with oxygen. Thus a chain reaction is initiated; (3) termination reactions which are essentially radical destroying reactions. The reaction of two peroxide radicals is one of the most important terminations.

a. The Initiation of Lipid Peroxidation

The initiation reaction is the least well-understood step in the sequence of lipid peroxidation. In general, the formation of initial free radicals is necessary to attack the allylic hydrogens of the unsaturated fatty acids.

The best candidate is the hydroxyl radical (OH·) which is a very strong oxidant. Especially from radiation studies it is known that hydroxyl radicals are very efficient initiators of lipid peroxidation.[17] Hydroxyl radicals may also be formed in the so-called Fenton reactions.[18-19] In this reaction trace amounts of ferrous ions (Fe^{2+}) and hydrogen peroxide (H_2O_2) can form hydroxyl radicals ($OH\cdot$). The hydrogen peroxide may have been formed[20] by dismutation of superoxide anions ($O_2^{-}\cdot$). The production of the latter compound is suggested to be formed by the action of ferrous ions on oxygen.[7,21] A sequence of reactions leading to hydrogen abstraction of unsaturated fatty acids in the presence of ferrous ions and oxygen may then be as follows:

$$Fe^{2+} + O_2 \rightarrow Fe^{3+} + O_2^{-}\cdot \quad (2)$$

$$2\,O_2^{-}\cdot + 2\,H^{+} \rightarrow H_2O_2 + O_2 \quad (3)$$

$$Fe^{2+} + H_2O_2 \rightarrow Fe^{3+} + OH\cdot + OH^{-} \quad (4)$$

$$LH + OH\cdot \rightarrow L\cdot + H_2O \quad (5)$$

When small amounts of hydroperoxides are present as impurities in the unsaturated fatty acid sample and also trace amounts of transition metal ions (Mn) are found, then lipid alkoxyl radicals ($LO\cdot$) and lipid peroxyl radicals ($LO_2\cdot$) are produced[21] which can initiate hydrogen abstraction on the unsaturated fatty acid (LH).

$$Mn^{n+} + LO_2H \rightarrow Mn^{(n+1)+} + LO\cdot + OH^{-} \quad (6)$$

$$Mn^{m+} + LO_2H \rightarrow Mn^{(m-1)+} + LO_2\cdot + H^{+} \quad (7)$$

Reaction 6 is reported[22] to be much more rapid than Reaction 7 and will be of importance when ferrous ions are present in a solution.

$$Fe^{2+} + LO_2H \rightarrow Fe^{3+} + LO\cdot + OH^{-} \quad (8)$$

$$LO\cdot + LH \rightarrow LOH + L\cdot \quad (9)$$

Pryor and co-workers[23] have recently shown that lipid diene hydroperoxides react with superoxide anion to produce alkoxyl radicals that subsequently initiate lipid peroxidation. If the production of the superoxide anions by ferrous ions and oxygen (Reaction 2) takes place in the presence of lipid hydroperoxides the following reaction occurs.

$$O_2^{-}\cdot + LO_2H \rightarrow O_2 + LO\cdot + OH^{-} \quad (10)$$

Pryor and co-workers showed that Reaction 10 is sufficiently fast to compete with Reaction 3 (dismutation of superoxide radicals). Radical formation by unimolar homolysis of lipid hydroperoxides:

$$LO_2H \rightarrow LO\cdot + OH\cdot \quad (11)$$

is very slow at moderate temperatures.[24] The bimolecular homolytic reaction:

$$LO_2H + LH \rightarrow LO\cdot + L\cdot + OH^{-} \quad (12)$$

is also slow and not significant at low temperatures.[25]

Autoxidation of unsaturated lipids will start rapidly when transition metal ions and lipid hydroperoxides are present. The transition-metal-catalyzed decomposition of hydroperoxides is more effective when the catalysts are in their lower valence states.

b. The Propagation of Lipid Peroxidation

As has been pointed out, the two most important initiation steps for the peroxidation of unsaturated fatty acids known at the moment are hydrogen abstraction by OH· radicals (Reaction 5) and by alkoxyl radicals (Reaction 9). The OH· radicals can be produced by radiation and in a Fenton- type reaction. The LO· radicals are derived from hydroperoxides after reaction with ferrous ions or superoxide anions.

Hydrogen abstraction from an unsaturated fatty acid produces the alkyl radical (L·) which reacts rapidly with oxygen to form the peroxyl radical ($LO_2\cdot$).

$$L\cdot + O_2 \rightarrow LO_2\cdot \quad (13)$$

This equation may be considered as the first step in the propagation of lipid peroxidation. When this peroxyl radical ($LO_2\cdot$) is in the proximity of unsaturated fatty acids or unsaturated acyl chains (*LH) and not sterically hindered, an allylic hydrogen will be abstracted thus yielding a hydroperoxide (LO_2H) and a new alkyl radical (*L·) which reacts with oxygen:

$$LO_2\cdot + {}^*LH \rightarrow LO_2H + {}^*L\cdot \quad (14)$$

$${}^*L\cdot + O_2 \rightarrow {}^*LO_2\cdot \quad (15)$$

The newly formed lipid peroxyl radical ($^*LO_2^{\cdot}$) may reaction with the next unsaturated fatty acid or acyl chain. In this way a chain reaction is started, the result of it being that a few initiations may produce propagation reactions in which a large number of unsaturated fatty acids are oxidized when sufficient oxygen is present. The reaction between the alkyl radical and oxygen is so rapid[26] that in many cases the overall peroxide yield is determined by the conformation of the lipids in the chain propagation reaction.[27,28]

c. The Termination of Lipid Peroxidation

When sufficient oxygen and suitable substrate (LH) is present one can assume that most of the free radicals are in the form of the peroxyl radical ($LO_2\cdot$).

The termination of chain reactions can take place by combination of radicals.[25] Under the conditions mentioned above most terminations will include peroxyl radicals

$$LO_2\cdot + LO_2\cdot \rightarrow (LO_2)_2 \quad (16)$$

$$(LO_2)_2 \rightarrow 2LO\cdot + O_2 \quad (17)$$

$$LO\cdot + LO\cdot \rightarrow (LO)_2 \quad (18)$$

The tetroxide formed in Reaction 16 will decompose to yield two alkoxyl radicals (Reaction 17) which combine to give LOOL (Reaction 18).

At low oxygen pressure ($<$ 10 mmHg) Reaction 16 will be less important than the two reactions depicted below.

$$L\cdot + LO_2\cdot \rightarrow (LO)_2 \quad (19)$$

$$L\cdot + L\cdot \rightarrow L_2 \quad (20)$$

Reaction 20 will be prominent when the oxygen pressure is below 1 mmHg.

Free-radical chain reactions can also be stopped by the presence of antioxidants (AH) resulting in by removal of either the peroxyl radicals or the alkyl radicals:

$$LO_2\cdot + AH \rightarrow LO_2H + A\cdot \quad (21)$$

$$L\cdot + AH \rightarrow LH + A\cdot \quad (22)$$

$$A\cdot + X \rightarrow \text{nonradicals} \quad (23)$$

When oxygen is not the limiting reaction, Reaction 22 can be neglected. Most antioxidant radicals (A·) produced are nonreactive and terminate in some nonradical-producing mechanisms.[22]

III. OXIDATIVE FREE-RADICAL DAMAGE TO LIPOSOMES AND MICELLES

The experiments described in this chapter were performed with three different types of liposomes. For reason of comparison also some micellar experiments are presented.

The liposomes are referred to by three letter acronyms, commonly used in present day liposomology: (1) multilamellar vesicles (MLV), (2) small unilamellar vesicles (SUV), (3) large unilamellar vesicles (LUV) formed from water in oil emulsions of phospholipids by a reverse-phase-evaporation technique, the reverse-phase-evaporation vesicles (REV).

A. Methods

1. Preparation of Liposomes and Micelles

All liposomes prepared, were made up of phospholipids isolated from rat liver. The extractions and separation procedures used are described elsewhere.[27] The approximate phospholipid composition was as follows: phosphatidylcholine (PC): 65%; phosphatidylethanolamine (PE): 20%; spingomyelin (SPH): 5%; phosphatidylinositol (PI) plus phosphatidylserine (PS): 5%; diphosphatidylglycerol (DPG): 5%. The average fatty acid composition of the phospholipid extract used was 16:0 (24%); 18:0 (18%); 18:1 (8%); 18:2 (17%); 20:4 (22%); 22:6 (11%).

Total phospholipid content in organic or aqueous solutions (liposomes) was measured by either evaporating aliquots of extracts under a stream of nitrogen gas and weighing the lipid on a Cahn (Gram) electrobalance or by a phosphorus determination.[27]

a. Preparation of MLV

Liver phospholipids (PL) stored under nitrogen at −20°C in methanol/dichlormethane (1:1) were warmed up to 0°C and the necessary volume pipetted into a round-bottom flask. When α-tocopherol (Sigma) or butylated hydroxytoluene (Koch Light) were used in the experiments, these antioxidants were solubilized in ethanol, pipetted into the flask, and mixed with the phospholipids. The organic solvents were evacuated at 30°C in a rotary evaporator. A thin film of lipids with or without antioxidants was deposited on the wall of the flask. Hereafter double distilled water with or without 5 μM $FeCl_2$ was added to obtain a final concentration of 1 mg PL per milliliter (pH about 5.5). From here on, two different conditions were employed. The suspensions were either purged with nitrogen gas or left in air and mechanically agitated on a Vortex mixer during about 30 sec. Hereafter the preparation received a short 30-sec sonication treatment with a Branson B12 probe sonicator (Danbury, Conn.) at 22°C. The sonication was performed either under a stream of nitrogen or in air. The suspension was centrifugated at 100 $\times g$ for 5 min and the supernatant adjusted to a lipid concentration of 0.5 mg/mℓ.

b. Preparation of SUV

The MLV preparation obtained just sprior to the step of low speed centrifugation served as starting material for SUV. The MLV were sonicated at 22°C in a bath-type sonicator (Megason, type T-240-80, 25 × 30 cm, Ultrasonics Instruments, Farmingdale, Mass.) for

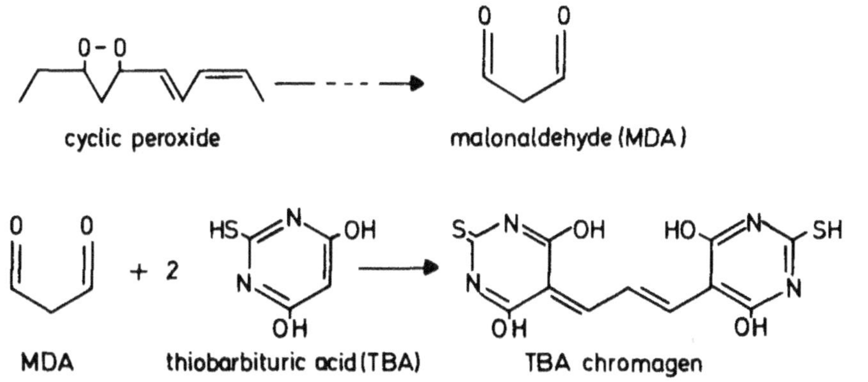

FIGURE 5. Formation of malonaldehyde and the TBA-malonaldehyde product.

2 hr either under nitrogen in a closed flask or in air in a closed flask. After the sonication procedure, the preparations were centrifigated at 105.000 × *g* at 4°C for 1 hr. The supernatant was adjusted to obtain a final PL concentration of 0.5 mg/mℓ.

c. Preparation of REV

The REV vesicles were prepared according to the method of Szoka and Papahadjopoulos.[29] About 85 mg of PL was dissolved in 6 mℓ diisopropylether/ dichlormethane (1:1) and 1.4 mℓ of double distilled water was added directly to this phospholipid/solvent mixture. The preparation was sonicated in the bath-type sonicator mentioned above for 5 min under nitrogen at 22°C. The mixture was then placed on a rotary evaporator and the organic solvents were removed under reduced pressure at 30°C. After about 30 min the emulsions became a viscous gel and about 10 min later an aqueous suspension was formed. At this point 1.4 mℓ double distilled water was added and the suspension was evaporated for anouther 15 min to remove more of the organic solvent. The aqueous suspension of vesicles thus obtained was extruded through an Unipore polycarbonate filter with a pore size of 0.2 μm.[30] Hereafter the REV preparation was diluted to a concentration of 0.5 mg PL per milliliter.

d. Preparation of Micelles

Concentrated solutions of fatty acids (Sigma) were freshly prepared in 0.5 *M* sodium phosphate buffer, pH 11.3 at room temperature, under continuous flushing of nitrogen. The solutions were adjusted to the desired pH by adding sodium dihydrogen phosphate and to the desired ionic strength (mostly 0.05 *M*) by dilution with double distilled water.

2. Autoxidation, Sonication, and Radiation

Incubations for the purpose of autoxidation experiments (with or without $FeCl_2$) were performed in a shaking waterbath at different temperatures in 25-mℓ Erlenmeyer flasks containing 5 to 15 mℓ of liposomes. The flasks were loosely covered with aluminum foil. For the determination of the "oxidation index" samples of 500 μℓ were taken after different incubation times.

Prolonged sonications of MLV were performed in test tubes with an inner diameter of 35 mm and a length of 10 cm. The volume of liposome suspensions was 5 to 12 mℓ. The tubes were closed. A part of the liposome suspension was bubbled with nitrogen and kept under a nitrogen atmosphere during the sonication. Most bath-type sonications were performed in the Megason 20 × 30 cm sonication bath. For reasons of comparison some sonications were done in a Bransonic 220, 14 × 24 cm bath.

Radiation of the lipid suspensions was performed in thin-walled glass tubes, with an inner

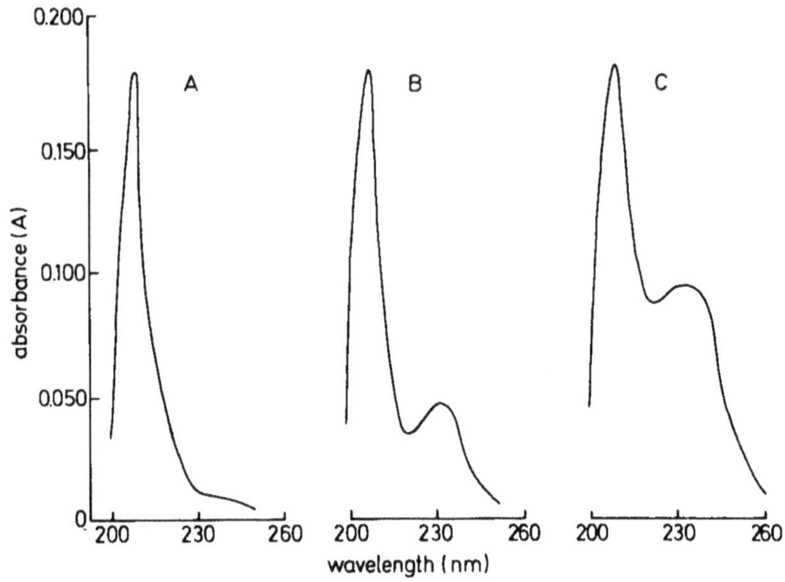

FIGURE 6. Part of the UV absorption spectrum of phospholipid preparations with different conjugation dienes. The oxidation index (A_{233} nm/A_{215} nm) is 0.054 (A), 0.270 (B) and 0.540 (C).

diameter of 15 mm and a length of 25 mm. The tubes were filled up to 3 mm from the top. Irradiation was performed with an X-ray tube (250 kV, h.v.l.: 1 mm Cu) positioned over the samples. Dose rates were monitored by the ferrous sulfate method (Fricke dosimetry).

3. Procedures for Assaying the Process of Lipid Peroxidation

Oxidation of different unsaturated fatty acids leads to a variety of hydroperoxides (Figures 1 to 4). Many analytical procedures to measure the peroxide value has been described in the past,[6,9,10,21,31,32] most of the methods being nonstoichiometric and nonspecific. In our laboratory three different methods have been used, one based on the disappearance of unsaturated fatty acids and the fatty acyl chains[27] and the other two on the production of conjugated dienes[27] and of malonaldehyde.[33,34]

As can be seen in Figures 1 to 4 conjugated dienes are formed when fatty acids with more than one double bond are oxidized. Malonaldehyde may be produced from cyclic peroxides (Figure 5) and therefore only lipid peroxidation is measured when fatty acids with three or more double bonds are present as a substrate. The latter is often the case when the phospholipid acyl chains originate from biological material.

Malonaldehyde has been identified as the lipid peroxidation product that reacts with thiobarbituric acid (TBA) to give red species absorbing at 535 nm,[35] as shown in Figure 5. The TBA assay is the most frequently used method for determining the extent of membrane lipid peroxidation in vitro. Gutteridge[36] standardized a TBA method for liposomes and showed that the degree of peroxidation could be related to the formation of TBA-reactive material.

In this chapter experiments will be discussed where lipid peroxidation is measured by either the disappearance of fatty acids (or acyl chains) or by the conjugated dienes produced.

a. Disappearance of Fatty Acids and Fatty Acyl Chains

The concentration of intact fatty acids was measured by gas-liquid chromatography (GLC). Heptadecanoic acid (17:0, margaric acid) was added to the samples as an internal standard.

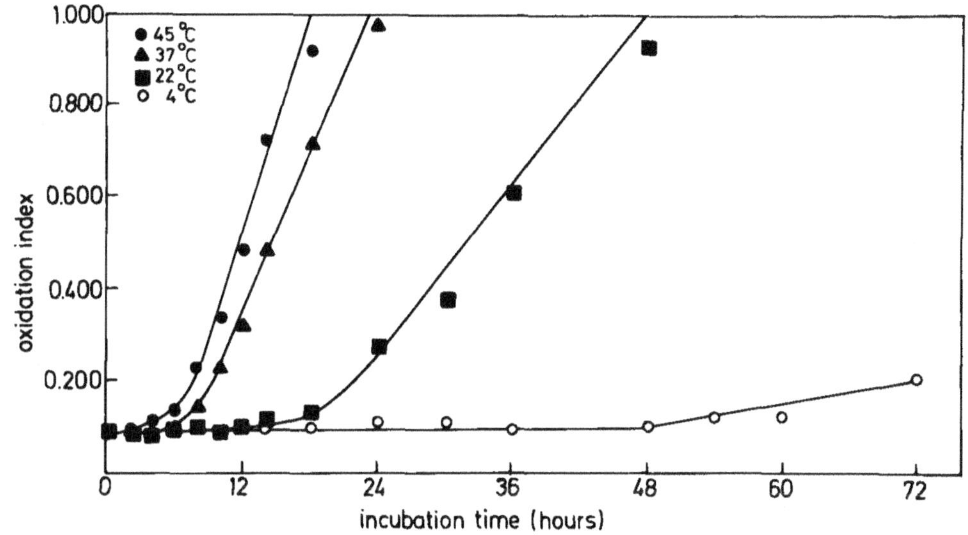

FIGURE 7. Rate of autoxidation of rat liver phospholipid MLV at different temperatures.

Samples of liposomal or micellar suspensions were subjected to lipid extraction.[27,37] The lipids were solubilized in methanol/sulfuric acid (95:5 v/v) in glass-stoppered tubes. Methyl esters were prepared and separated on a Becker Packard Gaschromatograph, model 29, equipped with a flame ionization detector and integrator system. The stationary phase used was a cyanoisilicone column consisting of 15% SP-2340 on Chromosorb® P AW-DMCS, 100 to 120 mesh (Supelco Inc.). The heating was carried out at an initial temperature of 200°C for 2 min, followed by an increase of 3°C per min up to 240°C.

b. Measurement of Conjugated Dienes

Oxidation of PUFA is accompanied by increased UV absorption. Changes in the 230 to 236 nm range are a measure of diene unsaturation and around 270 nm of triene unsaturation. These changes can be used as a relative measurement of lipid peroxidation.

The UV spectrum of lipids in the range of 200 to 260 nm is shown in Figure 6. The absorbance spectrum in Figure 6A is of phospholipids extracted from rat liver. Figures 6B and 6C show spectra of lightly and moderately peroxidized phospholipid samples. The increase of absorption at 233 nm is an indication of the appearance of conjugated dienes and thus of hydroperoxides. In accordance with the suggestion of Klein,[38] we have used 215 nm as a reference wavelength, which was the lowest wavelength where the absorption depended linearly on the amount of lipid used and was independent from the degree of peroxidation until a A233 nm/A215 nm of 0.8. The ratio of these absorptions is defined as the "oxidation index" and used in the figures presented in this chapter.

When the liposomes or micelles were dispersed in water, 0.5-mℓ aliquots were taken in each case and added to 4.5 mℓ acidified (HCl) ethanol. A blanc containing the same solution except for the lipids is substracted (double beam spectroscopy). The absorption at 300 nm is taken as zero. By measuring lipid peroxidation in this way the "oxidation index" of the lipid used in Figures 6A, B, and C is 0.054, 0.270, and 0.540, respectively. When water-soluble substances were present that could interfere with the UV absorbance of the lipids (e.g., $FeCl_2$) then the lipids were extracted from the liposomes and after evaporation of the organic solvents (under nitrogen) taken up in ethanol and measured against an ethanol blanc.

B. Experimental Results; Discussion

A selection of experiments will be given representing conditions under which lipid per-

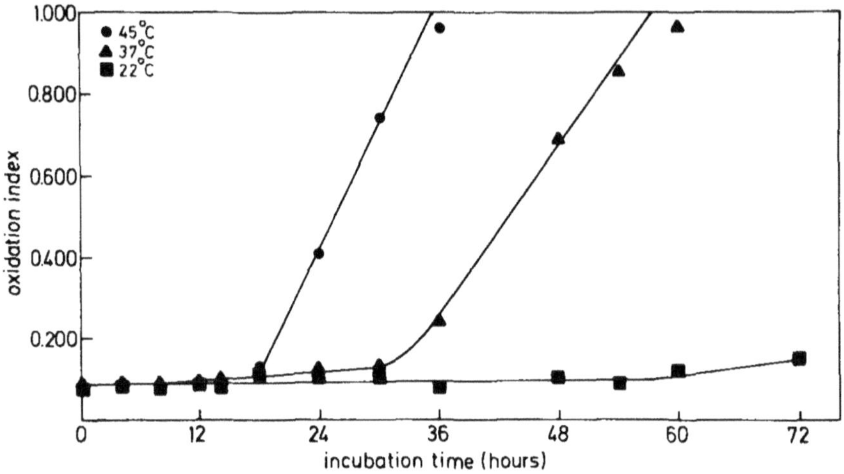

FIGURE 8. Rate of autoxidation of rat liver phospholipid REV at different temperatures.

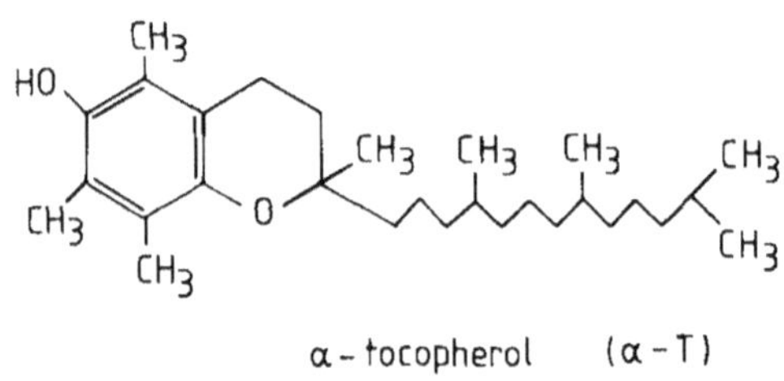

FIGURE 9. Structure of vitamin E.

oxidation reactions in rat liver phospholipids and in fatty acid preparations have been observed in our laboratory during the last 5 years. In many cases antioxidants have been used to prevent lipid peroxidation. Some results with α-tocopherol (α-T) and butylated hydroxytoluene (BHT) are reported.

1. Autoxidation of MLV and REV; Protection by Antioxidants

When MLV prepared in double distilled water are incubated under oxic conditions, as mentioned in the method section, lipid peroxidation will eventually take place. The temperature is very important, not only in determining the rate of lipid peroxidation but also in the duration of the induction period prior to the period of rapid lipid peroxidation. This is illustrated in Figure 7. Two hours after the start of the incubation no increase in the oxidation index could be observed in this experiment, not even at the relative high temperature of 45°C. After about 6 hr however an increase in conjugated dienes was observed and at about this time rapid lipid peroxidation started when the MLV were kept at 45°C. About 2 hr later lipid peroxidation at 37°C began and 18 hr after the onset of the incubation the MLV which were kept at 22°C started to produce increasing amounts of lipid peroxides. When REV were incubated at the same temperatures used earlier for MLV, lipid peroxidation occurred after a longer period of induction. After 12 hr almost no peroxides were formed in the REV preparations (while the oxidation index in MLV was around 0.5). This is shown in Figure 8. The REV start to peroxidize after about 20 hr at 45°C and after about 30 hr at 37°C. Up to 54 hr at 22°C the REV had an oxidation index equal to the controls. It may be concluded from Figures 7 and 8 that MLV do not peroxidize for 2 days when kept at 4°C while REV may be kept at 22°C without being oxidized during that period of time.

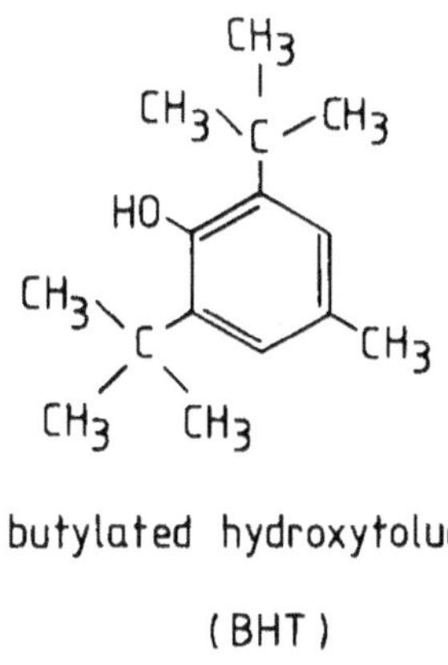

FIGURE 10. Structure of BHT.

Table 1
AUTOXIDATION OF NORMAL AND α-T-CONTAINING MLV, MEASURED BY THE OXIDATION INDEX AND THE DECREASE IN ARACHIDONIC ACYL CHAINS

Incubation time	Oxidation index (A_{233}/A_{215})	Decrease of 20:4 (%)
Normal–MLV after 12 hr	0.352	18
(α-T)-MLV after 12 hr	0.090	0
Normal-MLV after 24 hr	0.973	48
(α-T)-MLV after 24 hr	0.641	29
Normal-MLV after 48 hr	0.781	73
(α-T)-MLV after 48 hr	0.857	65

Note: The incubation was performed in a shaking waterbath at 37°C in 25 mℓ Erlenmeyer flasks containing 6 mg lipids in 12 mℓ H_2O. If present, α-T was at 0.5 mol %.

We have observed that the length of the lag period in autoxidation is dependent on the amount of lipid hydroperoxides present in the lipid sample (not shown). It is assumed that most of the initiations of lipid peroxidation take place by Reaction 9 where alkoxyl radicals abstract hydrogens from unsaturated fatty acids. Products are formed which themselves catalyze the reaction. A higher temperature will not only activate reacting molecules but also enhance decomposition of hydroperoxides and may shorten the induction period prior to the start of rapid lipid peroxidation. It is not clear at the moment why the induction period of REV preparations is generally found to be longer as compared to preparations of MLV. The same lipids and the same concentration (0.5 mg PL per milliliter) was used. The MLV preparation possibly has more prooxidant and less antioxidant compounds present. The REV were prepared in chloroform/dichloromethane. Trace amounts of these compounds are always present. However, addition of small amounts of these organic solvents to the MLV preparations did not lengthen the induction period (not shown). When sufficient hydroperoxides have been formed during the induction period a rapid peroxidation starts via the propagation steps mentioned earlier (Reactions 13 to 15).

The free-radical chain reactions may be inhibited by antioxidants as shown before. Basically, to be a good antioxidant, the compound must be able to delocalize effectively the unpaired electron produced in the reaction with the free radical. The more effective the

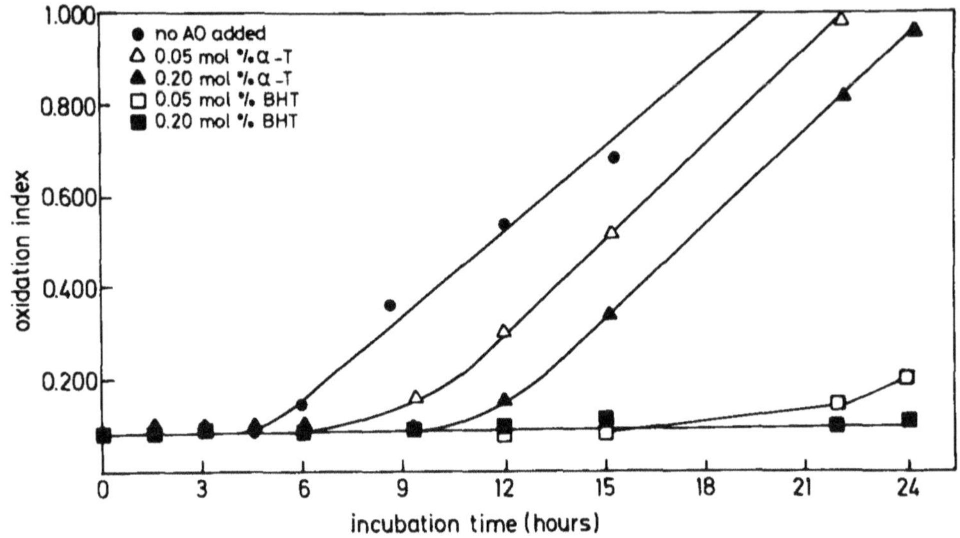

FIGURE 11. Effect of antioxidants (AO) on the autoxidation of rat liver phospholipid MLV.

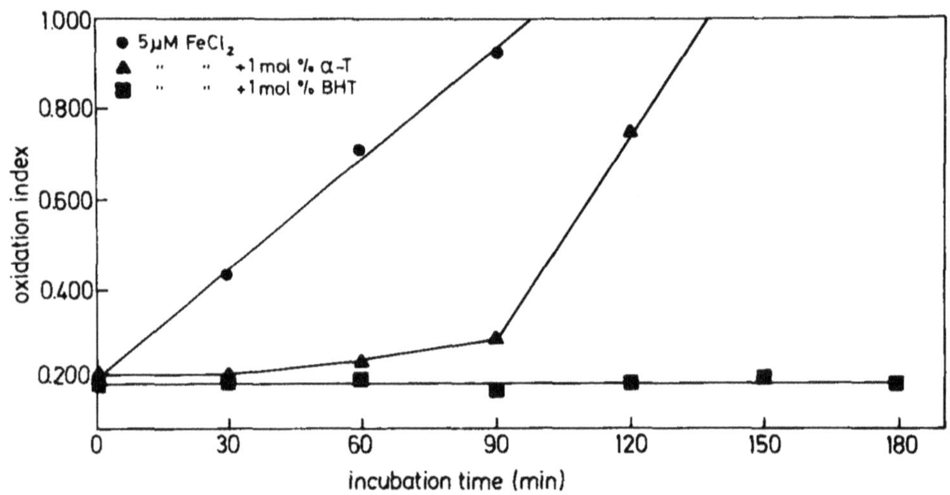

FIGURE 12. Effect of antioxidants on the ferrous ions catalyzed autoxidation of MLV.

resonance forms, the better the antioxidant. Many hydrogen donors of the phenolic type have proven to be good antioxidants and break chain reactions. The position and quality of the substituted groups in the phenols are important in determining the number of effective resonance forms. Two well-known phenolic antioxidants are vitamin E (α-T) and BHT. The chemical structure of α-T is depicted in Figure 9. The structure of BHT is shown in Figure 10.

Autoxidation of MLV containing 0.5 mol % α-T is clearly inhibited. After an incubation of MLV during 12 hr at 37°C in air the oxidation index of nonprotected lipids was 0.352 as compared to 0.090 when α-T was present (Table 1). Within the limits of detection no decrease of arachidonic acid (20:4) could be found when α-T was incorporated in the liposomes. Table 1 means to demonstrate that the oxidation index can only be used as a measure of lipid peroxidation when the A233/A215 is below 0.8 and during the initial state of rapid lipid peroxidation. Lipid peroxides will decompose after a certain time and the 233

Table 2
PREPARATION OF MLV AND SUV UNDER DIFFERENT EXPERIMENTAL CONDITIONS

Liposomes	no AO	1 mol % α-T	1 mol % BHT
MLV in air	0.093 ± 0.022	0.054 ± 0.018	0.051 ± 0.023
MLV in nitrogen	0.061 ± 0.018	0.055 ± 0.021	0.062 ± 0.017
SUV in air	0.205 ± 0.043	0.073 ± 0.019	0.067 ± 0.020
SUV in nitrogen	0.081 ± 0.025	0.070 ± 0.024	0.072 ± 0.018

Note: Both MLV and SUV were prepared at 22°C; the volume of MLV was 12 mℓ and of SUV 10 mℓ; the lipid concentration was 0.5 mg lipid per mℓ of H_2O. Data are expressed as the ratio: A_{233}/A_{215} (oxidation index) and given as the mean ± SD of four experiments.

nm absorption signal will gradually diminish. The disappearance of PUFAs as measured by GLC is a much better parameter. After 24 hr about 50% of the original amount of arachidonic acid has disappeared and 24 hr later about 75%. The oxidation index has dropped from 0.97 to 0.78.

In order to compare the protecting abilities of α-T and BHT these compounds were incorporated in the MLV at a mol % of 0.05 and 0.20. The liposomes were incubated in air at 44°C. The oxidation index was measured at different times after the start of the incubation. The result of this experiment is shown in Figure 11. As expected, the antioxidants prevent the lipids from being oxidized by a chain reaction and the induction period is enlarged. It is assumed that the peroxyl radicals formed during the propagation reactions (Reaction 13) are removed by the phenols. A hydrogen atom is donated to the peroxyl radical and a hydroperoxide is formed (Reaction 21). In this experiment BHT has a higher antioxidant capacity than α-T. As discussed earlier, the initiation of lipid peroxidation, and as a consequence the length of the induction period, will depend on the amount of lipid hydroperoxides in the sample. The presence of compounds with antioxidant and prooxidant characteristics will also influence the rate of the initiation reactions. Transition metals present as impurities may accelerate lipid peroxidation. When MLV are prepared in double distilled water to which 5 μM $FeCl_2$ is added, lipid peroxidation at 37°C will be very rapid. As shown in Figure 12, after 30 min an oxidation index of about 0.45 was reached. Without the additions of ferrous ions, it would take 12 hr (Figure 7). The ferrous-catalyzed lipid peroxidation could be inhibited by α-T and BHT. These compounds were added at 1 mol %. As in the autoxidation experiment without added ferrous ions, α-T was a less effective inhibitor when compared with BHT. It must be noticed however that the ferrous ions were present during the preparation (sonication) of the MLV. When $FeCl_2$ was added after the preparation, 1 mol % α-T could totally inhibit lipid peroxidation during that period of time (180 min), just like BHT.

2. *Sonication of MLV and SUV; Protection by Antioxidants*

It is of interest to know the effect of the presence of antioxidants during the preparation of MLV and SUV. Probe sonication is used during the MLV preparations. When the whole procedure is performed in a nitrogen atmosphere, then some protection against lipid peroxidation is seen. Table 2 shows that MLV lipids are slightly peroxidized when prepared in air. This peroxidation may completely be prevented by adding 1 mol % α-T or 1 mol % BHT (Figure 13). For the preparation of SUV it is very important to take precautions during the preparation. Sonicating in air will produce an appreciable amount of damage to the liposomes as is indicated by an oxidation index of 0.205. Lipid peroxidation was strongly inhibited when 1 mol % of α-T or BHT was present in the liposomes. Slight peroxidation

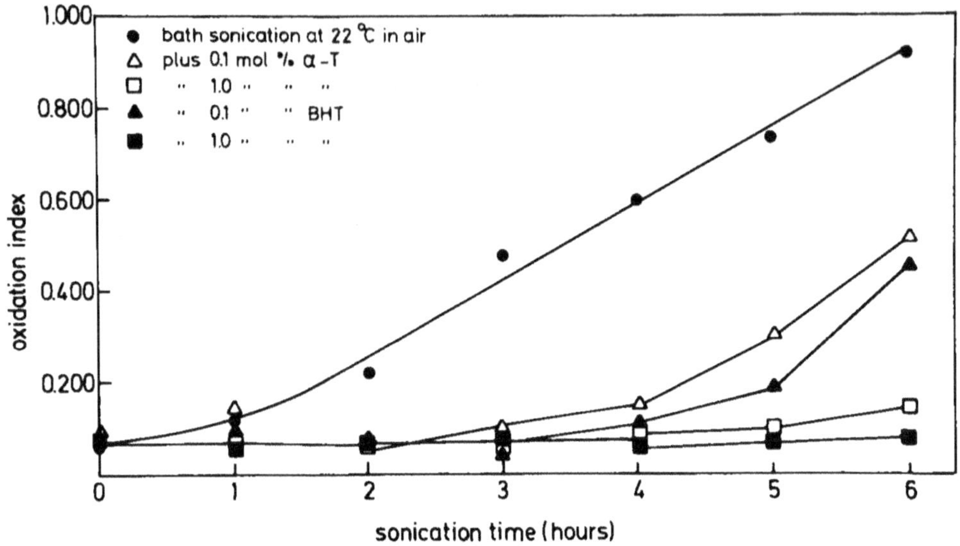

FIGURE 13. Effect of antioxidants on the rate of lipid peroxidation caused by sonication of MLV.

took place when the preparations were done under nitrogen gassing conditions. In order to have an indication of the protecting capacities of α-T and BHT a time course sonication experiment was performed with 0.1 mol % and 1 mol % of the antioxidants. The sonication time in this experiment was 6 hr, with a sample being taken every hour. The SUV without α-T or BHT oxidized rapidly and reached an oxidation index of 0.9 after 6 hr. Lipid peroxidation caused by sonication under aerobic conditions has been shown before.[38] It should be pointed out that the degree of lipid peroxidation depends on the type of sonication bath and on the place where the sample is positioned. Especially in large sonication baths the energy deposited varies enormously over the total area. We have performed some sonication experiments in a smaller bath (Bransonic 220) and found an average increase in oxidation index by a factor of two. It is shown in Figure 13 that with 0.1 mol % of antioxidants the liposomes were protected against 3 hr of sonication. The efficiency of α-T and BHT were comparable in this experiment. The sonication experiments performed in our laboratory indicate that 1 mol % α-T or BHT is sufficient to protect against lipid peroxidation when normal sonication times are used.

3. Radiation-Induced Lipid Peroxidation in MLV and Micelles; Protection by Antioxidants

Ionizing radiation generates free radicals. When the lipids are dispersed in an aqueous phase, water molecules will be the main source of the free radicals formed.

In the radiolysis of water, the princlipal reactive species are the hydrated electron e_{aq}^- the hydrogen atom H·, and the hydroxyl free radical OH·

$$H_2O \rightarrow e_{aq}^- , H\cdot , OH\cdot$$

In the presence of oxygen, e_{aq}^- and H· react at diffusion controlled rates to form univalently reduced oxygen, the superoxide anion ($O_2^-\cdot$) and its protonated form, the perhydroxyl radical ($HO_2\cdot$).

$$e_{aq}^- + O_2 \rightarrow O_2^-\cdot \quad (24)$$

$$H\cdot + O_2 \rightarrow HO_2\cdot \quad (25)$$

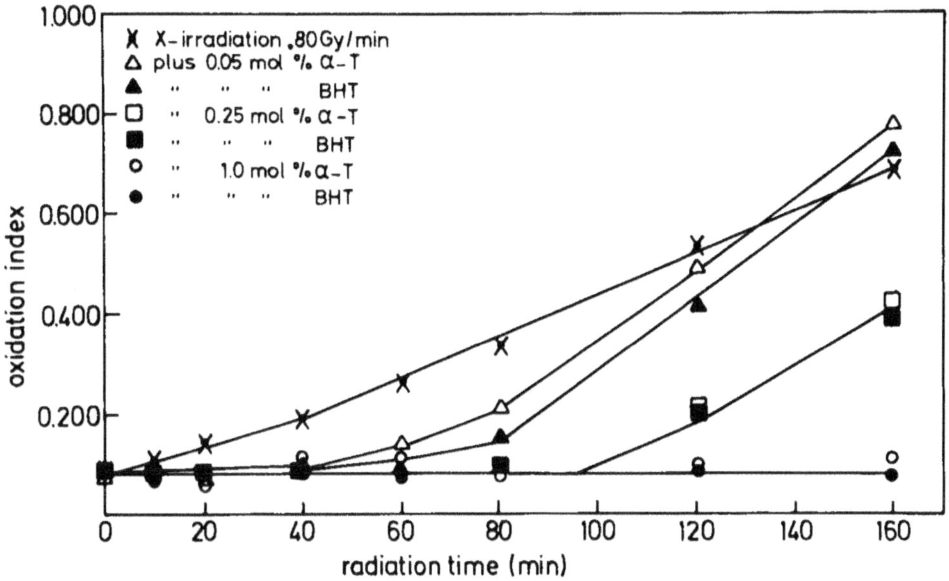

FIGURE 14. Effect of antioxidants on radiation induced lipid peroxidation in MLV at 22°C.

The time scale of the radiation action reaches from the earliest physical event followed by the transit of a quantum of energy through an atom in 10^{-18} sec. Reactions of radicals formed occur after 10^{-10} sec.

The energy absorbed from the radiation, the dose, is expressed in Gray (Gy.). 1 Gy = 1 J/kg = 100 rad. Chemical effects of radiation are often measured in G values (G). G_x is defined as the number of atoms or molecules of substance x produced or transformed in an irradiated system absorbing 100 eV of energy. The G values for most species produced are about 1 to 3, since it takes around 30 eV to produce an ion pair. High G values indicate the presence of chain reactions. Generally accepted values for radical yields in water irradiated with X-rays at neutral pH are as follows:

$$G_{(OH\cdot)} = 2.7 \qquad G_{(e^-_{aq})} = 2.7 \qquad G_{(H\cdot)} = 0.6$$

As mentioned earlier, the hydroxyl radicals have a powerful oxidative capacity. However, the hydrogen atom and its base the hydrated electron, have powerful reductive capacities. It has been recognized for a long time that hydroperoxides and their breakdown products are formed when emulsions of fats and oils are irradiated.[7,13,39-41]

When liposomes prepared from rat liver phospholipids are X-irradiated at a dose rate of 0.80 Gy/min the oxidation index increases to a value of about 0.6 in 160 min. This is shown in Figure 14. The lipids are protected against radiation-induced lipid peroxidation by 0.05 mol % α-T or BHT for about 60 min, by 0.25 mol % for about 100 min and for over 160 min by 1.0 mol %.

The protecting efficiency of both phenolic compounds is comparable when the same mole percent is used. As observed earlier [42] the rate of lipid peroxidation after the protected period (lag phase) is faster than observed in the MLV without the antioxidants. It may be that during the induction period there is accumulation of substances (lipid hydroperoxides) formed through Reaction 21. These provide the substrates for many initiation reactions leading to an increased propagation of lipid peroxidation (chain reactions) when the antioxidants are exhausted.

The rate of lipid peroxidation in liposomes is dependent on the dose rate applied (Table

Table 3
X-IRRADIATION OF LIPOSOMES

Dose rate (Gy/min)	Oxidation index (A_{233}/A_{215})
0.1	0.61
0.9	0.26
4.7	0.22
9.0	0.14

Note: 3 mℓ MLV (0.5 mg lipid/mℓ H_2O) was irradiated at 22°C under air with a total dose of 36 Gy.

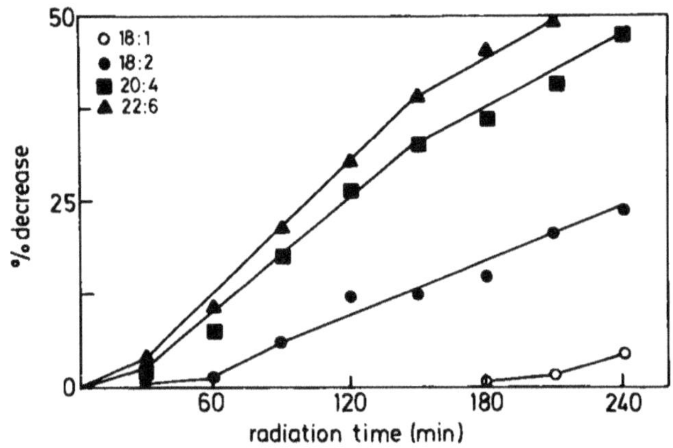

FIGURE 15. Loss of unsaturated fatty acyl chains in phospholipids of MLV after radiation in air at 22°C.

3). When a total dose of 36 Gy was given to MLV, the oxidation index was found to be much higher when low dose rates were applied. This has been observed when lipids are irradiated with different dose rates.[7,27,39,43-45]

When high dose rates are applied, more initiation reactions will take place and more lipid peroxyl radicals will be formed (Reaction 13). As a consequence of this, more termination reactions will occur when compared to low dose rate X-irradiation. At low dose rates less $LO_2\cdot$ will be formed in time and space and relatively more of these peroxyl radicals will react with a neighboring unsaturated fatty acid (Reaction 14). The extent of the chain reaction will be enhanced at lower dose rates.

When the disappearance of the unsaturated fatty acyl chains of the phospholipids was followed in time during X-irradiation, it was observed[27] that more clupanadonic and arachidonic acyl chains were destroyed (Figure 15). It should be mentioned that in the later stages of the radiation experiment also saturated fatty acids disappeared (not shown). The order of radiosensitivity of the unsaturated fatty acyl chains in the liposomes is the same as found when the liposomes were subjected to autoxidation with or without 5 μM $FeCl_2$ (not shown).

Experiments with fatty acid micelles performed by two different groups[17,46] indicated that the order of radiosensitivity was as follows: 18:2 > 18:3 ≥ 20.4. Therefore, our results with liposomes were different from the results of the above-mentioned investigators working with fatty acid micelles.

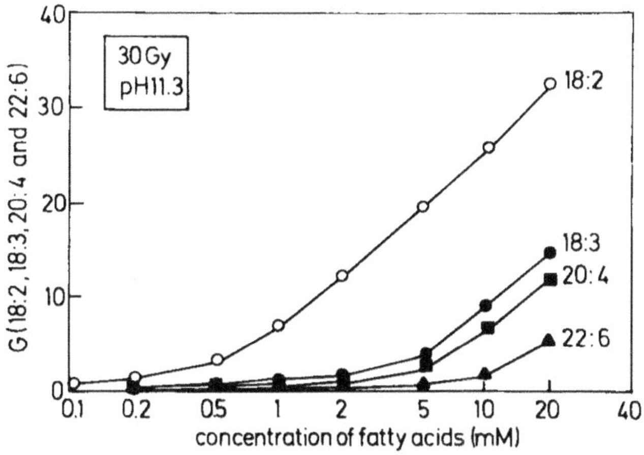

FIGURE 16. Dependence of G values for conjugated diene yield on the concentration of different fatty acids. (From Mooibroek, J., et al., *Int. J. Radiat. Biol.*, 42, 601, 1982. With permission.)

There are two important differences between the reported experiments with micelles and with liposomes:

1. The liposomal studies were performed at pH 7.4, while the studies on micelles were at pH 10 to 11.
2. The liposomes were composed of different acyl groups in a mixed configuration, while the micelles were single component fatty acid structures.

Experiments with micelles performed by us [28] led to the same conclusion: the more double bonds in the fatty acids the less radiosensitive these fatty acids were. This is shown in Figure 16. There is evidence for an inflection in each curve. The concentration region containing the inflection points coincides with the critical micelle concentration (CMC) for the different fatty acids. It can be seen that the fatty acids are less vulnerable for lipid peroxidation when present below CMC, where they are in the monomeric and dimeric form.

The experiments illustrated in Figure 16 were performed at pH 11.3. The influence of different pH values on the lipid peroxidation of 18:2, 18:3, 20:4, and 22:6 micelles was investigated. The concentration of the fatty acids in each micellar preparation was 2 m*M*. The order of radiosensitivity was found to be the same at all pH values explored, except for 18:2 around pH 10, where a drop in radiosensitivity was observed. As can be seen in Figure 17, a concentration of 2 m*M* is above the CMC for 18:2 and below the CMC for the other fatty acids tested. The drop in radiosensitivity for those fatty acids which are in the micellar form (18:2) at pH 10 can probably be explained by a critical value of surface hydrogen concentration at the micelle lipid-water interface, resulting in a concentration of undissociated fatty acids which is too high to maintain the original micellar structure. The micelles will break down and by lowering the pH, other structured aggregates will probably be formed, which again have a favorable alignment of the fatty acid hydrocarbon tails, enabling extended chain reactions to proceed.

When micelles of mixed fatty acid composition were prepared and irradiated, a different picture emerged (Figure 18). The four fatty acids were mixed in a 1:1 ratio. The experimental conditions were the same as for the experiment shown in Figure 17. The lipid concentration in each micellar preparation was again 2 m*M*.

When Figures 17 and 18 are compared it is clear that the radiosensitivity of single

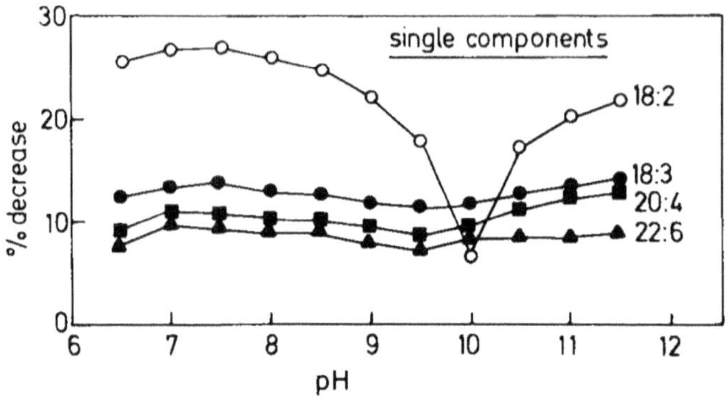

FIGURE 17. The effect of X-irradiation at 22°C on the disappearance of fatty acids in single component micelles. The concentration of the fatty acids was 2 m*M*. (From Mooibroek, J., et al., *Int. J. Radiat. Biol.*, 42, 601, 1982. With permission.)

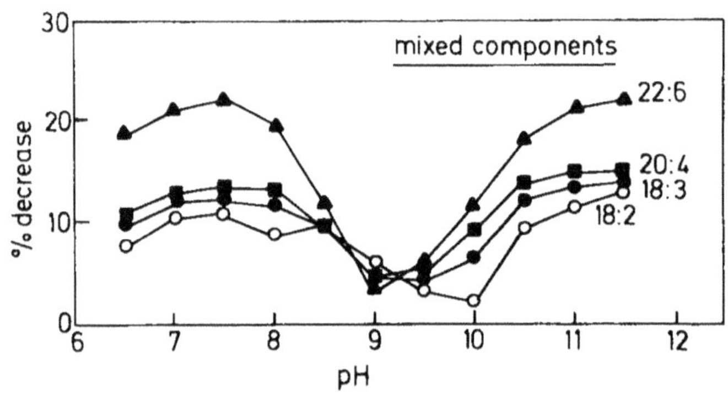

FIGURE 18. The effect of X-irradiation at 22°C on the disappearance of fatty acids in mixed component micelles. The total fatty acid concentration in each mixed preparation was 2 m*M*. (From Mooibroek, J., et al., *Int. J. Radiat. Biol.*, 42, 601, 1982. With permission.)

component micelles is different from the radiosensitivity of micelles consisting of mixed components. In both experimental systems 22:6 is present below CMC. In the mixed micelles the radiosensitivity of this fatty acid is much higher than when present in the single component micelles. For 18:2 the opposite observation is made. The mixed micelles have structures which are less favorable for chain reactions in which 18:2 can participate. The fact that the PUFAs in the mixed component experiments (Figure 18) are pH dependent indicate that these molecules are indeed part of the micellar structure. It is concluded that the efficiency of lipid peroxidation with respect to individual fatty acids or acyl chains depends primarily on the nature of the structural complex these lipids are in. Differences in the relative radiosensitivity arise from differences in the radical chain propagation steps of lipid peroxidation.

IV. CONCLUDING REMARKS

An attempt has been made to discuss shortly the present understanding of basic elements in lipid peroxidation of unsaturated fatty acids occurring in liposomes frequently used by many investigators today. A variety of experiments was introduced to illustrate the different ways lipid peroxidation can be initiated. Because of the involvement of our laboratory in lipid peroxidation induced by ionizing radiation, a substantial part of the experimental results was devoted to this aspect.

The most characteristic feature of lipid peroxidation under different experimental conditions is the radical chain propagation step, which occurs independently of the way the process is initiated.

Initiation of autoxidation in liposomes will probably not start when no transition metals are present in the system (glassware, water). The catalytic peroxidative capacity of these metals most probably has to be ascribed to their ability to accelerate the decomposition of existing lipid hydroperoxides. To a lesser extent the metals will activate oxygen leading to the generation of hydroxyl radicals. Hydroperoxide decomposition to alkoxyl radicals, might be the key step in the initiation of lipid autoxidation.

As has been shown in this chapter, the length of the lag phase (induction period) prior to the phase of rapid lipid peroxidation (chain reaction) is dependent on the concentration of pro- and antioxidants present. The antioxidants α-T and BHT have proven to be effective compounds in lengthening the induction period. From the experimental results presented it is suggested that BHT is a better protector against autoxidation of liposomes than α-T. Lipid peroxidation in biomembranes has also been shown to be dependent on antioxidants present.[47] Vitamin E-depleted membranes were peroxidized without any lag period.[34]

When lipids are sonicated, the extent of lipid peroxidation differs greatly according to the type of waterbath used or to the position in the waterbath. It is our experience that sonication of lipids with a titanium probe induces more oxidative damage than sonication in a bath-type sonicator. It has been shown[48] that appreciable nonoxidative damage can be caused by extensive sonication even in the presence of nitrogen gas. Especially hydrolysis of the phospholipids may be induced. Radiation-induced lipid peroxidation is not only dose dependent but also differs when different dose rates are used. Low dose rates induce more lipid peroxidation in the liposomal membranes. This dose rate dependency of lipid peroxidation has also been reported for biomembranes.[49]

After lipid peroxidation of liposomes, hydroperoxides present in the phospholipids will increase the degree of disorder in the bilayer and affect the permeability behavior of the vesicles. The aspect of a change in membrane permeability of the liposomes may be of importance especially when these vesicles are meant to be used as carriers for drugs or other substances. Although it is generally accepted that extensive lipid peroxidation in liposomes leads to an increased permeability of water-soluble compounds, no firm quantitative relationship has been established between the degree of lipid peroxidation and the leakiness of the vesicles for different classes of compounds.[38,50-54]

REFERENCES

1. **Gregoriadis, G.,** *Drug Carriers in Biology and Medicine,* Academic Press, London, 1979.
2. **Gregoriadis, G. and Allison, A.C.,** *Liposomes in Biological Systems,* J. Wiley & Sons, New York, 1980.
3. **Tom, B. H. and Six, H. R.,** *Liposomes and Immunobiology,* Elsevier/North-Holland, New York, 1980.
4. **Kaye, S. B.,** Liposomes — Problems and promise as selective drug carriers, *Cancer Treat. Rev.,* 8, 27, 1981.
5. **Knight, C. G.,** *Liposomes: From Physical Structure to Therapeutic Applications,* Elsevier/North Holland, Amsterdam, 1981.
6. **Uri, N.,** *Autoxidation and Antioxidants,* Vol. 1, Lundberg, W. E., Ed., Wiley Interscience, New York, 1961, 55.
7. **Mead, J. F.,** in: *Free Radicals in Biology,* Vol. 1, Pryor, W. A., Ed., Academic Press, New York, 1976, 55.

8. **Frankel, E. N.,** *Fatty Acids,* Pryde, E. H., Ed., American Chemist's Society, Champagin, Ill., 1979, 353.
9. **Schultz, H. W., Day, E. A., and Sinnhuber, R. O.,** *Symposium on Foods: Lipids and their Oxidation,* Avi Publishing, Westport, Conn., 1962.
10. **Dahle, L. K., Hill, E. G., and Holman, R. T.,** The thiobarbituric reaction and autoxidation of polyunsaturated fatty acid methylesters, *Arch. Biochem. Biophys.,* 98, 253, 1962.
11. **Pryor, W. A., Stanley, J. P., and Blair, E.,** Autoxidation of polyunsaturated fatty acids. II. A suggested mechanism for the formation of TBA-reactive materials from prostaglandin-like endoperoxides, *Lipids,* 11, 370, 1976.
12. **Frankel, E. N., Nef, W. E., Rohwedder, W. K., Khambay, B. S. P., Garwood, R. F., and Weedon, B. C.L.,** Analysis of autoxidized fats by gas chromotography-mass spectrometry. III. Methyl linolenate, *Lipids,* 12, 1055, 1977.
13. **Nawar, W. W.,** in *Radiation Chemistry of Major Food Components,* Elias, D. S. and Cohen, A. J., Eds., Elsevier, Amsterdam, 1977.
14. **Fong, K. L., McCay, P. B., Poyer, J. L., Keele, B. B., and Misra, H.,** Evidence that peroxidation of lysosomal membranes is initiated by hydroxyl free radicals produced during flavin enzyme activity, *J. Biol. Chem.,* 248, 7792, 1973.
15. **Tyler, D. D.,** Role of superoxide radicals in the lipid peroxidation of intracellular membranes, *FEBS Lett.,* 51, 180, 1975.
16. **Fong, K. L., McCay, P. B., and Poyer, J. L.,** Evidence from superoxide-dependent reduction of Fe^{3+} and its role in enzyme-generated hydroxyl radical formation, *Chem. Biol. Interact.,* 15, 77, 1976.
17. **Raleigh, J. A. and Kremers, W.,** Promotion of radiation peroxidation in models of lipid membranes by Caesium and Rubidium counter-ions; micellar linoleic and linolenic acids, *Int. J. Radiat. Biol.,* 34, 439, 1978.
18. **Walling, C.,** Fenton's reagent revisited, *Acc. Chem. Res.,* 8, 125, 1975.
19. **Pryor, W. A.,** *Free Radicals,* McGraw-Hill, New York, 1966.
20. **Fridovich, I.,** *Free Radicals in Biology,* Vol. 1, Pryor, W. A., Ed., Academic Press, New York, 1976, 239.
21. **Ingold, K. U.,** *Lipids and their Oxidation,* Schultz, H. W., Day, E. A., and Sinnhuber, R. O., Eds., Avi Publishing, Westport, Conn., 1962, 93.
22. **Reich, L. and Stivala, S. S.,** *Autoxidation of Hydrocarbons and Polyolefins,* Marcel Dekker, New York, 1969.
23. **Thomas, J. M., Mehl, K. S., and Pryor, W. A.,** The role of the superoxide anion in the xanthine oxidase-induced autoxidation of linoleic acid, *Biochem. Biophys. Res. Commun.,* 83, 927, 1978.
24. **Pryor, W. A.,** Free radical reactions and their importance in biochemical systems, *Fed. Proc. Fed. Am. Soc. Exp. Biol.,* 32, 1862, 1973.
25. **Pryor, W. A.,** *Free Radicals in Biology,* Vol. 1, Academic Press, New York, 1976.
26. **Ingold, K. U.,** Peroxy radicals, *Acc. Chem. Res.,* 2, 1, 1969.
27. **Konings, A. W. T., Damen, J., and Trieling, W. B.,** Protection of liposomal lipids against radiation-induced oxidative damage, *Int. J. Radiat. Biol.,*.35, 343, 1979.
28. **Mooibroek, J., Trieling, W. B., and Konings, A. W. T.** Comparison of the radiosensitivity of unsaturated fatty acids, structured as micelles or liposomes under different experimental conditions, *Int. J. Radiat. Biol.,* 42, 601, 1982.
29. **Szoka, F. and Papahadjopoulos, D.,** Procedure for preparation of liposomes with large internal aqueous space and high capture by reverse-phase evaporation, *Proc. Natl. Acad. Sci. U.S.A.,* 75, 4194, 1978.
30. **Olson, F., Hunt, C. A., Szoka, F. C., Vail, W. J., and Papahadjopoulos, D.,** Preparation of liposomes of defined size distribution by extrusion through polycarbonate membranes, *Biochim. Biophys. Acta,* 557, 9, 1979.
31. **Holman, R. T.,** *Progress in the Chemistry of Fats and Other Lipids,* Vol. 2, Holman, R. T., Lundberg, W. O., and Malkin, T., Eds., Pergamon Press, London, 1954, 51.
32. **Gray, J. I.,** Measurement of lipid oxidation: a review, *J. Am. Chem. Soc.,* 55, 539, 1978.
33. **Konings, A. W. T. and Trieling, W. B.,** The inhibition of DNA synthesis in vitamin-E-depleted lymphosarcoma cells by X-rays and cytostatics, *Int. J. Radiat. Biol.,* 31, 397, 1977.
34. **Konings, A. W. T. and Drijver, E. B.,** Radiation effects on membranes. I. Vitamin E deficiency and lipid peroxidation, *Radiat. Res.,* 80, 494, 1979.
35. **Wilburg, K. M., Bernheim, F., and Shapiro, O. W.,** The thiobarbiture acid reagent as a test for the oxidation of unsaturated fatty acids by various agents, *Arch. Biochem. Biophys.,* 24, 305, 1949.
36. **Gutteridge, J. M. C.,** The measurement of malondialdehyde in peroxidised ox-brain phospholipid liposomes, *Anal. Biochem.,* 69, 518, 1975.
37. **Konings, A. W. T.,** Substrate supply for nuclear oxidative phosphorylation, *Biochim. Biophys. Acta,* 223, 398, 1970.
38. **Klein, R. A.,** The detection of oxidation in liposome preparations, *Biochim. Biophys. Acta,* 210, 486, 1970.

39. **Mead, J. F.,** The irradiation-induced autoxidation of linoleic acid, *Science*, 115, 470, 1952.
40. **Mead, J. F.,** *Autoxidation and Antioxidants*, Vol. 1, Lundberg, W. O., Ed., Interscience, New York, 1961, 299.
41. **Wills, E. D. and Rotblad, J.,** The formation of peroxides in tissue lipids and unsaturated fatty acids by irradiation, *Int. J. Radiat. Biol.*, 8, 551, 1964.
42. **Konings, A. W. T.,** Radiation protection of membranes by α-tocopherol, *Int. J. Radiat. Biol.*, 38, 119, 1980.
43. **Raleigh, J. A., Kremers, W., and Gaboury, B.,** Dose rate and oxygen effects in models of lipid membranes: linoleic acid, *Int. J. Radiat. Biol.*, 31, 203, 1977.
44. **Gebicki, J. M.,** Biophysical Aspects of Radiation Quality, International Atomic Energy Agency, Vienna, 1971 229.
45. **Petkau, A. and Chelack, W. S.,** Radioprotective effect of superoxide dismutase on model phospholipid membranes, *Biochim. Biophys. Acta*, 433, 445, 1976.
46. **Patterson, L. K. and Repath, J. L.,** in *Micellation, Solubilization and Microemulsions*, Vol. 2, Mittal, K. S., Eds., Plenum Press, New York, 1977, 589.
47. **Konings, A. W. T. and Oosterloo, S. K.,** Radiation effects on membranes. II. A comparison of the effects of X-irradiation and ozon exposure with respect to the relation of antioxidant concentration and the capacity for lipid peroxidation, *Radiat. Res.*, 81, 200, 1980.
48. **Hauser, H. O.,** The effect of ultrasonic irradiation on the chemical structure of egg lecithin, *Biochem. Biophys. Res. Commun.*, 45, 1049, 1971.
49. **Konings, A. W. T.,** Dose rate effects in biomembranes, *J. Radiat. Res.*, 20, 259, 1979.
50. **Leibowitz, M. E. and Johnson, M. C.,** Relation of lipid peroxidation to loss of cations trapped in liposomes, *J. Lipid Res.*, 12, 662, 1971.
51. **Smolen, J. E. and Shohet, S. B.,** Permeability changes incuded by peroxidation in liposomes prepared from human erythrocyte lipids, *J. Lipid Res.*, 15, 273, 1974.
52. **Hicks, M. and Gebicki, J. M.,** A quantitative relationship between permeability and degree of peroxidation in UFA-some membranes, *Biochem. Biophys. Res. Commun.*, 80, 704, 1978.
53. **Mandal, T. K. and Chatterjee, S. N.,** Ultraviolet- and sunlight-induced lipid peroxidation in liposomal membrane, *Radiat. Res.*, 83, 290, 1980.
54. **Nakazawa, T. and Nagatsuka, S.,** Radiation-induced lipid peroxidation and membrane permeability in liposomes, *Int. J. Radiat. Biol.*, 38, 537, 1980.

Chapter 12

SIZE DISTRIBUTION ANALYSIS OF LIPOSOMES BY ELECTRON MICROSCOPY

Pierre Guiot and Pierre Baudhuin

TABLE OF CONTENTS

I. INTRODUCTION

The aim of the present chapter is to describe a methodology allowing an assessment of the heterogeneity of liposome suspensions. In many cases, it is assumed — at least implicitly — that this heterogeneity can be neglected and the control of the preparation is limited to the experimental determination of some average parameter of the particle population. However, techniques allowing a direct evaluation of the heterogeneity yield often results showing quite large variations in the geometric properties of the particles: in suspensions of monolamellar liposomes, standard deviations of the particle radius between 30 and 100% of the average radius, depending on the method of preparation, are not unusual. Hence, the measurement of bulk properties, such as the total mass of lipids, in liposome suspensions may give estimates reflecting essentially the occurrence of a small proportion of large particles. For example, let us assume that a preparation of unilamellar vesicles contains, on a number basis, 3% of particles with a diameter three times larger than that of the main (97%) component of the population. It can then be computed that these few large particles will contribute for 22% of the membrane area, and for 46% of the total volume of the particles in the suspension. Although it is clear that determination of bulk properties is important for the characterization of a preparation, the number of particles per unit volume and their true average size may be determinant parameters for the interactions of liposomes with cells.

Most of the techniques used for evaluating the geometric properties of particles, such as NMR[1] or classical light-scattering measurements, do not provide the true, statistical average size. These measurements depend indeed on the mean volume or the mean square volume. For example, let us suppose a hypothetical population with 97% of particles with a 15-nm radius, contaminated by 3% particles with 45-nm radius. The true, statistical, average radius is then 15.9 nm, but, according to Rayleigh's equation, light scattering will yield a value of 25.3 nm, corresponding to the average of r^6.

Electron microscopy provides a direct approach to the size of small particles. The technique of negative staining is particularly simple to apply, but unfortunately, as shown by Miyamoto and Stoeckenius,[2] it induces distortion of liposomes. Hence, its use is restricted to qualitative analysis of such particles. Direct evaluation of the distribution of geometrical properties of liposomes can however be obtained by electron microscopy of freeze-fracture replicas. Such analysis will provide correct statistical averages for particle radius, area, and volume. If the specific molecular area of the phospholipid headgroup is known, an evaluation of the number of particles per unit volume in the initial suspension can also be obtained, in the case of monolamellar liposomes.

The technique described here has been chosen essentially for two reasons: (1) it avoids the alteration of the suspension medium by fixatives or other reagents; (2) it yields representative samples of the suspension, required for quantitative estimates by stereological methods. Our own experience being essentially with monolamellar liposomes, we will deal here only with this type of particles.

II. PREPARATION OF SAMPLES FOR ELECTRON MICROSCOPY

The method was originally described by Bachmann and Schmitt[3] for the examination of particle suspensions. It involves the freezing of aerosol droplets (10 to 30 μm in diameter) in liquid propane, at −196°C, followed by recovery and embedding of the frozen droplets in butylbenzene, at −88°C. The preparation is then freeze fractured and a surface replica is examined at the electron microscope.

A. Applicability of the Spraying Technique to Liposome Suspensions

The general appearance of a preparation at the electron microscope is illustrated in Figure

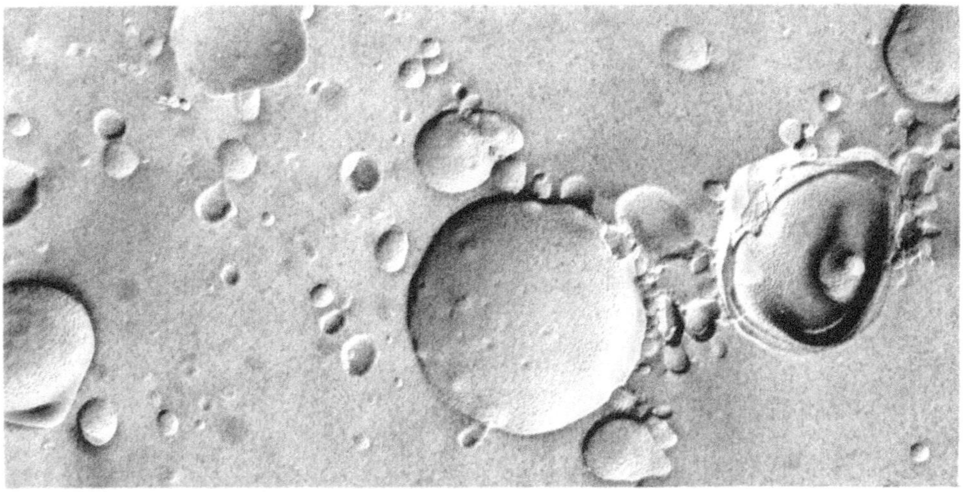

FIGURE 1. PS/PC (1:4) liposomes after suspension of phospholipids in phosphate-buffered saline, without sonication. (Magnification × 61,400.)

1. The suspension used in this case was rather heterogeneous: it was obtained from a phosphatidylserine/phosphatidylcholine (PS/PC) (1:4) mixture, by omitting the sonication step in the Johnson and Bangham[4] method. Particles display quite large variations in size and concentric lamellae of multilamellar liposomes are recognizable. As mentioned above, one of the main advantages of the method is that fixatives or cryoprotectants are not needed, owing to the rapid cooling of the sample. Electron micrographs, such as Figure 1, indeed suggest that the morphology of liposomes is well preserved. However, aerosol formation requires the rapid flow of the suspension, under gas pressure, through a narrow channel of a spraying gun. Associated turbulences can affect the properties of the suspension; hence the effect of the spraying procedure should be verified before attempting the examination of a preparation at the electron microscope. This can most easily be done by checking the light scattering at 90°C, immediately after spraying an aliquot of the suspension in a test tube, at room temperature. Since, as will be seen, the interest lays here essentially in relative modifications of light scattering, fluorescence spectrophotometers, without filter and with maximal slit, can be used for measuring the scattered light, for example at 650 nm.

Upon spraying of suspensions in phosphate-buffered saline, large (sixfold) modifications in the light-scattering properties can be observed for some liposome preparations, while for others the difference does not exceed 20% of the value before spraying (Figure 2). It should be noted that, in all cases, light scattering returned to the value before spraying after about 10 min. These changes in light-scattering properties are well related to changes in properties of liposomes, since they do not occur in the case of latex particles of 200 nm in diameter. That they are the consequence of turbulence is indicated by the fact that similar changes can be reproduced, to a smaller extent, by treatment of the suspensions with a laboratory mixer of the vortex type.

A preparation displaying large variations of light scattering upon spraying was examined at the electron microscope (Figure 3). It appeared heterogeneous and contained particles of largely different size and shape. Large (up to 1 μm) elements displayed a striated appearance, which could sometimes also be observed in smaller profiles. In our opinion, both the large modifications in light scattering and the striated appearance are due to the fact that in such preparation one component, PS, is fluid at room temperature, while the other, distearoylphosphatidylcholine (DSPC) is solid (transition temperature for distearoylphosphatidylcholine is 58°C[6]).

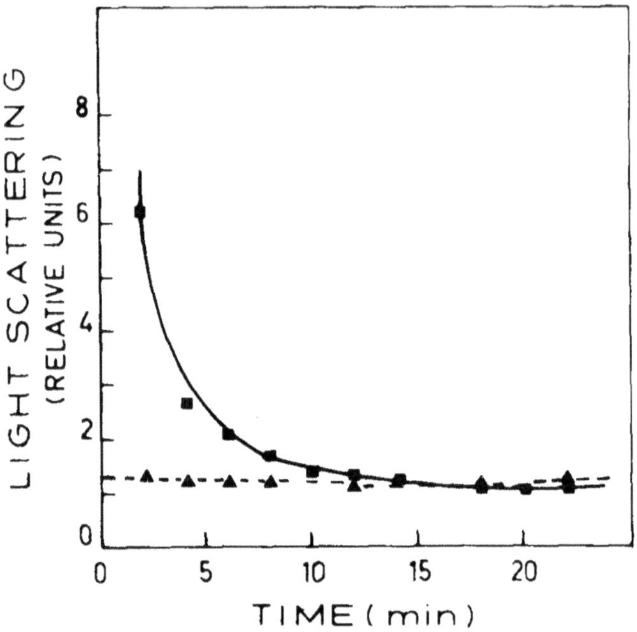

FIGURE 2. Light scattering of liposome suspensions after spraying. Values are expressed with respect to the light scattering of the suspension before spraying. Solid line: PS/DSPC (1:4) liposomes, prepared by a double emulsion technique.[5] Interrupted line: PS/PC (1:4) liposomes prepared by the same method.

The sensitivity of the light-scattering measurement is worth stressing. According to Rayleigh's equation, relative modifications in the intensity of scattered light (i.e., at constant wavelength and refraction index) is proportional to the number of particles and to the average squared particle volume. On the other hand, in a suspension of monolamellar liposomes, for a constant amount of phospholipid, the number of particles will be approximately inversely proportional to the squared radius. Hence, variations in the intensity of the scattered light will thus be roughly proportional to modifications of r.[4,7] In our work we have generally considered that a 5% change in the average size of particles could be tolerated; this corresponds approximately to a modification of 20% in light scattering.

The spraying procedure can also affect proteins which have been included in the suspension medium. Spraying of 20% serum in phosphate-buffered saline results in a large increase in the light scattering (Figure 4), which is progressively attenuated, although it does not return to the initial value. Moreover, as shown elsewhere,[8] addition of fetal calf serum to a liposome suspension results in the occurrence of numerous additional small profiles in the replicas of freeze-fractured suspension droplets. It must thus be concluded that the spraying procedure induces the formation of protein aggregates, 10 to 20 nm in diameter.

Measurement of light scattering as a function of time, immediately after spraying the suspension and over a period of 10 to 15 min, is thus a simple, but efficient control of the applicability of the spraying technique to a given preparation of liposomes. Except when otherwise stated, the effect of spraying has been tested and found negligible for all preparations presented as illustrations in this chapter.

B. Practical Procedures

The equipment for spray-freezing and for the recovery of frozen droplets is commercially available (Spray-freezing Unit, Balzers Union, Balzers, Lichtenstein). Liquid propane is

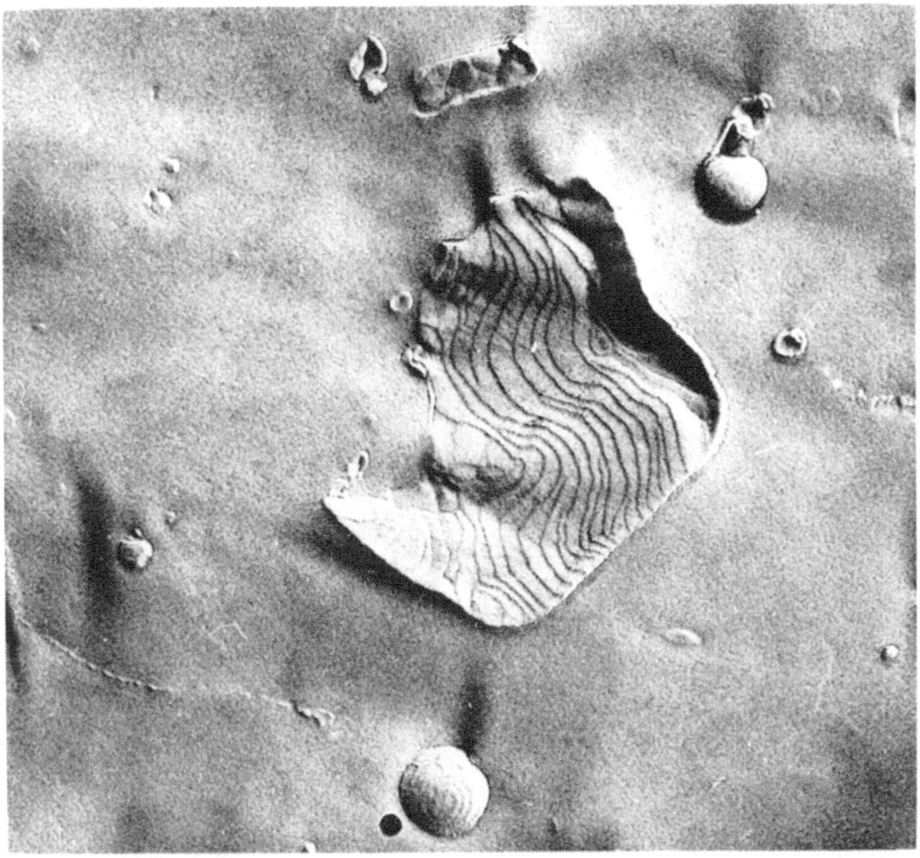

FIGURE 3. PS/DSPC (1:4) liposomes prepared by a double emulsion technique.[5] (Magnification × 60,000.)

produced by cooling gas in a glass condenser immersed in liquid nitrogen (−196°C). It is then poured in a small copper container, also cooled in liquid nitrogen. The spraying gun, equipped with a 150-μm nozzle and under a 0.5 kg/cm² pressure, is set perpendicular to the propane surface at a 10 cm distance. About 0.5 mℓ of sample is required for the spray; the concentration of phospholipid in the suspension should be 10 to 20 μmol/mℓ. As mentioned above, use of media containing significant amounts of protein may be a source of difficulty. Protein will not affect the preparation of the sample for electron microscopy, but stereological analysis will not be possible if protein aggregates have dimensions comparable to the average particle size, and hence cannot be differentiated from fractured liposomes. After spraying, the copper container is transferred to a metallic block cooled at −88°C, in a closed cabinet equipped with a perspex cover and manipulation gloves. An inverted glass funnel, connected to a vacuum pump, allows the evaporation of liquid propane. Thereafter, the frozen droplets are embedded in butylbenzene, previously cooled at −88°C, and transferred to small gold disks, which are then immersed and kept in liquid nitrogen. Samples are fractured at −110°C, at a pressure of 2.10^{-6} mmHg, with a knife at −150°C, according to the procedure of Moor et al.[9] The freeze-etching equipment used in our laboratory is model BA 360 M from Balzers A. G., Balzers, Lichtenstein. The replica is obtained by evaporation under vacuum of 4 nm platinum, at a 45° angle, followed by 18 nm carbon, at an angle of 90°. The replicas are then washed for 2 hr in 40% chromic acid at room temperature, rinsed in distilled water, and placed on a copper grid.

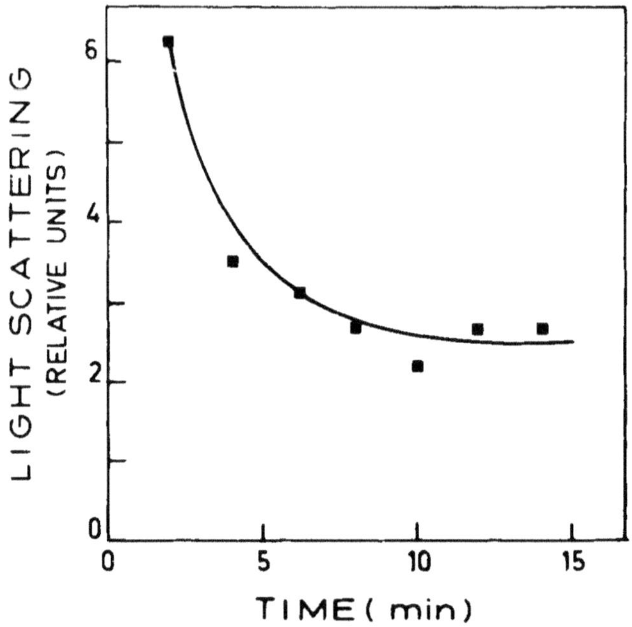

FIGURE 4. Light scattering of a 20% decomplemented fetal calf serum in phosphate-buffered saline, after spraying. Values are expressed with respect to the light scattering of the suspension before spraying.

FIGURE 5. Small unilamellar vesicles (SUV) made from PS/PC (1:4) and prepared by the Johnson and Bangham[4] procedure. (Magnification × 137,000.)

III. ELECTRON MICROSCOPY

In view of the subsequent application of stereological methods, a few points deserve special attention at the electron microscope. Micrographs should be taken at random: this means that any selection eventually introduced should be on a technical basis, independent of the particles. Overlapping of micrographs should be avoided. Since each droplet is representative of the sample and since the fracture level is essentially determined by the

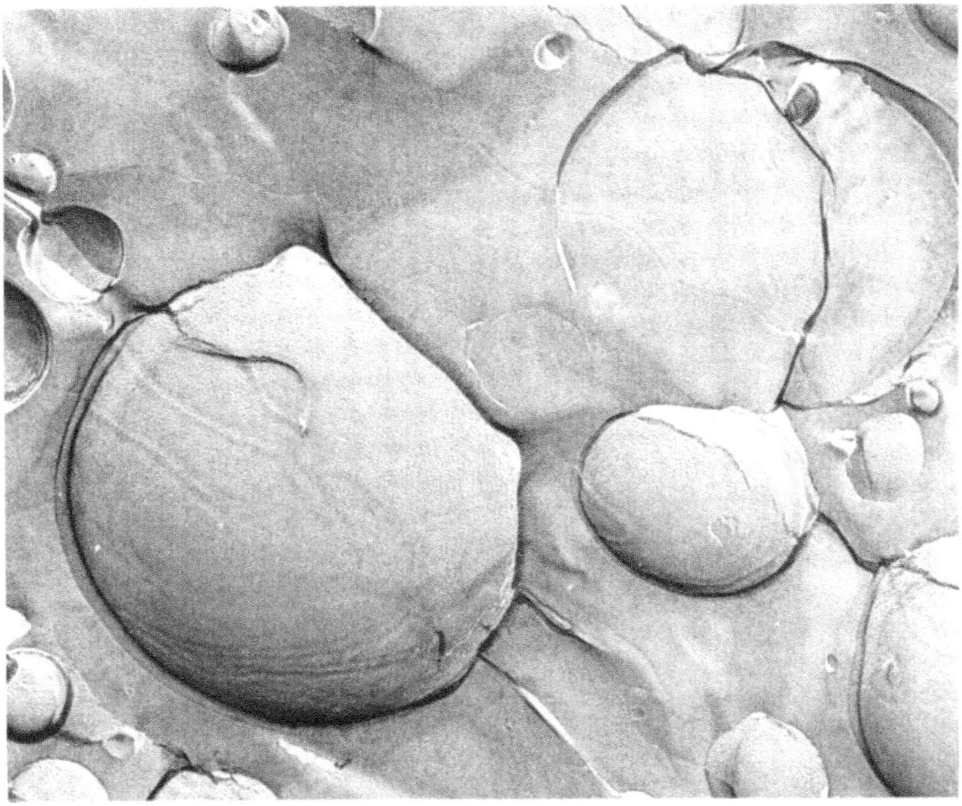

FIGURE 6. Large unilamellar vesicles (LUV) made from PS/PC (1:4) and prepared by the REV method of Szoka and Papahadjopoulos.[10] (Magnification × 64,000.)

level of the ice surrounding the particles, the micrographs obtained in these conditions may be considered as random samples.

Since size distribution analysis will be performed, it is essential to have an accurate estimate of the final magnification on the photographic prints. Possible source of errors are not only instrumental, but may also be related to modifications of the area of photographic papers, during processing. The simplest control here is to use a calibrated grating replica, which is photographed at the electron microscope after each set of micrographs of the sample and processed simultaneously. Moreover, micrographs of the calibration grating replicas can be used to reproduce all micrographs at the same final magnification and this will simplify the subsequent analysis.

Figures 5 and 6 illustrate typical micrographs, such as those used for the stereological analysis of small or large unilamellar vesicles. The sample size will of course depend on the accuracy required; generally a set of micrographs for size distribution analysis includes a minimum of 400 profiles.

IV. STEREOLOGICAL ANALYSIS

A. Data Collection

For the determination of size, the boundary of a profile should be taken at the level of the surrounding ice, so that the exact fracture path in the membrane becomes irrelevant. The definition of the particle size as the radius of the circle with equivalent area is generally convenient. If fracture profiles are not circular, it is very easy to match visually two areas

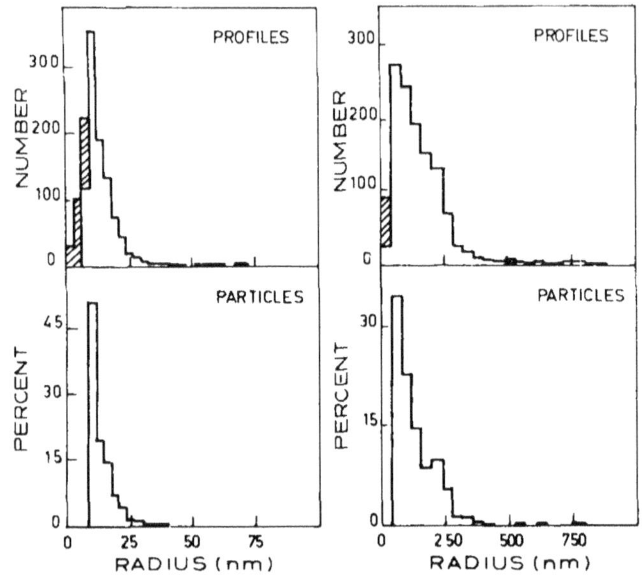

FIGURE 7. Size distribution analysis. Left: PS/PC (1:4) liposomes, prepared by the Johnson and Bangham[4] method; right: PS/PC/Chol (1:4:5) liposomes, prepared by the Szoka and Papahadjopoulos[10] method. Upper graphs: profile size distribution. The open part of the histogram is the experimental profile distribution. The hatched area represents the minimal number of small polar profiles which were missed. Lower graphs: particle size distribution obtained by Wicksell's transformation (see text).

of equal size but different shape and, even for nearly circular profiles, problems arising from the orientation of the measuring system are avoided. Several devices, of various sophistication levels, are commercially available for such measurements. The simplest is a set of concentric circles, with linear increments in radius and starting with zero radius, on a transparency, which can easily be made and reproduced photographically. About 15 to 25 circles will provide a convenient test system (for a review on profile size measurement, see Weibel[11]). Other systems are based on digitizing tablets or on image analyzers connected on line to a small computer. Whatever the method used, for the simplification of subsequent computations, classes of the experimental profile size histogram should have same width and the first class should start at the origin. Particles of a given class are considered as having all the radius corresponding to the middle of the class interval.

Obviously, the sample size will determine the statistical accuracy of statistical evaluations. It is useful to remember here that a Poisson distribution is generally followed in counting procedures and that the coefficient of variation (100 × SD/mean) of a count is equal to $100/\sqrt{N}$, where N stands for the number particle counted. As mentioned above, a minimum of 400 profiles is generally photographed at the electron microscope. This will thus give a ±5% accuracy on the average size, but in a class containing 50 profiles the coefficient of variation of the count will be 14%.

B. Particle Size Distribution

It is clear that the profile size in replicas depends both on the dimension of the particle and on the level at which a particle is fractured with respect to its center. In the case of spherical particles observed in sections, mathematical procedures have been developed by Wicksell[12] and later by Bach,[13] which take into account the effect of the fracture level on the size of the profiles. Briefly, the method is based on the fact that the class with largest

Table 1
STATISTICAL PARAMETERS DERIVED FROM SIZE DISTRIBUTION ANALYSIS

Parameter	Value	Radius of corresponding particle
Average radius (nm)	130	130
SD of radius (nm)	83	—
Average area (nm^2)		
External	298,000	154
Internal	285,000	
Average volume (nm^3)		
Total	26.0×10^6	184
Internal	24.7×10^6	

Note: Values given in this example pertain to liposomes prepared by REV[10] and containing PS, PC, and Chol (1:4:5 molar ratio).

radius is entirely made of sections passing through the center of the largest particles. Since these largest particles form an homogeneous subpopulation, of known radius, their abundance in the original suspension can be easily estimated from the abundance of their profiles in section. Moreover, contribution of the largest particles to classes of profiles with smaller radius can also be evaluated and subtracted from the experimental histogram. One obtains then a new histogram with one class less, on which the same computation can be applied. The procedure is repeated until all the profiles seen in section have been attributed either to equatorial or to nonequatorial sections.

In practice, however, these repetitive subtractions yield negative values for the abundance of particles with small radius. As discussed in detail for an application on isolated subcellular particles observed in ultrathin sections,[14] this is due to the fact that small polar profiles are not identified exhaustively. As a consequence, there are not enough small profiles to account for the nonequatorial fractures through particles of larger radius. The negative frequencies of the particle distribution have of course no physical meaning and should be set to zero. It is thus important to realize that the lower limit of the first nonzero class of the histogram does not necessarily represent the lower end of the particle size distribution, but may simply result from the truncation of the distribution of profiles. The importance of this truncation will depend on the possibility of exhaustive identification of the profile fractures and hence on the characteristics of the particles. Although it is not possible to have information on the abundance of particles below the truncation limit, a minimal evaluation of the number of missed profiles can be obtained by calculating backward the profile size distribution from the particle size distribution, after the negative frequencies have been set to zero. It is not uncommon to find that the missed fracture profiles may represent up to 25% of the total number of profiles counted. Examples of stereological analysis are presented in Figure 7.

The computations just outlined are rather lengthy and complex. Several graphical shortcuts have been proposed (for a review, see Weibel[11]). Another solution is to perform the analysis by computer and a Fortran IV program for this purpose has been published.[15] Moreover, microcomputers are presently largely sufficient to manage easily the computation, within reasonable (± 2 min) execution time. A Pascal version of a program for these calculations is available upon request. It has been written specifically to run under the Apple II Pascal Operating System (Apple Computers Inc., Calif.). Care has been taken to avoid as much as possible extensions of the language specific of Apple Pascal and to maintain maximum compatibility with the UCSD® version of Pascal; the program should thus be adaptable to other microcomputers with minor modifications only.

C. Statistical Parameters

A complete characterization of the geometric properties of a suspension of monolamellar liposomes is presented in Table 1. Similar data for particles with different lipid composition or prepared by different methods have been published elsewhere.[7,8] The obtention of most of these parameters from the size distribution analysis is straightforward. The estimation of the internal area is made by taking into account the membrane thickness, t, and calculating $4\ \pi\ (r-t)^2$ for the population. From data of the literature, it can be inferred that t is equal to 3.8 or 4.1 nm for particles without and with 50% cholesterol, respectively. A detailed discussion of this point and of the value adopted for the thickness can be found in Guiot et al.[8] For large unilamellar liposomes such as those from Table 1, the difference between external and internal area of the membrane is negligible, but for small particles prepared by sonication, a large (up to 50%) difference may exist. Similarly, the volume of the internal aqueous phase of monolamellar liposomes can be estimated by computing $4/3\ \pi\ (r-t)^3$. In the example of Table 1, the membrane represents 5% of the total particle volume, but for smaller particles, values up to 50% can be found.

In some cases, it may be useful to convert the discontinuous particle size distribution into a continuous one. The fitting of a distribution function should take into account the truncation affecting the experimental distribution and — especially for small unilamellar liposomes — the thickness of the membrane, which defines the minimum possible size of a particle. A mathematical procedure for fitting a log-normal distribution to liposome populations is described by Guiot.[7] The choice of a log-normal distribution rests on the following two properties of this function. First, the distribution is restricted to positive values of the independent variable, $x = r-t$, which is obviously the case in a particle population. Secondly, the shape of the distribution remains the same for x, x^2, and x^3.

D. Limitations of the Analysis

Size distribution analysis by stereological methods can only be performed on populations of spherical particles or, within certain limits, of revolution ellipsoids. In the latter case, it can be shown that, as long as the ratio between long and short axes is between 1 and 1.5, the error related to the nonspherical shape of the particles will be smaller than 3% for the average radius and 5% for the average volume.[15] In the case of ellipsoidal particles, it is mandatory to use as "radius" of profiles the radius of the circle with equivalent area.

In principle, stereological methods could be used for determination of "total" area or volume of the particles per unit volume of suspension. Dividing these parameters by the average particle area or volume would then yield the number of particles per unit volume in the sample. However, stereological analysis assumes that a plane section or a polished surface is available for measurements and this is obviously not the case for a fracture surface. For the evaluation of the number of particles, replacement of a plane section by an irregular fracture face will lead to an overestimation of the number of profiles present in a given plane. According to Guiot et al.,[8] the overestimation may be of the order of 60%. Hence, evaluation of the abundance of particles based on the area occupied by polar groups (see below), as determined by X-ray diffraction analysis, should be preferred.

V. ABUNDANCE AND MOLECULAR COMPOSITION OF PARTICLES

The total area of the monomolecular phospholipid leaflets present in a preparation can be obtained by multiplying the total amount of phospholipid present in the preparation by the area occupied by a molecule of phospholipid, taking into account other associated components if necessary. For example, we have used for the headgroup of PS/PC (1:4) liposomes, a value of 0.7 nm^2 per molecule, which is a rounded average of the values reported in the literature, from X-ray diffraction studies.[16,17] For PS/PC/Chol (1:4:5) liposomes, a value of

Table 2
ABUNDANCE AND MOLECULAR COMPOSITION OF LIPOSOME POPULATION

(1) Number of molecules in the membrane of the average particle 583,000 $nm^2/0.45\ nm^2$	1.30×10^6
(2) Mass of the membrane, daltons, (1) × 595	774×10^6
(3) Concentration of phospholipids in suspension, μmol/mℓ	10
(4) Number of particles per unit volume, $m\ell^{-1}$ N × (3) × 2 × 10^{-6}/(1)	0.93×10^{13}
(5) Number of particles per unit phospholipid mass, μmol^{-1} (4)/(3)	0.93×10^{12}

Note: Same preparation as in Table 1, from which the evaluation of 583,000 nm^2 of total (external + internal) area is used. Taking a molecular weight of 800 and 390, for phospholipids and cholesterol, respectively, an average molecular weight of 595 was deduced from the composition of the suspension. A specific molecular area of 0.45 nm^2 is assumed (see text). N is Avogadro's number.

0.19 nm^2 per cholesterol molecule[18] should be taken into account, yielding in this case a value of 0.45 nm^2 as the area associated to a single headgroup. An example of application to a liposome preparation is given in Table 2. The parameters listed for the molecular composition pertain to an average particle, defined as the particle with the average membrane area. The heterogeneity of the preparation should, however, not be overlooked. In the preparation of Table 2, size distribution analysis shows that the range for total membrane area extends from 84,500 to 18,500,000 nm^2, for particles having a radius of 60 or 860 nm, respectively. Hence, the number of molecules in the membrane of a particle will vary between 187,800 and 41,100,000 in such a population.

VI. EXAMPLE OF APPLICATION: ADSORPTION OF LARGE UNILAMELLAR LIPOSOMES TO MACROPHAGES IN CULTURE

This last section will be devoted to a brief description of a study of the adsorption of liposomes, prepared by the reverse-phase-evaporation (REV) method[10] on cultured macrophages. The aim is to illustrate how this first step of the interactions between liposomes and cells can be characterized on a quantitative basis at the cellular level.

A. Experimental System

The cells were peritoneal macrophages from C57 black mice, cultured at 37°C, for 3 to 10 days, in Eagle Dulbecco's medium, with addition of 20% decomplemented fetal calf serum, under an atmosphere enriched in CO_2 (10%). Two types of liposomes were used. The first were made of PS and PC with a 1:4 molar ratio. The characteristics of these liposomes, as obtained from stereological analyses were as follows: average radius 115 nm; average external area of membrane 291,000 nm^2; average volume $32.7 \times 10^6\ nm^3$. By the method described above, it was evaluated that the preparation contained 0.738×10^{12} liposomes per μmole phospholipid. The second type of liposomes contained 50% Chol in addition to the phospholipids; their composition and characteristics were those listed in Tables 1 and 2.

B. Mathematical Model

The time course of the association of liposomes to peritoneal macrophages cultured for 7 days is illustrated in Figure 8. It has been shown[7] that in such experiments the association of the particles with the cells is essentially reversible and hence corresponds to adsorption.

A simple model can be fitted to the experimental data, if it is assumed that the surface available for adsorption is limiting and that liposomes are present in large excess in the

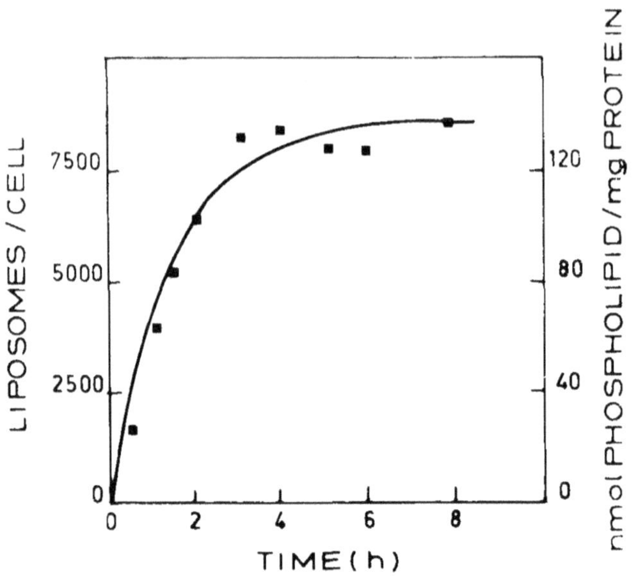

FIGURE 8. Kinetics of the adsorption of large unilamellar liposomes to cultured peritoneal macrophages. About 10^6 macrophages, cultured for 7 days, were incubated at 37°C in presence of PS/PC (1:4) liposomes (0.738 × 10^{12} liposomes/mℓ), in a medium containing 20% fetal calf serum. The curve fitted corresponds to Equation 2, with n_{max} = 8500 and kα = 0.69/hr.

extracellular medium. Let S be the surface of the cell available for adsorption and α the area occupied on average by a liposome on the cell membrane. If n is the number of liposomes adsorbed to the cells at time t, then

$$\frac{dn}{dt} = k\,(S - \alpha n) \quad (1)$$

where k is the rate constant for adsorption. Upon integration, one obtains:

$$n = n_{max}\,(1 - e^{-k\alpha t}) \quad (2)$$

where n_{max}, the maximum number of particles which can be adsorbed, is equal to the ratio S/α. As can be seen in Figure 8, a good agreement between this model and the experimental data is obtained with n_{max} = 8500 and kα = 0.69/hr in the case of PS/PC (1:4) liposomes. The last value corresponds to a half-life of 1 hr for the exponential term.

At the cellular level, the significance of these parameters can best be assessed from the value reported in the literature for the area of the plasma membrane. From a stereological analysis of peritoneal macrophages in culture, Steinman et al.[19] have reported the area of 825 μm^2 for the plasma membrane. Since α represents the cross-section of the particle, it is equal to 1/4 of the average area of the liposomes (291,000/4 = 72,750 nm^2) if a spherical shape can be assumed as it is the case here. Hence, if the whole plasma membrane is available for adsorption, n_{max} should be equal to 825/0.07275 = 11,300 liposomes. Our value (8500) assumes thus that a maximum of 75% of the plasma membrane is available for adsorption of the particles. This seems quite reasonable, especially since we are dealing with adherent cells.

Similar experiments were also performed in order to study the effect of liposome con-

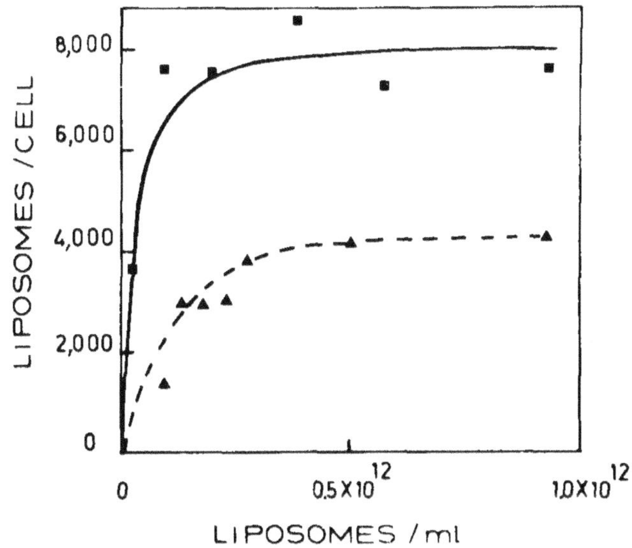

FIGURE 9. Effect of liposome concentration of large unilamellar liposomes to cultured peritoneal macrophages. About 10^6 macrophages, cultured for 10 days were incubated for 3 hr at 37°C in the presence of liposomes, in a medium containing 20% fetal calf serum. The curves fitted correspond to Equation 4. Continuous line: PS/PC (1:4) liposomes; $n_{max} = 8000$ and $K_d = 22 \times 10^9$ liposomes/mℓ. Interrupted line: PS/PC/Chol (1:4:5) liposomes; $n_{max} = 4500$ and $K_d = 87 \times 10^9$ liposomes/mℓ.

centration. Exposure of the cells was then prolonged to a time much longer than the half-life of the exponential term of Equation 2, which could thus be neglected. In this case, a dissociation constant, K_d, can be defined by the equation:

$$K_d = \frac{C\,(S - \alpha n)}{\alpha n} \tag{3}$$

where C is the concentration of liposomes, and the other symbols have the same meaning as defined before. By rearranging, Equation 3 becomes:

$$n = n_{max} \frac{C}{K_d + C} \tag{4}$$

Fitting this function to the results of Figure 9 gives a reasonable concordance with the experimental data for $n_{max} = 8000$ and $K_d = 22 \times 10^9$ liposomes/mℓ for liposomes without cholesterol and $n_{max} = 4500$ and $K_d = 87 \times 10^9$ liposomes/mℓ for liposomes with 50% Chol.

C. Summary of Data

The association of liposomes with peritoneal macrophages can thus be satisfactorily described by an adsorptive model based on an equation of the form:

$$n = n_{max}\,(1 - e^{-k\alpha t}) \frac{C}{K_d + C} \tag{5}$$

which combines Equations 2 and 4. Table 3 lists the values of the parameters observed for two types of liposomes. It should be, however, mentioned that the kinetics described by

Table 3
PARAMETERS OF THE ADSORPTION OF LARGE MONOLAMELLAR LIPOSOMES ON CULTURED PERITONEAL MACROPHAGES

Composition	Average radius (nm)	Age of culture (days)	n_{max} ($cell^{-1}$)	% of plasma membrane area covered	Half-life of exponential term (Eq. 2) (hr)	K_d (liposomes/mℓ)	K_d (μmole phospholipid/mℓ)
PS/PC (1:4)	115	3	6100	54	0.96	—	—
		7	8500	75	1.00	—	—
		10	8000	71	—	22×10^9	0.030
PS/PC/Chol (1:4:5)	130	3	1950	17	1.18	—	—
		10	4500	40	—	87×10^9	0.094

Note: Liposomes were prepared by the REV method.[10] For significance of parameters, see Equations 2, 4 and 5.

Equation 5 is applicable only when 20% fetal calf serum is present in the extracellular medium.[7] Furthermore, as it is apparent from the data of Table 3, the value of the parameters, and more specifically the maximal number of liposomes that can be adsorbed on a cell, can be modified appreciably as a function of the age of the culture. It is not known if this difference reflects a change in the total area of the plasma membrane per cell, which was considered as constant here, or some modification in the spatial arrangement (microvilli, attachment to the culture dish, . . .) of the plasma membrane. It should, however, be noted that the data presented in Table 3 show that inclusion of cholesterol in the liposomes decreases appreciably the number of particles which can be adsorbed on the membrane, while the half-life of the adsorptive process is slightly increased. Other factors than modification of the plasma membrane are thus here certainly at play. Furthermore, it is obvious that inclusion of a protein in the liposome membrane may also change radically the affinity of liposomes for the cell membrane.

ACKNOWLEDGMENTS

This work was supported by the Belgian Fonds de la Recherche Scientifique Médicale (grant 3.4547.79) and by the Volkswagenstiftung (FRG).

REFERENCES

1. **Dekruijff, B., Cullis, P. R., and Radda, G. R.,** Outside-inside distributions and sizes of mixed phosphatidylcholine=cholesterol vesicles, *Biochim. Biophys. Acta,* 436, 729, 1976.
2. **Miyamoto, V. K. and Stoeckenius, W.,** Preparation and characteristics of lipid vesicles, *J. Membr. Biol.,* 4, 252, 1971.
3. **Bachmann, L. and Schmitt, W.,** Improved cryofixation applicable to freeze-etching, *Proc. Natl. Acad. Sci. U.S.A.,* 68, 2149, 1971.
4. **Johnson, S. M. and Bangham, A. D.,** Potassium permeability of single compartment liposomes with and without valinomycin, *Biochim. Biophys. Acta,* 193, 82, 1969.
5. **Gotfredsen, C. F.,** Cellular and Tissular Uptake Distribution and Metabolism of Liposomes, Doctoral thesis, Université Catholique de Louvain, Brussels, 1982.
6. **Pagano, R. E. and Weinstein, J. N.,** Interactions of liposomes with mammalian cells, *Ann. Rev. Biophys. Bioeng.,* 7, 435, 1978.
7. **Guiot, P.,** Aspects Quantitatifs de l'Interaction entre Vecteurs Particulaires et Cellules, Thèse de Doctorat, Université Catholique de Louvain, Brussels, 1981.
8. **Guiot, P., Baudhuin, P., and Gotfredsen, C. F.,** Morphological characterization of liposome suspensions by stereological analysis of freeze-fracture replicas from spray-frozen samples, *J. Microsc.,* 120, 159, 1980.
9. **Moor, H., Muhlethaler, K., Waldner, H., and Frey-Wyssling, A.,** A new freezing-ultramicrotome, *J. Biophys. Biochem. Cytol.,* 10, 1, 1961.
10. **Szoka, F. and Papahadjopoulos, D.,** Procedure for preparation of liposomes with large internal aqueous space and high capture by reverse-phase evaporation, *Proc. Natl. Acad. Sci. U.S.A.,* 75, 4194, 1978.
11. **Weibel, E. R.,** *Practical Methods for Biological Morphometry,* Vol. 1, Academic Press, New York, 1979.
12. **Wicksell, S. D.,** The corpuscle problem. A mathematical study of a biometric problem, *Biometrika,* 18, 151, 1925.
13. **Bach, G.,** Kugelgrössenverteilung und Verteilung der Schnittkreise; ihre Wechselseitingen Beziehungen und Verfahren zur Bestimmung der Einen aus der Anderen, in *Quantitative Methods in Morphology,* Weibel, E. R. and Elias, H., Eds., Springer-Verlag, Berlin, 1967, 23.
14. **Baudhuin, P. and Berthet, J.,** Electron microscopic examination of subcellular fractions. II. Quantitative analysis of the mitochondrial populations isolated from rat liver, *J. Cell Biol.,* 35, 631, 1967.
15. **Baudhuin, P.,** L'Analyse Morphométrique Quantitative de Fractions Subcellulaires, Thèse d'Agrégation de l'Enseignement Supérieur, Université Catholique de Louvain, Brussels, 1968.
16. **Lecuyer, H. and Dervichian, D. G.,** Structure of aqueous mixtures of lecithin and cholesterol, *J. Mol. Biol.,* 45, 39, 1969.

17. **Chrzeszczyk, A., Wishnia, A., and Springer, C. S.,** The intrinsic structural asymmetry of highly curved phospholipid bilayer membranes, *Biochim. Biophys. Acta,* 470, 161, 1977.
18. **Israelachvili, J. N. and Mitchell, D. J.,** A model for the packing of lipids in bilayer membranes, *Biochim. Biophys. Acta,* 389, 13, 1975.
19. **Steinman, R. M., Brodie, S. E., and Cohn, Z. A.,** Membrane flow during pinocytosis. A stereological analysis, *J. Cell Biol.,* 68, 665, 1976.

Chapter 13

HYDROPHOBIC AFFINITY PARTITION OF LIPOSOMES IN AQUEOUS TWO-PHASE SYSTEMS

Eva Eriksson

TABLE OF CONTENTS

I. INTRODUCTION

Aqueous two-phase systems consisting of the two polymers dextran and poly(ethylene glycol) (PEG), have been widely used in the separation and study of cells and membrane particles.[1-5] Both the upper, PEG-rich phase and the lower, dextran-rich phase have a high content of water, 80 to 95%. Salts and sugars may be added at suitable concentrations. Partition in two-phase systems is based on differences in surface properties, in contrast to several other separation methods which utilize differences in size or density.

The specificity for membranes with the phase partition method is improved by using polymer-bound hydrophobic ligands which selectively bind to membrane surfaces. Hydrophobic affinity partition with fatty acids covalently attached to PEG has been used for selective partition of proteins[6,7] and membranes.[8-10] Materials having affinity for the PEG-bound fatty acids will be partitioned from the lower phase to the upper phase.

Liposomes of different lipid compositions have been partitioned in phase systems.[11,12] The effect of variations in the polar head group region of the liposomes was studied by partition of liposomes consisting of phosphatidylcholine (PC) phosphatidylserine (PS) phosphatidylinositol (PI), or sphingomyelin (SPH). The influence of the hydrophobic part of the bilayer on the partition behavior was investigated using PC liposomes in which the degree of unsaturation of the fatty acids was varied. It was concluded from these investigations that the polar head group of the phospholipid plays a dominant role in determining the partition of liposomes while the degree of unsaturation of the acyl groups is of less importance.

Hydrophobic affinity partition of liposomes shows that PEG-bound fatty acids have affinity for lipid bilayers and can change the partition of liposomes from the lower phase to the upper phase. The longer the PEG-bound fatty acid, the lower is the amount of it required for extraction of liposomes from the lower phase. Also, in hydrophobic affinity partition, differences in partition behavior were found when the polar head groups of the phospholipids were varied but not when the hydrophobic part of the bilayer was varied. Neither differences in length nor in degree of unsaturation of the liposomal phospholipids give partition differences with PEG-bound fatty acids of various lengths. The presence of cholesterol in the bilayer does not influence the partition behavior.

II. METHODOLOGY

A. Phase Systems

When aqueous solutions of dextran and PEG are mixed above certain concentrations, aqueous two-phase systems are formed. The phase diagram of a water-dextran-PEG phase system is shown in Figure 1.

Particles distribute between the two phases and the interface. Partition of particles is dependent of the molecular weight of the polymers, polymer concentration, temperature, and salt composition.[1] If the molecular weight of one polymer is reduced, the particles will have more affinity for the phase rich in that polymer. If the polymer concentration is increased the number of particles adsorbed to the interface will also increase.

Partition of particles often depends strongly on the type of salt present in the system.[1] This directing effect of salts on partition increases with the electrical charge of the particle. When the anion and the cation of the salt differ in their relative affinity for the phases, a different distribution of the ions will occur at the interface, giving rise to a potential difference between the phases, a so-called interfacial potential.[13]

The partition of charged particles can also be strongly influenced by introducing charged groups on a fraction of PEG.[14,15] If PEG is positively charged, e.g., trimethylamino-PEG, it will attract the negative particles into the upper phase. Negatively charged PEG such as PEG-sulfonate or carboxymethyl-PEG will attract positively charged particles. When systems

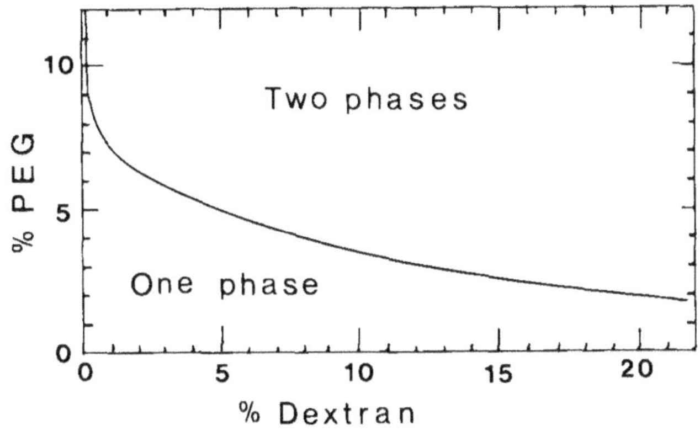

FIGURE 1. Phase diagram of the system water-dextran mol wt 500,000 — PEG mol wt 4000 at 0°C.

with PEG of mol wt 4000 and at low ionic strength (< 5 m*M*) are used, little (< 5%) of the total PEG has to be replaced by its charged form, to extract all the material from one phase to another.

B. Affinity Partition

Specific ligands covalently attached to one of the polymers can selectively change the partition of membrane particles. Hydrophobic ligands or ligands with high specificity for membranes can be used.[6-10,16] Fatty acids esterified with PEG have affinity for lipid bilayers. The carboxylic acids can be bound directly to one of the two terminal hydroxyl groups of PEG with the aid of carbodiimides[6] or by using the acid chloride[17] or anhydride.[18]

C. Preparation of Phase Systems

When choosing a phase system of a certain composition, the temperature, tonicity, salt composition, and pH should be suitable for the material to be partitioned. An example of a specific system (the percent values being w/w) is the following: a phase system (5 g) at 4°C containing 5.6% dextran of mol wt 500,000, 5.6% PEG of mol wt 5000, 5 m*M* NaCl, 10 m*M* sodium phosphate buffer, pH = 7.4, and 20 m*M* sucrose. Dextran is obtainable from Pharmacia Fine Chemicals, Uppsala, Sweden and PEG (Carbowax) 4000 and 6000 (recently renamed as PEG 3350 and 8000, respectively) from Union Carbide, New York. The following stock solutions are prepared (the first two to be stored in the cold):

1. 20% dextran. Dextran powder normally contains 3 to 5% water and therefore the concentration of the solution must be determined. Layer 220 g of dextran on 780 g of water. Heat with gentle stirring and keep the mixture boiling until all the dextran is dissolved. This procedure also minimizes microbiological contamination. To determine the dextran concentration, dilute 5 g of the solution to 25 mℓ with water and measure the optical rotation of this solution. The specific rotation (for 1 dm, 1 g/mℓ) is $[\alpha]_D^{25} = +199° \cdot$ mℓ/g · dm. Adjust the concentration of the stock solution to 20% with water.
2. 40% PEG. This polymer contains only traces of water. Mix 400 g of PEG with 600 g of water at room temperature. Mechanical stirring is advisable.
3. Salt and buffer solutions. Prepare these 10 times stronger than the final concentrations to be used in the phase systems.

Example — The amounts of stock solutions necessary for the phase system given above

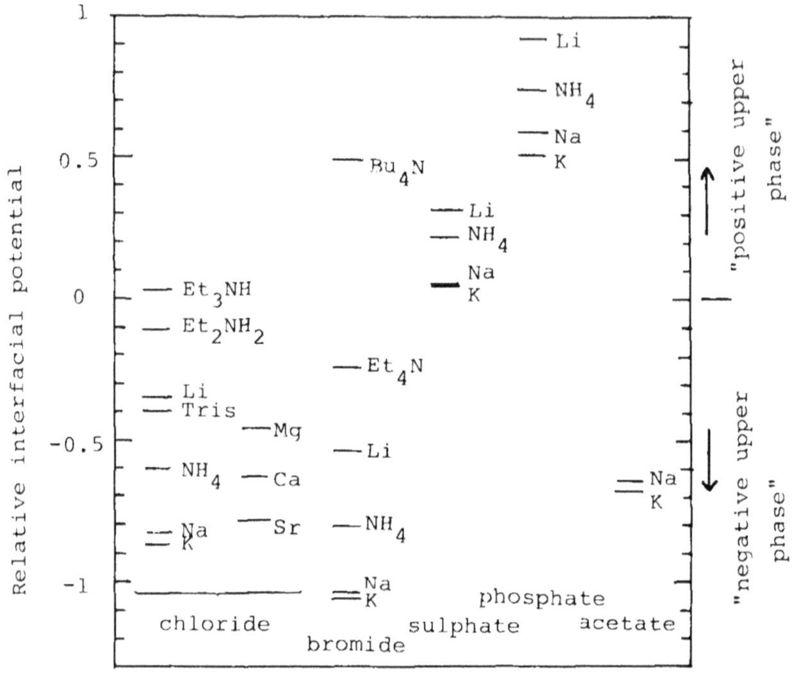

FIGURE 2. Relative interfacial potential obtained with various salts in systems consisting of 8% dextran mol wt 500,000 and 8% PEG mol wt 4000 at 25°C. Salt concentration 25 m*M*. (From Eriksson, E. and Johansson, G., *Methodol. Surv. (B) Biochem.*, 8, 81, 1979. With permission.)

are 20% dextran, 1.4 g; 40% PEG, 0.7 g; 0.05 *M* NaCl, 0.4 mℓ; 0.1 *M* sodium phosphate buffer, pH = 7.4, 0.4 mℓ; 0.2 *M* sucrose, 0.4 mℓ. Mix polymer and salt solutions in a test tube. Add water to 4 g total weight and bring the tubes to the desired temperature. Suspend a sample of the particulate test material in a solution containing 5 m*M* NaCl, 10 m*M* sodium phosphate buffer, pH = 7.4, and 20 m*M* sucrose. Add 1 g of this suspension to the phase system. Mix the phases by several (10 to 40) inversions and allow them to settle. The settling takes 5 to 30 min at unit gravity but can be facilitated by a low speed centrifugation. The centrifugation must be carried out with care since the sedimentation of the particles is increased. Keep the tubes at a constant temperature during the whole procedure. To determine the partition, withdraw an aliquot from each phase with a pipette and determine the percentage of material in the phases. The amount at the interface is taken as the difference between added material and material in the two phases.

To obtain more material in the upper phase:

1. Choose a salt according to Figure 2.
2. Choose a phase system of lower polymer concentrations.
3. Use PEG with charge or with ligand that will attract the material.
4. Choose a phase system with PEG of lower molecular weight, e.g., PEG of mol wt 4000 instead of PEG of mol wt 6000.

To obtain more material in the lower phase:

1. Choose a salt according to Figure 2.
2. Choose a phase system of lower polymer concentrations.
3. Use PEG with a charge or dextran with bound ligands.
4. Choose a phase system with dextran of lower molecular weight.

Table 1
THE DISTRIBUTION OF UNILAMELLAR LIPOSOMES BETWEEN THE TWO PHASES IN 5% (W/W) DEXTRAN MOL WT 500,000, 4% (W/W) PEG MOL WT 6000, 0.03 *M* NaCl AND 0.11 *M* SODIUM PHOSPHATE BUFFER, pH = 7.0, AT 22°C

	Percentage of liposomes		
Liposome composition	In the upper phase	At interface	In the lower phase
PS	2	25	73
PI	47	6	47

From Eriksson, E. and Albertsson, P.-Å., *Biochim. Biophys. Acta,* 507, 425, 1978. With permission.

D. Partition of Liposomes

1. Polar Head Groups

The partition of liposomes composed of phospholipids with different polar groups has been described.[11] Large differences in partition of liposomes were found when the polar head groups were varied.

Example — In a phase system containing 5.6% dextran mol wt 500,000, 5.6% PEG mol wt 4000, 5 m*M* NaCl, 10 m*M* sodium phosphate buffer, pH = 7.4, and 20 m*M* sucrose at 4°C more than 90% of egg PC liposomes partition in the upper phase, while more than 80% of PS liposomes are in the lower phase. The remaining ones are the interface.[19]

The charge of the liposome is of importance for the partition. It is not only the charge that is of importance, however. For instance PI and PS liposomes have the same net charge but not the same partition behavior (Table 1).

A further example is partition of liposomes composed of phospholipids where the methyl groups of PC have been substituted with ethyl groups. The exchange of three methyl groups with three ethyl groups results in a higher partition of both unilamellar and multilayered liposomes (Figure 3).

2. The Hydrophobic Part of the Membrane

PC liposomes with different degrees of unsaturation in the phospholipid fatty acyl chains have been partitioned in two-phase systems.[11] Soya bean PC was the most unsaturated phospholipid used, egg PC was intermediate, and dipalmitoyl phosphatidylcholine (DPPC) was entirely saturated. There were no differences in partition of multilayered liposomes with different degrees of unsaturation. In contrast, small bilayer liposomes prepared from soya bean PC had a higher partition than liposomes with more saturated acyl chains (Table 2).

Incorporation of cholesterol in molar ratio 1:1 into small liposomes of soya bean PC and DPPC equalizes the partition of these liposomes, (Table 2).

3. Hydrophobic Ligands Attached to PEG

Fatty acids esterified with PEG can change the partition of liposomes from the lower phase via the interface to the upper phase,[11,12] (Figure 4). For bilayer vesicles the amount of PEG-palmitate that is required to obtain 50% of the liposomes in the upper phase is in the order of five PEG palmitate molecules per liposome. The corresponding number for erythrocytes was 5×10^6 molecules per erythrocyte. The amount of PEG palmitate per mole of membrane lipid is ten times higher for erythrocytes than for bilayer liposomes.

The length of the fatty acid is of importance. The longer the PEG-bound fatty acid, the lower is the amount of it required for extraction of PC and PS liposomes from the lower

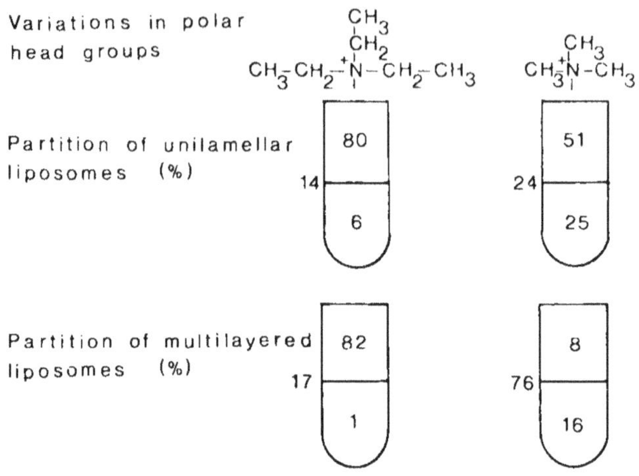

FIGURE 3. Partition of liposomes composed of soya bean PC and its modified analog in 5% dextran mol wt 500,000, 4% PEG mol wt 6000, 0.15 *M* NaCl and 0.01 *M* sodium phosphate buffer, pH = 7.0 at 22°C. The modified phospholipid was a generous gift from Dr. Björn Åkesson.[19]

Table 2
THE DISTRIBUTION OF UNILAMELLAR LIPOSOMES BETWEEN THE TWO PHASES IN THE SAME PHASE SYSTEM AS IN TABLE 1

	Percentage of liposomes		
Liposome composition	**In the upper phase**	**At interface**	**In the lower phase**
Soya bean PC	41	23	36
Egg PC	25	42	33
DPPC	4	55	41
Soya bean PC:Chol 1:1	1	45	54
DPPC:Chol 1:1	1	53	46

From Eriksson, E. and Albertsson, P.-Å., *Biochim. Biophys. Acta*, 507, 425, 1978. With permission.

phase (Figure 5). Differences in partition of PC and PS liposomes were more pronounced when the PEG-bound fatty acid was decreased in length from stearic to capric acid.

PC liposomes require lower concentrations of PEG-palmitate in order to be extracted into the upper phase than SPH liposomes.[11] PS and PI liposomes are transferred to the upper phase at even higher concentration of PEG-ester (Figure 6).

Variations in the hydrophobic part of the membrane do not influence the hydrophobic affinity partition. Liposomes composed of dilauroyl-, dimyristoyl-, dipalmitoyl-; or distearoyl PC (DLPC, DMPC, DPPC, DSPC) as well as PC from soya bean or egg are all transferred to the upper phase at the same concentration of PEG-ester and independently of the length of the PEG-bound fatty acid.

Incorporation of cholesterol in soya bean and DPPC liposomes in molar ratios 1:1 has no effect on the partition behavior in phase systems containing PEG-palmitate.

III. DISCUSSION

Partition of liposomes in aqueous two-phase systems shows that the polar head group of the phospholipid plays a dominant role in determining the partition, while the degree of

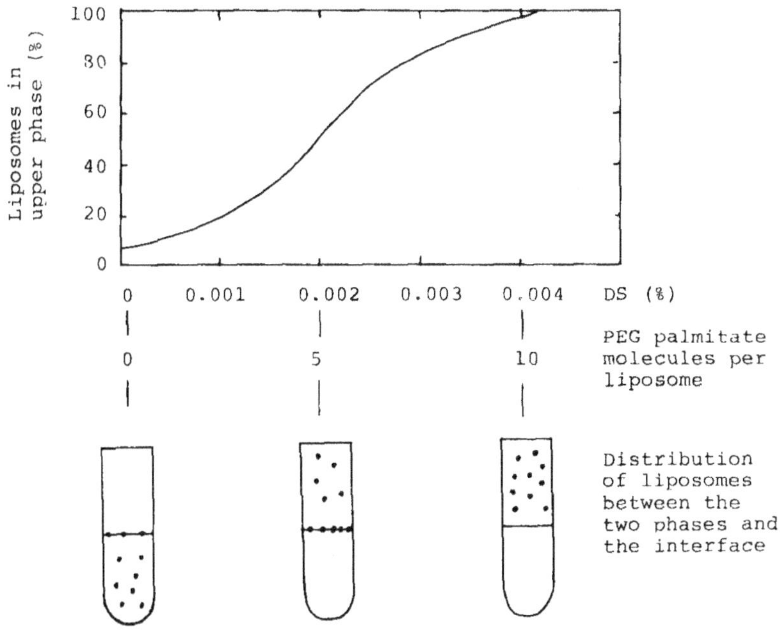

FIGURE 4. Unilamellar PC liposomes in phase systems containing PEG palmitate. The percentage of liposomes in the upper phase as function of the percentage PEG end groups esterified with palmitic acid; DS, degree of substitution. The phase systems were composed of 8% dextran mol wt 40,000, 4% PEG mol wt 6000, 0.15 *M* NaCl and 0.01 *M* sodium phosphate buffer, pH = 7.0. Each 5 g phase system contains 0.35 μmol phospholipid. (From Eriksson, E. and Albertsson, P.-Å., *Biochim. Biophys. Acta*, 507, 425, 1978. With permission.)

unsaturation is of less importance. The only time a partition difference was found when the degree of unsaturation was varied, was between unilamellar (but not between multilayered) soya bean PC and DPPC liposomes. This might be explained by the higher degree of curvature for unilamellar liposomes, thereby exposing more of the hydrophobic part of the bilayer. This partition difference was eliminated by the presence of cholesterol in molar ratio 1:1.

By using PEG with hydrophobic ligands the partition of lipid bilayers is changed from the lower phase via the interface to the upper phase. Only a very small amount of the terminal hydroxyl groups of PEG has to be esterified with fatty acids to change the partition of liposomes to the upper phase.

The PEG-bound hydrocarbon chain is supposedly inserted into the membranes and, in that way, interacts with the hydrophobic interior of the bilayer. Owing to a weaker interaction of the shorter fatty acids, higher concentrations of these are required for extraction of liposomes from the lower phase. The deeper the PEG-bound fatty acids are inserted into the liposomal bilayer, the stronger are the hydrophobic interactions between the fatty acids and the hydrophobic part of the bilayer. This interaction is independent of the nature of the hydrophobic part of the liposomes, since there were no differences in the partition of liposomes with different degrees of unsaturation in their phospholipid fatty acid chains.

The polar head group of the phospholipid is a dominant factor in determining the partition of liposomes when PEG-bound fatty acids are present. Since there were no partition differences when the hydrophobic part was varied, it is the accessibility to the hydrophobic part that is important for the interaction. Both electrostatic and steric factors may contribute to this effect.

The fact that SPH-rich membranes require higher concentrations of PEG-palmitate than PC-rich membranes for extraction into the upper phase (shown with liposomes[11] and

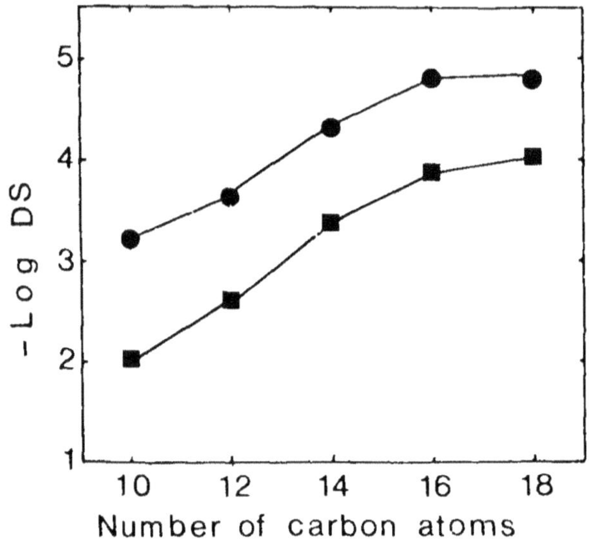

FIGURE 5. The fraction of PEG end groups esterified with fatty acids (DS) for extraction of 50% of the liposomes from the lower phase as a function of the number of carbon atoms of the fatty acids. ●, PC; ■, PS liposomes. (From Eriksson, E. and Albertsson, P.-Å., *Biochim. Biophys. Acta*, 507, 425, 1978. With permission.)

erythrocytes[8]) may be explained by considering the "hydrogen belts" of the membranes.[20] These regions consist of hydrogen bond acceptors (e.g., carbonyl groups of glycero- and sphingolipids) and hydrogen bond donors (e.g., sphingolipid hydroxyl groups). In the case of SPH liposomes there are several opportunities for lipid-lipid hydrogen binding since SPH contains both donor and acceptor. In contrast, PC only contains acceptor. Thus, there is likely to be a network of hydrogen bonds in SPH liposomes which might be difficult to penetrate for PEG-palmitate molecules.

The phase systems are very useful for separation purposes, as exemplified by the phase system where PC liposomes are in the upper phase and PS liposomes in the lower phase after 10 min separation time. The usefulness of the phase systems in the separation and study of biological membranes has been shown several times.[1-5] Inside-out and right-side-out membrane vesicles from chloroplast thylakoid membranes,[21] from plant mitochondrial membranes,[22] and also from red blood cells[23] have been separated using phase partition. It is possible that the lipid composition also of biological membranes is an important factor in determining their partition.

By using hydrophobic affinity partition, the separation capacity of the phase systems is improved. Membrane particles with affinity for PEG-bound hydrophobic ligands can easily be separated from other material.

Although shorter PEG-bound fatty acids have to be used in higher concentrations they are still very useful for separation purposes. This is reflected in the greater extraction of PC liposomes compared to PS liposomes from the lower phase as the PEG-bound fatty acid decreases in length. Similar behavior has been observed with chloroplast membranes where intact and broken chloroplasts were separated when capric acid was attached to PEG but not when longer fatty acids were used.[10] Thus, several different lengths of PEG-bound fatty acids should be tried in order to achieve the best separation.

Hydrophobic affinity partition is useful for binding studies between liposomes and a ligand attached to PEG. The PEG-bound ligand must be rather small, however, in order to partition

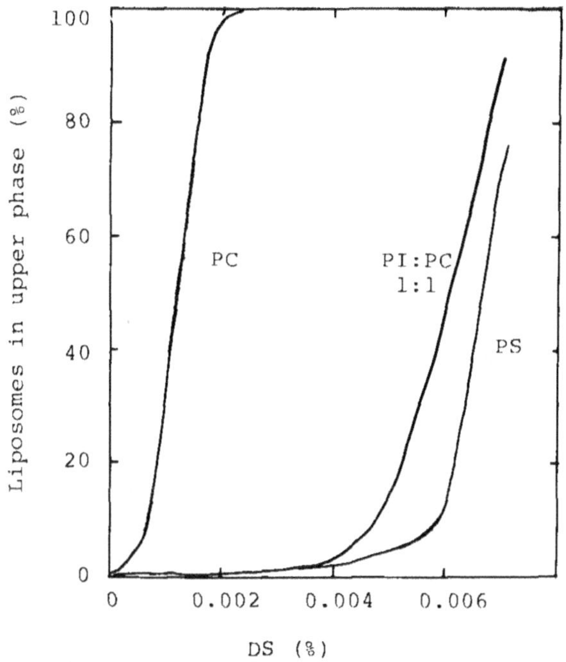

FIGURE 6. Multilamellar liposomes in phase systems containing PEG palmitate. The percentage of liposomes in the upper phase as a function of the percentage PEG end groups esterified with palmitic acid (DS). The phase systems contain 8% dextran mol wt 40,000, 4% PEG mol wt 6000, 0.15 *M* NaCl and 0.01 *M* sodium phosphate buffer, pH = 7.0. Each 5 g phase system contain 0.35 μmol phospholipid. (From Eriksson, E. and Albertsson, P.-Å., *Biochim. Biophys. Acta*, 507, 425, 1978. With permission.)

in the PEG-rich phase. A phase system is then chosen to contain all the material at the interface or in the lower phase. By replacing a part of the PEG with PEG-ligand, only particles with affinity for the ligand will be extracted into the upper phase.

Phase systems are also useful for studies of interactions between liposomes and cells or organelles. A phase system is chosen where all liposomes partition in one phase and the cells in the other phase. The partition of a mixture of liposomes and cells is then studied. It is also possible to study the interaction between cells and liposomes by comparing the partition of the cells before and after incubation with liposomes as has been shown for polymorphonuclear leukocytes and liposomes.[24]

The interaction of small water-soluble molecules and liposomes can be studied in phase systems. Small molecules distribute almost 1:1 between the two phases (they do not adsorb to the interface). By studying the concentration of molecules in the phase which does not contain liposomes, the interaction can be followed.

REFERENCES

1. **Albertsson, P.-Å.,** *Partition of Cell Particles and Macromolecules*, 2nd ed., Almqvist & Wiksell, Stockholm, 1971, 118.
2. **Albertsson, P.-Å.,** Separation of particles and macromolecules by phase partition, *Endeavour*, 1, 69, 1977.

3. **Albertsson, P.-Å.,** Partition between polymer phases, *J. Chromatogr.*, 159, 111, 1978.
4. **Albertsson, P.-Å., Andersson, B., Larsson, C., and Åkerlund, H.-E.,** Phase partition — a method for purification and analysis of cell organelles and membrane vesicles, *Meth. Biochem. Anal.*, 28, 115, 1982.
5. **Eriksson, E. and Johansson, G.,** Partition of cells between aqueous phases, *Methodol. Surv. (B) Biochem.*, 8, 81, 1979.
6. **Johansson, G.,** The effect of poly(ethylene glycol) esters on the partition of proteins and fragmented membranes in aqueous biphasic systems, *Biochim. Biophys. Acta*, 451, 517, 1976.
7. **Shanbhag, V. P. and Axelsson, C. G.,** Hydrophobic interaction determined by partition in aqueous two-phase systems. Partition of proteins in systems containing fatty-acid esters of poly(ethylene glycol), *Eur. J. Biochem.*, 60, 17, 1975.
8. **Eriksson, E., Albertsson, P.-Å., and Johansson, G.,** Hydrophobic surface properties of erythrocytes studied by affinity partition in aqueous two-phase systems, *Mol. Cell. Biochem.*, 10, 123, 1976.
9. **Westrin, H., Albertsson, P.-Å., and Johansson, G.,** Hydrophobic affinity partition of spinach chloroplasts in aqueous two-phase systems, *Biochim. Biophys. Acta*, 436, 696, 1976.
10. **Johansson, G. and Westrin, H.,** Specific extraction of intact chloroplasts using aqueous biphasic systems, *Plant. Sci. Lett.*, 13, 201, 1978.
11. **Eriksson, E. and Albertsson, P.-Å.,** The effect of the lipid composition on the partition of liposomes in aqueous two-phase systems, *Biochim. Biophys. Acta*, 507, 425, 1978.
12. **Eriksson, E.,** Hydrophobic affinity partition of liposomes in aqueous two-phase systems, *J. Chromatogr.*, 205, 189, 1981.
13. **Johansson, G.,** Effects of salts on the partition of proteins in aqueous polymeric biphasic systems, *Acta Chem. Scand.*, B 28, 873, 1974.
14. **Johansson, G., Hartman, A., and Albertsson, P.-Å.,** Partition of proteins in 2-phase systems containing charged poly(ethylene glycol), *Eur. J. Biochem.*, 33, 379, 1973.
15. **Johansson, G.,** Partition of proteins and microorganisms in aqueous biphasic systems. *Mol. Cell. Biochem.*, 4, 169, 1974.
16. **Flanagan, S. D., Barondes, S. H., and Taylor, P.,** Affinity partitioning of membranes. Cholinergic receptor-containing membranes from torpedo californica, *J. Biol. Chem.*, 251, 858, 1976.
17. **Shanbhag, V. P. and Johansson, G.,** Specific extraction of human serum albumin by partition in aqueous biphasic systems containing poly(ethylene glycol) bound ligand, *Biochem. Biophys. Res. Commun.*, 61, 1141, 1974.
18. **Bückmann, A. F., Morr, M., and Johansson, G.,** Functionalization of poly(ethylene glycol) and monomethoxy-poly(ethylene glycol), *Makromol. Chem.*, 182, 1379, 1981.
19. **Eriksson, E.,** Partition of Membranes in Aqueous Two-Phase Systems. A study with Liposomes as a Model Membrane, Ph.D. thesis, University of Lund, Sweden, 1981.
20. **Fong, J. W., Tirri, L. J., Deshmukh, D. S., and Brockerhoff, H.,** Studies on the hydrogen belts of membranes. I. Diester, diether, and dialkyl phosphatidylcholines and polyoxyethylene glycerides in monolayers with cholesterol, *Lipids*, 12, 857, 1977.
21. **Andersson, B. and Åkerlund, H.-E.,** Inside-out membrane vesicles isolated from spinach thylakoids, *Biochim. Biophys. Acta*, 503, 462, 1978.
22. **Möller, I. M., Bergman, A., Gardeström, P., Ericson, I., and Palmer, J. M.,** Characterisation and purification of inside-out submitochondrial particles obtained from plant mitochondria, *FEBS Lett.*, 126, 13, 1981.
23. **Steck, T. L.,** Preparation of impermeable inside-out and right-side-out vesicles from erythrocyte membranes, *Meth. Memb. Biol.*, 2, 245, 1974.
24. **Dahlgren, C., Kihlström, E., Magnusson, K.-E., Stendahl, O., and Tagesson, C.,** Interaction of liposomes with polymorphonuclear leukocytes, *Exp. Cell Res.*, 108, 175, 1977.

Chapter 14

APPLICATIONS OF A TECHNIQUE FOR RAPID SEPARATION OF LIPOSOMES FROM CELLS IN CULTURE

Michael W. Fountain

TABLE OF CONTENTS

I. INTRODUCTION

Evaluation of experiments involving interactions of liposomes with cultured cells must include methods for quantitative separation of liposomes from cells at the termination of the experiments.[1-6] As many experiments involve interactions of liposomes with cells other than fibroblastic cells, techniques for gentle, rapid separation of cells from liposomes, preserving the integrity of both liposomes and cells, are of importance in areas of molecular biology,[3,5] immunology,[1,2,4] and pharmacology.[6-8] The contents of this chapter are designed to include information for the researcher associated with cell culture and the investigator who wishes to use liposomes as a model for modifying cellular behavior.

Information is presented for cells derived from primary isolation from bovine,[3,9] channel catfish,[10] and canine peripheral blood and from continuous L-1210[1,2] cell cultures. There are specific physiologic differences between cells obtained from primary cell culture and continuous culture, but these differences may be overcome in liposome-cell-separations if one adheres to basic knowledge of cell and liposome densities.[2]

Although the animal models most commonly used in our laboratories have been cattle and dogs, other mammalian cells and circulating fish cells have been used successfully in our hands because of their similarities in size and densities. The bovine and canine cells discussed in this chapter are employed for specific liposome-cell interactions which may be applied to numerous other liposome-cell studies involving nonfibroblastic cells capable of being sustained in suspension culture.

The following sections are designed to contain specific examples of applications of liposome-cell separation technology with references from which investigators can locate in-depth studies of information not specifically narrated in the text. Because all applications involve techniques originally developed using L-1210 cell-liposome separations, an in-depth explanation of the usage of discontinuous gradient for separation of liposome and cells is first described followed by specific applications in which we have employed the techniques for a variety of types of experiments involving cells and liposomes.

II. TECHNIQUE FOR RAPID SEPARATION OF LIPOSOMES AND CELLS

Our method of separation of liposomes from cells in suspension cultures provides a rapid and gentle technique which overcomes many of the problems in repeated centrifugation of cells to remove noncell-associated liposomes, i.e., decreases in cellular viability, damages by sheer forces on liposomes, and length of time needed to enact separation of liposomes from cells and so reducing impositions normally imposed on frequency of sampling and processing of multiple samples. The following example is reproduced from an earlier study.[1,2]

Murine leukemia (L-1210) cells were maintained in spinner culture with RPMI 1640 medium containing 10% fetal calf serum and glutamine (2 m*M*). The cells were harvested from log-phase growth and washed three times by gentle (500 $\times$ *g*, 10 min) centrifugation and resuspension in Ca^{2+} Mg^{2+} free Hanks buffer, pH 7.4. After the final washing, the cells were adjusted to 5 $\times$ 10^6/mℓ and were held at 37°C for a maximum of 15 min. Viability was monitored by trypan blue exclusion.

Unlabeled egg PC in ethanol was obtained from Avanti Biochemicals, Birmingham, Ala.; choline-methyl-[^{14}C] PC in ethanol/toluene (1:1) (spec. act., 50 Ci/mol) was obtained from New England Nuclear, Boston, Mass.; 6-carboxyfluorescein (6-CF) was purchased from Eastman Chemicals, Rochester, N.Y.

10 mg PC in ethanol were rotoevaporated to dryness and lyophilized overnight. The dried lipid was rehydrated in 1 mℓ of the above-mentioned buffer, allowed to swell at room temperature for 20 min, and sonicated in a bath-type sonicator (Branson Model B-12) for 35 min. The sonicated suspension was spun at 10,000 rev/min (8500 $\times$ *g*) for 60 min in a

Beckman J-21 B centrifuge, and the supernatant was retained. Radiolabeled liposomes were prepared by adding choline-methyl-[^{14}C]-PC ($5.0 \times 10^5 - 1 \times 10^6$ cpm) to the unlabeled PC before rotoevaporation. The fluorescent dye 6-CF was entrapped in liposomes by sonicating the lipid in the buffer containing 200 m*M* 6-CF. In all instances, the supernatant from the centrifugation was passed over a 1 cm × 40 cm Sephadex® G-50 column equilibrated with the buffer. The void volume (containing small unilamellar liposomes) was retained.

Ficoll-Paque was purchased from Pharmacia, Inc., Piscataway, N.J., as a sterile 5.7% (w/v) solution. For clarity, the dilutions mentioned hereafter refer to v/v dilutions of the original 5.7% Ficoll-Paque solution with the buffer. All gradients were prepared in 5-mℓ cellulose nitrate centrifuge tubes (No. 305050, Beckman Instruments, Palo Alto, Calif.). Gradients of various dilutions of Ficoll-Paque (70, 60, 50, 35, and 30%) were prepared, and each tube received 2.5 mℓ of one of the gradients. A 1-mℓ cushion undiluted Ficoll-Paque was underlaid by placing a capillary pipette through the gradient to the bottom of the tube. The pipette was connected to a peristaltic pump, and the cushion was added at a flow rate of 0.3 mℓ/min.

For the purpose of this study, we wished to demonstrate the rapidity and the efficacy of the technique. Cells were therefore incubated with liposomes containing either ^{14}C-labeled phospholipid or quenched (200 m*M*) 6-CF and were immediately centrifuged and fractionated. Equal volumes of cells and liposomes (1 mℓ of each, to yield 2.5×10^6 cells and approx. 3.0 mg phospholipid) were mixed, and 0.75 mℓ of the mixture was gently layered on top of the Ficoll gradient. The gradients were spun at 1200 rev/min (300 × *g*) for 1 min in a table-top centrifuge (IEC Model HNS) equipped with a swinging-bucket rotor. As controls, cells at the same density but without liposomes were also spun through gradients.

Immediately after centrifugation, the tubes were punctured and fractionated in a downward-flow direction, with 0.25-mℓ fractions being collected. Each fraction was examined for cells by counting in triplicate in a Neubauer hemacytometer. Radioactivity was determined by liquid-scintillation counting and for studies utilizing 6-CF, each fraction was lysed by the addition of 1 mℓ of 10% Triton® X-100 (Rohm and Haas, Philadelphia, Pa.) in the buffer. The relative fluorescence of each fraction was obtained on a Perkin–Elmer Model MPF3 spectrofluorometer with $\lambda_{ex} = 480$ nm, $\lambda_{em} = 520$ nm, excitation and emission bandpass of 4 nm, and a vertically polarized excitation beam.

Trypan blue exclusion was performed before and after washing and resuspension of cells in the buffer; before and after incubation with liposomes and after centrifugation and fractionation. At no time did the viability fall below 96%. We have incidentally noted an artifactual increase in the viability of the cells after centrifugation (e.g., from 96% to 98 to 99%). This is apparently related to the properties of the gradient and the migration of dead cells through the gradient; cells which are dead, swollen, and lysed remain at the top of the gradient, and those cells which take up trypan blue but appear intact and of the same size as viable cells migrate with the viable cells through the gradient. The centrifugation therefore removes the grossly aberrant dead cells from the general population.

The results obtained using the discontinuous-gradient centrifugation technique demonstrates that the distribution of cells shows a marked dependence on the concentration of the Ficoll-Paque which overlays the cushion. In a 70% gradient, the cells do not band at the gradient-cushion interface, but as the concentration of the gradient is lowered a larger proportion of the cells accumulate at the interface. Maximal banding is observed on 30 to 35% gradients, in which 80 to 90% of the cells are recovered in fractions 4 to 6. This area corresponds to the cushion-gradient interface.

Although cells will band on a 30% gradient, in the presence of liposomes there is an incomplete separation. We believe this phenomenon is due to turbulence as a result of the cells migrating through the gradient. The liposomes are apparently carried through the

Table 1
UPTAKE OF ^{125}I-TRIIODOTHYRONINE (t_3) FROM LIPOSOMES[a] BY CANINE PERIPHERAL BLOOD LYMPHOCYTES

Cells	Initial ^{125}I-T_3 (CPM)	Cell-associated ^{125}I-T_3 (CPM)	Cell efficiency of ^{125}I-T_3 uptake (%)	Free ^{125}I-T_3 (CPM)	Retention in incubation medium (%)
Lymphocytes	7.51×10^3	0.13×10^3	1.7	7.35×10^3	98.3

[a] Multilamellar liposomes composed of DMPC:Chol:DCP (7:2:1), 50 μmol, containing 10 μg T_3/mg phospholipid.

Table 2
VIABILITY OF CANINE LYMPHOCYTES DURING INCUBATION WITH TRIIODOTHYRONINE, LIPOSOMES, AND LIPOSOMES CONTAINING TRIIODOTHYRONINE AFTER SEPARATION USING DISCONTINUOUS-GRADIENT CENTRIFUGATION

	Minutes						
Culture	0	30	60	90	120	180	240
Lymphocytes	95[a]	95	95	93	92	91	90
Lymphocytes + T_3[b]	95	95	93	91	90	90	88
Lymphocytes + liposomes	95	95	94	94	93	90	90
Lymphocytes + liposomes containing T_3[b]	95	95	93	91	90	90	89

[a] Percentage of viable cells after separation over discontinuous-gradient centrifugation; means of three replicates per sample.

[b] Concentration of triiodothyronine as monitored by trypan blue exclusion.

gradient behind the cells. The problem is easily rectified by increasing the gradient to 35%. There are minor variations among gradients, but parallel experiments demonstrate equivalent separations of cells from both [^{14}C] PC liposomes and fluorescent liposomes containing 6-CF; the cells are separated from free liposomes by at least seven fractions.

III. APPLICATIONS OF DISCONTINUOUS-GRADIENT CENTRIFUGATION TECHNIQUES

A. Interactions of Liposome Drug Carriers and Canine Leukocytes

The technique of discontinuous-gradient centrifugation has been employed in our laboratory for isolation of canine peripheral blood leukocytes after incubation with liposomes containing entrapped therapeutic agents including antibiotics; (amikacin, gentamicin, kanamycin, and tobramycin)[8] and thyroid hormones[7] thyroxine and triiodothyronine. The technique has been used successfully in ascertaining uptake of liposome-entrapped drugs (Table 1) and for determination of toxic effects of liposome-encapsulated therapeutic agents on canine leukocytes (Table 2). As the technique is gentle, accurate measurements of toxicities to leukocytes by liposomes or liposome-entrapped drugs may be determined with great reliability and reproducibility than with other conventional techniques involving repeated washings. In addition, several replicates may be run simultaneously to ensure accuracy between experimental groupings.

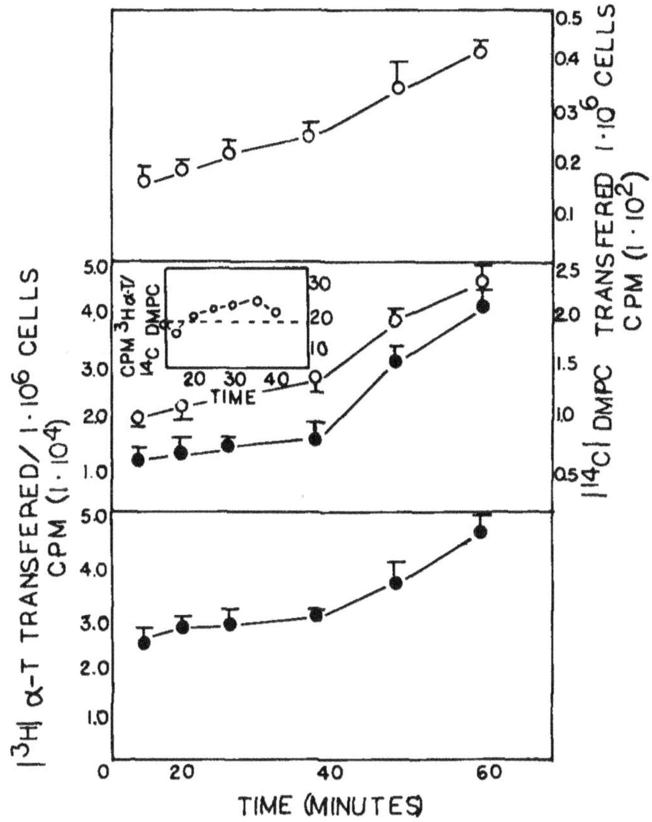

FIGURE 1. Use of discontinuous-gradient centrifugation in separation of bovine lymphocytes and liposomes during lipid exchange studies. Transfer of DMPC and α-T from vesicles to BPBL at 39°C. Vesicles containing isotopic amounts of [^{14}C]DMPC and/or [^{3}H]α-T were incubated with BPBL and the amount of radiolabeled α-T (●) and DMPC (○) were determined. Insert, Ratio of [^{3}H]α-T to [^{14}C]DMPC transferred to cells, dotted line, ratio of ^{3}H α-T to [^{14}C]DMPC in starting vesicles.

B. Studies of Lipid Transfer from Liposomes to Bovine Peripheral Blood Lymphocytes

The technique of discontinuous-gradient centrifugation has been used in our studies of effects of various phospholipids, cholesterol, and tocopherol on bovine lymphocyte function and in particular in studies involving lipid transfer from liposomes to bovine lymphocytes.[3] The use of a technique for separation of noncell-associated liposomes provided a means of effectively removing cell-adherent liposomes which might otherwise have been believed to have transferred lipids to cells. As shown in Figure 1, discrete differences in exchange rates between two lipid components of the same liposome can be determined easily using the gradient centrifugation technique.

C. Studies of Bovine Polymorphonuclear Leukocytes and Liposome Interactions

In addition to separation of lymphocytes and liposome after interaction, the discontinuous-gradient centrifugation technique has recently been used to separate liposomes from bovine polymorphonuclear leukocytes (PMN) after interactions in vitro.[9] As PMN exhibit phagocytic activities in suspension culture and exhibit chemoluminescence in such activities, the rapid separation of liposomes from PMN cells after interactions is necessary to ensure accurate measurements of alteration of PMN cellular activities by liposome-cell interactions. Our

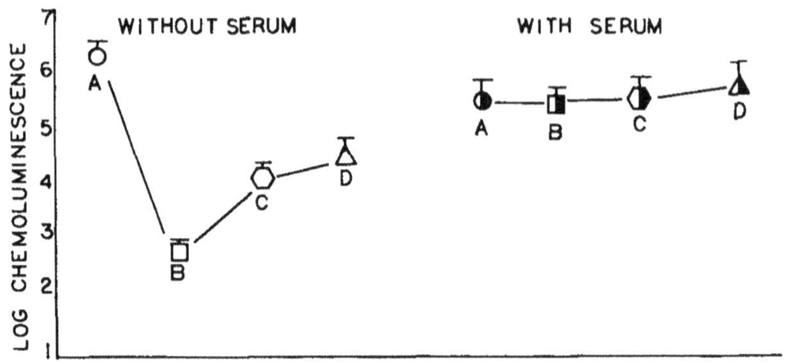

FIGURE 2. The effect of α-T in liposomes on phagocytosis of opsonized zymosan by PMN as measured by luminol-dependent PMN chemoluminescence after preincubation with liposomes. Peak chemoluminescence response from luminol-dependent chemofluorescence. Bovine polymorphonuclear leukocytes (5 × 10^5 cells/mℓ) were incubated with liposomes for 1 hr at 39°C either with or without 10% autologous bovine serum. Zymosan which had been opsonized with bovine serum (0.1 mℓ) was added to cultures along with luminol (1:125 in Hanks' without phenol red, pH 7.2). Luminol was used to monitor peak chemoluminescence response indicative of PMN phagocytosis of opsonized zymosan. Peaks occurred between 7.5 and 13 min. Symbols represent means of triplicate determinations ± SD of the mean. Cells were separated from liposomes by density centrifugation prior to addition of zymosan. A, Controls without liposomes + 0.1 mℓ Hanks' BSS + 0.05 mℓ luminol + 0.1 mℓ zymosan. B, Liposomes (DMPC: α-T; 5:5 molar ratio) + 0.05 mℓ luminol + zymosan. C, Liposomes (DMPC: α-T; 5:1 molar ratio) + 0.05 mℓ luminol + zymosan. D, liposomes (DMPC: α-T) + 0.05 mℓ luminol + zymosan.

results indicate that PMN separated from liposomes by discontinuous-gradient centrifugation are not impaired physiologically, as monitored by their phagocytic activities, for instance the uptake of opsonized zymosan particles in subsequent incubations following cell-liposome separations (Figure 2).

D. Studies of Peripheral Blood Cells from Channel Catfish (*Ictalurus punctatius*) and Liposome Interactions

Primary isolations of circulating blood cells from channel catfish have been separated from liposomes using the discontinuous-gradient centrifugation technique[10] as shown in Table 3. This example illustrates the usefulness of this technique for experiments involving interactions of liposomes and cells in suspension culture derived from lower vertebrates, i.e., channel catfish. No reduction in cell viability in catfish cells was observed following separations involving density-gradient centrifugation. The application of the discontinuous-density centrifugation technique for cells derived from sources other than mammals opens avenues for exploration of interactions of liposomes or liposome drug carriers in species not previously deemed accessible for conventional assay techniques.

IV. DISCUSSION

In general, the use of a single-step density-gradient centrifugation of liposome/cell suspension is advantageous over more commonly used methods of repeated washings in the following ways:

1. It is more rapid, requiring only 1 min centrifugation. Previous experimental designs for evaluating the introduction of lipids into mammalian cell surfaces using liposomes have employed multiple washings/centrifugations[4-6] or use of rapid centrifugation through

Table 3
UPTAKE OF LIPOSOMES CONTAINING ENTRAPPED ^{51}Cr BY PERIPHERAL BLOOD CELLS OF CHANNEL CATFISH *(ICTALURUS PUNCTATIUS)*

	Minutes				
	0	15	30	45	60
Liposome entrapped ^{51}Cr[a]	0[b] (0)[c]	2 (0)	45 (1.0)	160 (2.9)	285 (5.2)
Free ^{51}Cr[a]	0 (0)	5 (0)	6 (0)	18 (0.3)	21 (0.3)

[a] Na ^{51}Cr (5455 cpm) added per culture, total volume 0.3 mℓ.
[b] CPM in cells after incubation and separation using discontinuous-gradient centrifugation; means of four culture per sampling.
[c] Percent uptake of ^{51}Cr as a function of time.

silicone.[11] Although these techniques may provide separation of cells from liposomes, they, in general, do not provide as clean and rapid separations as those obtained using the single-step density-gradient centrifugation.

2. The discontinuous-gradient centrifugation technique is performed in centrifuge tubes which are nonwetting (cellulose nitrate). The use of tubes of this construction provides a vessel which may be used without any need for siliconizing prior to separation of liposomes from cells. In addition the cellulose nitrate tubes are easily punctured for separation of cells and liposomes in a downward flow fractionation procedure.[2]
3. Since discontinuous-gradient centrifugation requires only 1 min centrifugation the procedure may be preformed in a table-top centrifuge, accessible to virtually any investigation. Such gentle separations, as the ones employed in our technique, minimize possible damage to cells from shear force commonly encountered in separation techniques involving nonsiliconized glass centrifuge tubes.[3,7]
4. The cell viability after separation of liposomes and cells is as good as viability of cells prior to separation. In our original study involving L-1210 cells[2,3] and in subsequent studies involving bovine,[3,9] catfish,[10] and canine cells,[7,8] no decrease in viability of cells at any point during the separation procedure has been observed.
5. The functional integrity of cells separated using discontinuous-gradient centrifugation is preserved following this procedure. As demonstrated in studies involving bovine PMN separated from liposomes and subsequently assayed for phagocytic activities indicate no impairment of cellular function following separation using our technique.[9]
6. As the discontinuous-gradient centrifugation technique is rapid and complete in its ability to separate cells from noncell-associated liposomes,[2,3] cells may be taken directly from the gradient and either examined by epifluorescence (as there is no detectable background fluorescence to interfere with photomicrography following separation[2]), or assayed for uptake of radioactive liposome-entrapped[2,7] water- or lipid-soluble markers which may have been transferred to cells. Further, cells may be used[3] in other biological assays such as lymphocyte transformation assays,[12,13] phagocytic assays,[9] or bacterial assays[8] following liposome-cell interactions, and liposomes may be examined for alterations in membrane ingredients, loss of internal contents, or transfer of membrane components during liposome-cell interaction studies.
7. The discontinuous-gradient separation technique may be used with cells in suspension culture from a variety of mammals, namely, mouse, cattle, dogs, and lower vertebrates such as fish. It is our belief that this technique may be applied to any number of cells from a variety of hosts, provided the cells to be separated are maintained in suspension culture and do not adhere to the walls of incubation container.

ACKNOWLEDGMENTS

The author gratefully acknowledges the collaborations of Drs. J. P. Segrest and R. Chiovetti, Jr. in original development of this technique and Dr. R. D. Schultz and Mr. C. Dees in development of applications using this technique.

REFERENCES

1. **Fountain, M. W. and Segrest, J. P.,** A rapid assay for cells in suspension culture, *Ala. J. Med. Sci.*, 15, 273, 1978.
2. **Fountain, M. W., Chiovetti, R., Jr., Kerceret, M., Parrish, D. O., and Segrest, J. P.,** Liposome-cell interactions: a rapid assay for cells in suspension culture, *Biochim. Biophys. Acta*, 596, 420, 1980.
3. **Fountain, M. W. and Schultz, R. D.,** Introduction of α-tocopherol into bovine lymphocytes surfaces by lipid vesicles: effects of phospholipid composition on α-tocopherol transfer, *Immunobiology*, in press, 1982.
4. **Ostro, M. J., Lavelle, D., Paxton, W., Matthews, B., and Giacomoni, D.,** Parameters affecting the liposome mediated insertion of RNA into eukaryotic cells *in vitro*, *Arch. Biochem. Biophys.*, 210, 1980.
5. **Sandra, A. and Pagano, R. E.,** Liposome-cell interactions. Studies of lipid transfer using isotopically asymmetric vesicles, *J. Biol. Chem.*, 254, 2244, 1979.
6. **MacDonald, R. C., Carr, V. M., MacDonald, R. I., Carbello, P. P., and Fisher, J. A.,** Interactions between lipid vesicles and cell membranes, *Ann. N.Y. Acad. Sci.*, 308, 200, 1978.
7. **Fountain, M. W., Ganjam, V. K., and Schultz, R. D.,** Interactions of canine leukocytes with a canine vehicle for triiodothryonine, *Vet. Immunol. Immunopathol.*, 2, 401, 1981.
8. **Fountain, M. W., Dees, C., and Schultz, R. D.,** Enhanced intracellular killing of *Staphylococcus aureus* by canine monocytes with liposomes containing amikacin, gentamicin, kanamycin, and tobramycin, *Curr. Microbiol.*, 6, 373, 1981.
9. **Fountain, M. W., Dees, C., Chambers, W., Klesius, P., and Schultz, R. D.,** unpublished data, 1982.
10. **Fountain, M. W., Dees, C., and Schultz, R. D.,** unpublished data, 1981.
11. **Finkelstein, M. C. and Adelberg, E. A.,** Neutral amino acid transport in an established mouse lymphocyte cell line, *J. Biol. Chem.*, 252, 7101, 1977.
12. **Fountain, M. W. and Schultz, R. D.,** Effects of nitrogen and liposome phospholipid concentration on the blastogenic response of bovine lymphocytes to phytohemagglutinin, *Immunobiology*, 161, 409, 1982.
13. **Fountain, M. W. and Schultz, R. D.,** Effects of enrichment of phosphatidylcholine liposomes with cholesterol or α-tocopherol on the response of lymphocytes to phytohemagglutinin, *Mol. Immunol.*, 19, 59, 1982.

Chapter 15

FREEZING AND THAWING OF LIPOSOME SUSPENSIONS

George Strauss

TABLE OF CONTENTS

I. INTRODUCTION

This chapter discusses the role of freeze-thaw (F/T) processes in liposome research and technology. It surveys available investigative techniques, as used in our laboratory and elsewhere. The general nature of the F/T process is described by giving illustrative results.

Freezing and thawing of liposomes has important applications for investigative and preparative purposes. F/T processes give insight into the stability of liposomes having different structures, the transformation of one structural type into another, including the key process of membrane fusion, and the ability of liposomes to encapsulate and release solutes. Such F/T studies supplement liposome investigations by other means[1] and serve as an important aid in the investigation of the F/T response of cells and tissues.

Results of such studies can be applied to the design of liposomes prepared for specific purposes, such as their use as drug carriers,[2,3] as energy transducers,[4] and as catalytic sites for chemical reactions.[5]

Additional uses of F/T processes are found in liposome technology. These include the encapsulation of drugs and other solutes by taking advantage of the transient destabilization of the bilayer structure during freezing. The preservation of drug-carrying liposomes by long-term storage in the frozen or lyophilized state is part of this technology. Here, freeze-destabilization must be avoided by use of suitable additives.

Freezing of liposomes is also part of freeze-fracture and freeze-etch procedures used prior to electron microscopic examination of liposomes. This application lies outside the present discussion, however, since it is not designed to produce, or to monitor, any freezing effects but is intended to immobilize structures in unchanged form.

A. Freezing of Whole Cells

Before the development of liposome technology as a research tool the effect of F/T on lipid bilayers was studied mostly with whole cells or natural cell membranes, such as erythrocyte ghosts.[6] From such studies the principal causes of cell damage and death due to freezing were identified as (1) mechanical damage by intracellular ice crystals which can distort and pierce the cell membrane and (2) dehydration, which occurs when extracellular water freezes before the cell contents do, causing an efflux of water from the cell.[7] This creates high salt concentrations which denature and inactivate cellular enzymes. As a further consequence, cells become leaky, as manifested by the loss of hemoglobin from erythrocytes.[8] Cell destruction also depends in part on the rate of cooling and subsequent warming. The fraction of cells damaged is large at slow and fast freezing and thawing, and is minimal at intermediate rates. Moreover, the optimum rate depends on the type of cell.[9] This rate dependence has been correlated to the rate of water transport through the lipid bilayer[10] which in turn determines the extent of dehydration. The water permeability is a function of phase equilibria of the membrane lipids.

B. Response of Liposomes to Freezing

F/T effects quite similar to those in cells are observed in liposomes. Very large multilamellar liposomes were found to lose water progressively upon repeated F/T cycling.[11] Leakage of glucose and various inorganic ions occurred after F/T treatment.[12] The loss of glucose was minimal at intermediate cooling rates, and liposomes became distorted, possibly due to the extrusion of lipid from the membrane during freezing.[13]

Freeze-fracture electron microscopy studies of large multilamellar liposomes showed the formation of small nonbilayer particles upon F/T treatment.[14] Such intermembrane particles appear to act as "rivets" between adjacent bilayers. According to a model formulated by Hui et al.,[15] fusion is initiated by local dehydration or ice formation (the same initiating effects deduced from whole cell studies), causing high local curvature of the bilayers, and their eventual fusion. The small nonbilayer aggregates which accompany fusion are regarded as secondary products of membrane distortion. Such aggregates may be identical to the hexagonal lipid phase, designated H_{II}, which has been observed when liposomes consisting of phosphatidylethanolamine (PE) are heated beyond their gel-liquid crystal transition temperature.[16]

The foregoing results refer to multilamellar liposomes large enough to be seen under the light microscope, i.e., of the order of 10,000 Å. The behavior during F/T of very much smaller liposomes, of 250 to 500 Å diameter, was studied by light scattering to monitor aggregation and by fluorescent probes to observe structural changes.[17] Such liposomes, which were mostly unilamellar, increased in size due to aggregation or fusion and suffered losses in membrane integrity. These effects were essentially independent of cooling and warming rates, but did increase with decreasing temperature level at which the liposomes were held while frozen.

Phospholipid bilayers undergo a phase transition from the liquid crystal to the gel phase at a characteristic transition temperature, T_c. For many naturally occurring phospholipids with unsaturated alkyl chains T_c lies close to the freezing point. Freezing of cells and liposomes composed of such lipids thus unavoidably also involves passage through the transition temperature, so that the effect of freezing and thawing per se is obscured. It is possible to separate these effects by working with liposomes composed of saturated lipids of sufficiently high T_c, such as dimyristoyl phosphatidylcholine (DMPC) and dipalmitoyl phosphatidylcholine (DPPC). When small unilamellar vesicles (SUV) of these lipids are heated or cooled through their T_c, or preferably held at T_c, they rapidly aggregate and form larger multilamellar vesicles, and exhibit enhanced membrane permeability.[18-20]

Small micellar particles are observed in the electron microscope when fusion takes place at T_c.[21] These effects thus are quite similar to those caused by F/T treatment. One may conclude that similar initiating conditions, namely a destabilization of adjoining bilayers, exist in both cases.

C. Techniques for Investigation of F/T Effects

The following techniques have been found effective in our investigations to study various aspects of F/T-induced changes in liposomes:

1. Electron microscopy of negatively stained liposome suspensions, to observe changes in size, cluster formation, and conversion of uni- to multilamellar vesicles (MLV)
2. Light-scattering measurements to determine changes in liposome size
3. Fluorescent energy transfer between membrane-embedded probes to recognize and measure formation of liposome aggregates, or fusion
4. Fluorescence increase or loss of membrane-embedded probes to measure loss of membrane integrity
5. Fluorescence changes in water-soluble probes encapsulated in the inner aqueous volume of liposomes to measure leakage into or out of vesicles

II. PREPARATION AND FREEZING OF LIPOSOMES

A. Types of Liposomes

A variety of techniques is available for the preparation of liposomes,[22] and methods can be chosen to make liposomes of several specific structures. The structural types of interest for F/T studies are SUV, defined as those under 1000 Å in size; large unilamellar vesicles (LUV), those over 1000 Å; and MLV which may range in size from 1000 Å to several microns.

For studies of structural transitions during F/T cycles, SUV provide the most information since they are least stable and are transformed into one of the other types under F/T conditions. The other types, LUV and MLV, are preferred for the preparation and study of solute-carrying liposomes, because of their greater stability and much greater capacity per unit weight of lipid to encapsulate solutes.

B. Lipids

The techniques described here refer to liposomes in which the major constitutents are one or more of the following: phosphatidylcholine (PC) from egg yolk; phosphatidylserine (PS) from bovine brain; cardiolipin (CL) from bovine heart; and synthetic DMPC and DPPC.

For studies of F/T-induced changes it is important to ensure that single-chain lipids, such as lysolecithin and free fatty acids, are absent since these compounds act as fusogenic agents, even in small concentrations. (It may be noted that lysophosphatides are toxic to cell membranes.) Care must be taken to avoid formation of these hydrolysis products from the parent phospholipids. Since negatively charged lipids such as PS and CL behave differently from neutral ones such as PC, it is also necessary to guard against contamination of one type of lipid by another.

Phospholipids preferably are stored in hexane, chloroform, or ethanol solution, at −20°C, under nitrogen. Bottles with serum caps are convenient. In use, aliquots are withdrawn from the storage bottle with a hypodermic syringe.

C. Preparative Methods

For the preparation of SUV, the best-known techniques are the injection method and the sonication method. These are widely used, with various modifications. Specific adaptations of these techniques, as used in our F/T studies, are as follows.

1. The Injection Method

In this technique, introduced by Batzri and Korn,[23] an ethanolic solution of the phospholipids, and of any other desired membrane constituents, such as cholesterol, lipid-soluble probes, etc., was injected into an aqueous solution of sufficient volume so that the final alcohol concentration did not exceed 5%. If the lipid is supplied in a hydrocarbon solvent this must first be removed by evaporation. To remove the last traces of hydrocarbon the lipid was redissolved in ether, reevaporated, and kept under vacuum until solvent free. A rotary evaporator was used so that the lipid could be deposited as a thin film coating the inside of a round-bottom flask for complete removal of solvent. The lipid was dissolved in ethanol to give a concentration of 5 to 50 mg/mℓ. The lipid concentration in ethanol is reported to affect the size of the resulting liposomes.[24] The aqueous solution was buffered at pH 7 to 7.5, using, for example, 5 to 50 m*M* Tris. It additionally contained NaCl or KCl at up to 1.0 *M*, and/or other water-soluble constituents which were to be trapped in the internal aqueous compartment of the liposomes.

The injection was performed with a Hamilton syringe fitted with a 22-gauge needle 10 cm long, and was done rapidly, in 1 to 2 sec, while the aqueous solution was vigorously agitated by a Vortex mixer or magnetic stirrer. The solution volume should be kept small;

typical volumes were 5 mℓ aqueous solution and 0.1 to 0.25 mℓ alchoholic solution. The shape of the container may affect the mixing efficiency. We found flat-bottomed vials, about 2 cm diameter and 8 cm high, ideal for this purpose. It is not recommended to scale up the volumes unduly. This procedure gave a fairly uniform population of liposomes 200 to 400 Å in diameter, without further treatment.

2. The Sonication Method

This is the original technique for the preparation of SUV, as described by Huang.[25] A solution of the lipid, together with any other membrane constituents, in hexane or ethanol was evaporated to dryness on a rotary evaporator so as to deposit the lipid as a thin film on the walls of the container. Here again it is essential, perhaps even more so than in the injection method, to remove all traces of hydrocarbon solvent by reevaporating once or twice from ether, and thorough drying under vacuum. Failure to remove all hydrocarbon may result in formation of microemulsions and give nonuniform liposomes. The dry lipid and any other coevaporated constituents were dispersed in an aqueous solution, of composition as described under the injection method. The dispersion can contain up to 20 mg lipid per milliliter of aqueous solution. It was first agitated on a Vortex mixer which produced a milky appearance. This was followed by sonication with a bath-type or probe-type sonicator. Good results were obtained with a Branson probe-type sonicator fitted with microtip. A maximum of 2 mℓ of dispersion was placed into a 2-cm-diameter tube with a conical bottom. The probe tip was positioned about 5 mm from the bottom and the conical sides. Sonication at power level 3, for 10 min, produced an almost clear, only slightly opalescent dispersion. The suspension next was centrifuged at 15,000 × *g* to remove oversized lipid aggregates and metal particles from the tip. An Eppendorf microcentrifuge was convenient for this purpose. This procedure yielded a population of mostly unilamellar vesicles 300 to 500 Å in size, with a few larger multilamellar ones.

3. Comparative Advantages of the Injection and Sonication Methods

The injection method avoids possible air oxidation or degradation of the lipids and dyes used as probes during sonication. It is also faster. The sonication method, on the other hand, allows the preparation of much higher lipid concentrations in the final dispersion, up to about 20 mg/mℓ, and for some purposes it is the only suitable method. The injection method is limited to about 2.5 mg/mℓ due to the limited solubility of lipids in ethanol and the need to limit the concentration of the latter in water.

An important difference between the two methods is that sonicated liposomes are much less "reactive" than those made by injection, possibly because they are well annealed. For example, they aggregate to a lesser extent upon F/T treatment. They also elute reasonably freely from a gel filtration chromatographic column, whereas liposomes made by injection tend to stick to the packing material despite their exclusion from the gel due to their size. For this reason it is advantageous to give a brief (10 to 30 sec) sonication treatment to injection-prepared liposomes.

4. Gel Filtration Chromatography

To obtain liposomes which carry solutes in their interior aqueous compartments, they are prepared by injection into, or sonication with, an aqueous solution of the solute to be trapped. The solute then is present inside and outside the liposomes. The external solute is removed by gel filtration chromatography, i.e., by size exclusion. Sephadex® G-50 or G-75 media are suitable for removal of most types of inorganic or organic compounds. For separation of unbound proteins from liposomes, Sephadex® G-200 is effective. For size-fractionation of liposomes, such as separation of large MLV from SUV, agarose gels such as Sepharose® 4B or 6B can be used.

To separate organic dyes, a 1 × 15 cm Sephadex® G-75 column was used. The column was first equilibrated with the buffer to be used for elution. It was found that liposomes made by injection could not be chromatographed since they were bound to the column so strongly that they could not be eluted. Sonicated liposomes eluted quite well. However, it is advisable to preequilibrate and saturate the column with the liposomes to be chromatographed. A 1-mℓ sample of the liposome suspension was carefully allowed to flow into the column material, followed by small aliquots of buffer, to establish an even band. Elution then followed by downward flow. The liposomes, being very much larger than the pore size of this gel, eluted immediately after the void volume. The first appearance of lipid in the eluate, and its complete elution could be monitored by turbidity.

5. Preparation of Protein-Containing Liposomes

The incorporation of proteins into liposomes can be accomplished by simply incubating the liposomes with an aqueous protein solution. More rapid and more complete protein uptake can be achieved by the technique described below, using a protein such as serum glycoprotein (Cohn Fraction VI). This protein has a molecular weight of approximately 50,000.

A solution of 8 mg glycoprotein in 0.1 mℓ of an aqueous solution 100 m*M* in KCl and 50 m*M* in pH 7.5 Tris buffer was prepared. To this was added 0.9 mℓ of ethanol containing 8 mg of PC. The mixed solution, initially turbid, was agitated on a Vortex mixer for 20 min and in that time became almost clear. It is probably a colloidal protein suspension, with the phospholid bound to the protein and aiding its dispersion. A 0.25 mℓ aliquot of the protein/PC dispersion was injected into 5 mℓ of KCl-Tris buffer (same composition as used to make the dispersion). This caused liposomes to form, but only part of the protein became incorporated. To remove free protein the liposome suspension was passed through a Sephadex® G-200 gel filtration column. The liposomes were eluted immediately after the void volume. Absorption spectra at 280 nm taken before and after the chromatographic separation, when corrected for the dilution effect of the column, showed that about 25% of the total protein, or 0.25 mg protein per milligram lipid, had become bound. Alternatively, the liposome suspension was passed under pressure through a Millipore PSJM Pellicon ultrafiltration membrane. This has a nominal molecular weight cut off of 100,000 which allows passage of free protein but retains liposomes. Spectroscopic analyses of the original suspension and the filtrate gave the percentage of bound protein. In both methods the "bound" protein includes the amount dissolved in the inner aqueous compartment. However, the SUV have very small internal volumes which contain only a small fraction the membrane-bound protein.

D. Cooling and Warming Procedures

Investigation of F/T effects requires that liposome suspensions be subjected to various cooling and warming rates and be held at various cold storage temperatures. These objectives are easily achieved with relatively simple equipment. For temperatures of 0 to −50°C, a refrigerated constant temperature bath is convenient, using a 60:40 (v/v) ethylene glycol/water mixture as the bath fluid. This mixture has an eutectic temperature of −52°C. Dewar flasks filled with various cooling mixtures can be used for lower temperatures. These include dry ice/acetone which gives a temperature of −86°C, and liquid nitrogen which maintains a temperature of −196°C, its boiling point. Sample volumes should be small, e.g., 3 mℓ in a 13 mm × 100 mm test tube, so as to minimize temperature variations within the sample. Cooling rates, i.e., the rates of heat removal, obviously depend both on the heat conductivity of the container wall and on the temperature difference between the sample and bath. The sample temperature thus approaches the bath temperature asymptotically, and the cooling rate is not constant. The rates given below for certain arrangements are only approximations. Sample temperatures were measured with a remote thermistor thermometer.

For slow cooling, the sample tube was surrounded by an empty larger tube (50 mm diameter) serving as an air jacket. For reproducible conditions, the jacket should be immersed and temperature equilibrated before insertion of the sample tube. This arrangement gave a cooling rate of the order of 0.3°C/min with bath temperatures down to −86°C. Faster cooling at about 4°C/min was obtained by direct immersion of the sample tube in the bath. In order to maintain an approximately even cooling rate, such direct cooling should be done in stages by succesive immersion in different baths at, say, −25, −50, and −70°C. For very fast cooling one may use the large temperature differential provided by a liquid nitrogen bath. The use of a narrow (5 mm diameter), thin-walled aluminum sample tube provides particularly fast heat transfer. Such an arrangement gave a cooling rate of 750°C/min.

Warming the frozen sample in air at room temperature gave a warming rate of about 0.3°C/min. Warming in a 25°C water bath produced a warming rate of about 5°C/min.

As pointed out by Meryman,[26] it is possible to maintain a fairly steady cooling rate, with only a brief halt at the freezing point. During thawing, however, the sample temperature quickly rises to 0°C before any significant melting occurs. The thaw line then advances, comparatively slowly, from the sample wall to the interior. Thus, thawing proceeds isothermally, over a relatively long time. The rate of melting can be varied only slightly. Use of a warming bath much above room temperature does somewhat increase the thawing rate, but also subjects the melted portion of the sample to undesirably high temperatures.

III. INVESTIGATIVE TECHNIQUES

This section describes several experimental approaches used to study F/T effects on the structure of liposomes, their stability, and their solute-retaining ability. As seen later in Section IV, these methods gain in usefulness when several techniques are used in conjunction.

A. Electron Microscopy

Direct visualization in the electron microscope of changes in liposome size and structure effected by F/T treatment is extremely useful, provided one can be sure that no artifacts are introduced during evaporation of the liposome suspension on the sample grid. Negative staining is suitable for liposomes up to 5 μm in size; larger vesicles become distorted during the evaporation of the sample.[22,27] Only volatile salts, e.g., ammonium acetate, can be present in appreciable concentrations. Following a procedure similar to that of Olson et al.,[28] a liposome suspension was prepared in 100 m*M* ammonium acetate and 50 m*M* Tris at pH 7.5 at a lipid concentration of 0.3 to 0.6 mg/mℓ. After F/T or other treatment, if any, the sample was applied to the grid, then drawn off with filter paper. A drop of 2% (w/v) ammonium molybdate was applied and drawn off, then allowed to dry for 30 min. The buffer and staining solution were prefiltered through an ultrafilter membrane of 0.1 μm pore size (Millipore or BioRad® polycarbonate filter). The sample did not spread uniformly over the grid; however, very satisfactory images were obtained by seeking out areas that were covered. Precoating the grid with a 0.1 mg/mℓ solution of bacitracin improved the spreading of the sample but often caused too much staining which made it difficult to see details. Staining with uranyl acetate was unsatisfactory as it produced long chains of clustered liposomes, probably due to electrostatic interaction.

B. Turbidity and Light Scattering

1. Theoretical Background

Changes in liposome size due to aggregation and fusion can be followed by measuring the intensity of light scattered by liposome suspensions. Such Rayleigh scattering can be related to the size, and sometimes shape, of the solute particles. The particle size is a function of the so-called Rayleigh ratio, R_θ, defined as $\frac{i_\theta r}{I_o}$, where i_θ is the intensity of light scattered at an angle θ to the incident beam, I_o is the incident light intensity, and r is the distance of the detector from the sample.

For the purpose of finding relative, rather than absolute, particle sizes it is sufficient to calculate the Rayleigh ratio R_{90} for scattering at 90°. One method of finding R_{90} is to measure the loss in transmittance with a spectrophotometer. An instrument of high sensitivity is required since the fractional transmittance loss by typical liposome suspensions is small. From the attenuation of transmitted light one calculates the turbidity, τ, defined by

$$\frac{I_o - I}{I_o} = 1 - e^{-\tau\ell} \approx \tau\ell = 2.303 \text{ OD} \tag{1}$$

where I is the intensity of transmitted light, ℓ is the pathlength through the sample, and OD is the optical density, equal to $\log_{10}(I_o/I)$. The value of R_{90} is then found from the relation

$$R_{90} = \frac{3}{16\pi} \cdot \tau \tag{2}$$

Alternatively, and preferably, one can obtain R_{90} directly by measuring i_{90}, the scattered light intensity at 90°, with a fluorescence spectrophotometer, using the same wavelength in the excitation and emission beams. This gives $R_{90} = i_{90}\, r/I_o$.

For particles smaller than about 1/20 of the wavelength of the scattered light, the value of $(R_{90}/c)_o$, where the subscript indicates extrapolation to zero concentration, is proportional to the molecular weight of the particles, that is to the cube of the diameter. For liposomes which are of the same order of magnitude as the wavelength used (3500 to 4000 Å), the extrapolated specific turbidity $(\tau/c)_o$ and thus also the extrapolated value of $(R_{90}/c)_o$, is an oscillatory function of the particle diameter.[29] However, for particles up to approximately 5000 Å, this function varies approximately linearly with particle diameter. For comparative measurements, values of R_{90}/c can be replaced by i_{90}/c since the instrumental constants r and I_o cancel out.

In the case of liposome aggregation and fusion by F/T or other treatment, the relation between external diameter and scattering is further complicated if unilamellar vesicles change into multilamellar ones, so that the wall thickness of the scattering shell is altered.[30] Our scattering data (see Section IV and Table 2) were found to be linearly proportional to the liposome diameters, within the very approximate size estimates from electron micrographs.

To determine absolute, rather than relative liposome sizes, the refractive index increment of the suspension must be determined, and scattering intensities must be measured at several scattering angles and several concentrations. This is done with a light-scattering photometer in which the detector is mounted on a turntable. A double extrapolation to zero concentration and zero scattering angle of i_θ/c then gives the particle size, following the method of Zimm.[24,31] More recently, another method for particle size determination, known as quasielastic or dynamic light scattering has come into use. Here the time scale of fluctuations of scattered light intensity, due to Brownian motion, is determined and used to compute the translational diffusion coefficient. The latter, together with the physical parameters of the solvent, gives the particle size.[32] A variant of this method is the analysis of the spectral distribution of scattered light arising from a narrowly monochromatic light source, due to Doppler shifts created by Brownian motion.[33] The data from these two methods are the Fourier transforms of each other and provide the same information. Both these methods are rapid, do not require a knowledge of the lipid concentration, and, unlike the classical Zimm method, give the distribution of particle sizes in the sample, rather than just an average size.

2. *Light-Scattering Measurements*

Effects of F/T cycling on liposomes were followed by light-scattering measurements at 90°, using a Hitachi-Perkin Elmer MPF-3 fluorescence spectrophotometer. Since scattering is proportional to $1/\lambda^4$, short wavelengths of 350 or 400 nm were generally used. When the liposomes contain fluorescent probes one must be careful to avoid wavelengths where light absorption or emission could occur. Scattering intensities were measured at several concentrations, and the ratio i_{90}/c was extrapolated to zero concentration. The relative change in particle size during a F/T cycle or other treatment was obtained as the ratio of the $(i_{90}/c)_o$ values measured before and after the treatment.

C. Energy Transfer Between Membrane-Embedded Probes

Electron microscopy and light-scattering methods are effective for monitoring size changes in liposomes during F/T treatment, but cannot give information about the structural nature of such changes. Clearly, there must be a breakdown and reformation of bilayer structures whenever liposomes increase in size or change from a unilamellar to multilamellar structure. Such rearrangement may happen even when a MLV undergoes no outwardly apparent change during a F/T cycle. These processes can be investigated in further detail by the use of energy transfer between fluorescent probes.

1. *Theory of Energy Transfer Measurements*

Energy transfer occurs when a fluorescent molecule (the donor) is excited and transfers some or all of the electronic excitation energy to another molecule (the acceptor) which in turn emits this energy as fluorescence. Such energy transfer is demonstrated by the appearance of acceptor fluorescence as a result of irradiating the system at a wavelength absorbed only by the donor. The essential characteristic of this process for the present purpose is that it is strongly dependent on the distance R between donor and acceptor, decreasing with $1/R^6$. If two liposome populations are mixed which have been separately labeled with energy donor and acceptor probes, respectively, no energy transfer is seen at first since the average distance between liposomes is large. The appearance of energy transfer following F/T or some other treatment shows that donor-labeled and acceptor-labeled liposomes are at close proximity. In the absence of other information this means that the liposomes have formed clusters, undergone disproportionation through exchange of monomeric lipid (and of probes), or fused.

The efficiency of energy transfer, ϕ_{DA}, is defined as the fraction of excitation energy of the donor which appears as fluorescence in the acceptor. At donor-acceptor distances of 20 to 100Å which are typical for liposome clustering or fusion, such transfer occurs by dipole-dipole interaction, the so-called Förster mechanism. In practice, no energy transfer occurs unless the liposomes are in contact. The fraction of liposomes in a mixed population of donor- and acceptor-carriers which has become aggregated thus is given by the value of $(\phi_{DA})_{mix}$, relative to the value of $(\phi_{DA})_{double}$ for doubly labeled liposomes each of which carries both donor and acceptor.

Several methods for measuring ϕ_{DA} have been described. One way to demonstrate energy transfer is to compare the additive fluorescence emission spectra of donor and acceptor in a system before and after a reaction which produces aggregation.[34] Another technique is to measure the fluorescence yield (or fluorescence intensity) of the donor in presence of the acceptor, ϕ_f, and also in its absence, ϕ_f^o.[35] The transfer efficiency then is given by

$$\phi_{DA} = 1 - \frac{\phi}{\phi_f^o} \tag{3}$$

This method assumes that energy transfer is the only mechanism by which the donor is quenched on addition of acceptor. However, there may also be quenching by internal con-

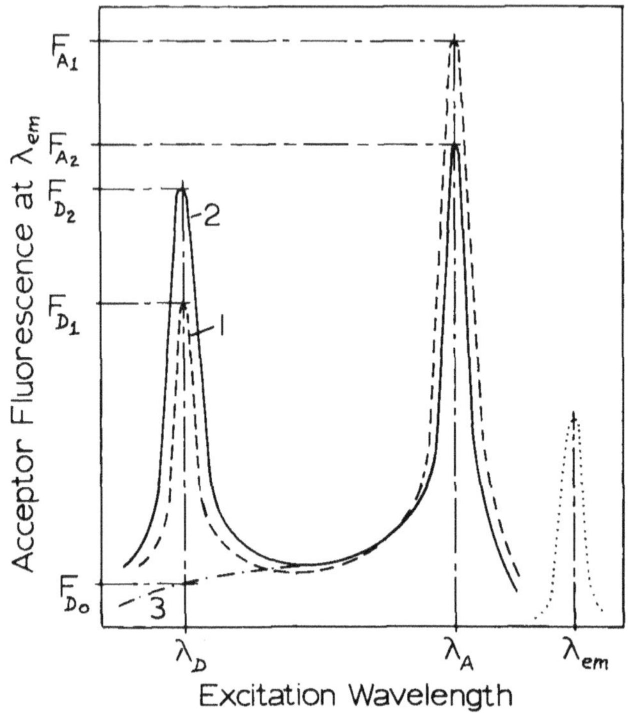

FIGURE 1. Determination of energy transfer efficiencies, ϕ_{DA} from fluorescence excitation spectra. Fluorescence emission of the acceptor is monitored at λ_{em} as a function of excitation wavelength. Wavelengths λ_D and λ_A are the excitation maxima of the donor and acceptor. Curves 1 and 2 represent two test systems; curve 3 represents a control system containing only acceptor. System 2 has higher ϕ_{DA} but also greater quenching by internal conversion, than system 1.

version and static quenching by complex formation. When using this method for comparative measurements, e.g., before and after a F/T cycle, one must assume that these processes are absent or at least occur to the same extent. Such an assumption is not justified, especially since internal conversion is sensitive to changes in rigidity and polarity which are likely to accompany structural changes in liposomes.

Values of ϕ_{DA} can be obtained without making the above assumptions by measuring the fluorescence of the acceptor induced by excitation of the donor and comparing it with the fluorescence of the acceptor generated by its direct excitation.[36] The latter measurement serves as internal standard, and adjusts the induced fluorescence reading for any quenching effects not due to energy transfer. This method does make the assumption that the donor and acceptor are equally affected by such environmental quenching.

To make a measurement, excitation spectra, i.e., plots of acceptor fluorescence intensity at a fixed emission wavelength vs. wavelength of excitation, are taken for the system under test and also for a control system. In the latter, energy transfer is eliminated either by adding a detergent, e.g., Triton® X-100, which destroys the liposomes and makes the donor-acceptor distance very large, or by using a system containing only the acceptor.

A representative set of fluorescence excitation spectra is shown in Figure 1. The acceptor fluorescence is monitored at λ_{em}, while the excitation wavelength is scanned over the region containing λ_D and λ_A, the absorption maxima (and therefore also the excitation maxima) of the donor and acceptor, respectively. The spectra of two systems being measured (curves 1 and 2) and of a control system (curve 3) are shown. Peak heights F_{D_1} and F_{D_2} at λ_D represent

Table 1
FLUORESCENT PROBES FOR ENERGY TRANSFER MEASUREMENT

Compound	Ionic charge	λ_{ex}	λ_{em}	Used as
Dansyl DPPE	−	335	515	Donor
Dansyl DPPS	−	350	515	Donor
Pyrene DPPE	−	350	420	Donor
Dioctadecyloxacarbocyanine	+	486	504	Acceptor or donor
Dioctadecylindocarbocyanine	+	540	556	Acceptor
Chl *a*	0	420,660, and intermed.	672	Acceptor
Octadecanoylaminofluorescein	−	495	530	Donor or acceptor
Octadecylrhodamine B	+	540	580	Acceptor

the fluorescence induced in the acceptor by energy transfer, plus any small amount, indicated by F_{D_0}, due to direct excitation of the acceptor at λ_D. For simplicity this is assumed to be the same for both test systems. Peak heights F_{A_1} and F_{A_2} represent the direct excitation of the acceptor at λ_A. These heights will be different if the two systems experience different degrees of quenching, e.g., by internal conversion.

The energy transfer efficiencies, on an arbitrary scale, are given by

$$\phi_{DA_1} = (F_{D_1} - F_{D_0})/F_{A_1} \text{ and } \phi_{DA_2} = (F_{D_2} - F_{D_0})/F_{A_2} \quad (4)$$

from which the ratio of transfer efficiencies for the two systems is obtained. If the two systems are (1) a mixture of separately labeled liposomes and (2) doubly labeled liposomes, this ratio represents the fraction of liposomes which have aggregated. The quantity F_{A_1}/F_{A_2} is the ratio of the fluorescence quantum yields of the acceptor (and, by assumption also of the donor) in the two systems. This provides useful information about the relative environments of the probes in the two systems.

2. *Choice of Fluorescent Probes*

Donor and acceptor compounds must be chosen so that the spectral region of donor emission overlaps an absorption region of the acceptor, in order to have good energy transfer. Additionally, there should be no donor emission in the spectral region where the acceptor emission is measured. A further requirement of these probes is that they must be firmly anchored in the lipid bilayer, as long as it is intact, and should as nearly as possible have the same amphiphilic characteristics as the phospholipids themselves. The presence of one or preferably two, hydrocarbon chains of 16 or 18 carbons ensures such attachment. As an incidental advantage, such long alkyl chains cause the probe to aggregate and become nonfluorescent in water. Such probes can therefore also be used to measure solute uptake or loss during F/T, as discussed in another section. Suitable probes meeting these requirements are listed in Table 1.

Probes lacking proper amphiphilicity, such as the completely nonpolar donor-acceptor pair *p*-terphenyl and diphenyloctatetraene, when incorporated into two liposome populations, gave energy transfer even in the absence of aggregation or fusion, since they exchanged between liposomes.

Chlorophyll *a* (Chl *a*) is unique as energy acceptor. It has two principal absorption maxima (see Table 1), either of which can be used as reference excitation peak (F_A in Figure 1). The peak at 420 nm can serve as acceptor peak lying at a wavelength *shorter* than the excitation peak for most donors.

3. Incorporation of Probes into Liposomes

For liposomes prepared by the injection method, the probes were codissolved in the ethanolic injection solution which was then injected as usual. For the sonication method, the probes were codissolved in the organic solvent and lyophilized together with the lipids. Sonication may degrade some of the probe compounds. Suitable molar ratios of probe to phospholipid ranged from 1:100 to 1:1000. This corresponds to 3 to 30 probe molecules per liposome of 3000 lipid molecules.

4. Fluorescence Measurements

A fluorescence spectrophotometer of sufficient sensitivity is required to permit the use of low probe concentrations. Since the energy transfer method involves comparison of spectral peaks at different wavelengths the instrument should give an excitation spectrum which is automatically corrected for change in instrumental response with wavelength. A Hitachi-Perkin Elmer MPF-3 fluorescence spectrometer equipped with a corrected spectra accessory performed satisfactorily.

D. Uptake and Release of Liposome Constituents

The structural rearrangement of lipid bilayers during F/T can be further explored by measuring the loss of membrane constituents, as well as their uptake by the membrane from the external solution.

1. Uptake and Loss of Membrane-Soluble Probes

Dyes and other compounds carrying long hydrocarbon chains, such as dioctadecyloxacarbocyanine (DCY), Chl *a*, etc. (see Table 1), when dissolved in the lipid membrane are highly fluorescent and largely monomeric, as indicated by their intense and well-resolved spectra. In water they are practically nonfluorescent and form aggregates, as indicated by their lowered and broadened absorption spectra. It is quite easy therefore to monitor the movement of such a probe into or out of the membrane as a result of F/T. To measure dye uptake, liposomes were prepared by one of the standard methods, and a small volume of an ethanolic solution of the dye was added to the aqueous suspension, e.g., 10 $\mu\ell$ of dye solution to 3 mℓ of liposome suspension. Essentially no fluorescence was observed since most of the dye remained outside the liposomes. Any subsequent increase in fluorescence as a result of F/T cycling indicated that some of the dye had entered the lipid bilayer. Fractional uptake could be calculated by comparing the result with a liposome preparation in which the same total amount of dye had been incorporated into the membrane during preparation. This was done as described under energy transfer techniques by injecting or sonicating the dye together with the lipid.

To measure movement of the dye towards equilibrium in the opposite direction, liposomes were prepared with the dye incorporated into the vesicles. Migration of dye into the outer solution following F/T treatment was registered as a loss of fluorescence. A variant of this technique was to incorporate an energy donor and acceptor into liposomes and to measure the energy transfer before and after the F/T process. A loss of probes after F/T then was indicated by a drop in energy transfer.

2. Uptake and Loss of Inner-Compartment Solutes

The trapping of solutes in the inner aqueous volume of liposomes, or conversely, their leakage from the interior into the external solution, is more difficult to measure than the movement of membrane-bound probes. Here, no spectral change occurs as the probe moves from one aqueous environment to another. The probe selected should not bind to the lipid membrane and must not leak through the membrane in its intact state, i.e., before and after the F/T cycle.

Carboxyfluorescein (CF) is a probe which meets these requirements reasonably well. It is highly fluorescent when dilute, with $\lambda_{ex} = 490$ nm and $\lambda_{em} = 518$ nm. Its fairly large size and negative charge give it a low membrane permeability, especially through negatively charged lipids such as PS and CL.[37]

To measure CF uptake, liposomes were prepared with CF in the external solution at a concentration of 10 m*M* or less. The liposomes were next put through a F/T cycle, after which the external CF was removed by passing the liposomes through a Sephadex® G-50 or G-75 chromatographic column. The amount of trapped CF per unit weight of lipid was determined from the fluorescence of the eluate and that of the original external CF solution.

Release of CF from vesicles during F/T can be demonstrated by taking advantage of its self-quenching behavior. At concentrations of the order of 200 m*M*, CF is essentially nonfluorescent. When liposomes containing CF at this concentration in their internal volume are subjected to a F/T cycle, the fluorescence strongly increases as the CF is released into the external aqueous phase and becomes highly diluted.[38]

Another technique to determine solute release during F/T treatment makes use of a highly fluorescent complex formed by Tb^{+++} with dipicolinic acid (DPA). This complexation reaction was used previously by Wilschut and Papahadjopoulos[39] to study the integrity of Ca^{++}-induced liposome fusion. For the present purpose, the decomposition of this complex, and the consequent quenching of its fluorescence, by EDTA was utilized.

Liposomes were prepared in presence of 15 m*M* $TbCl_3$, 150 m*M* DPA, and 2 m*M* Tris at pH 7.5. The liposome suspension was placed onto a 1 cm × 15 cm Sephadex® G-75 column, and eluted with a solution containing 100 m*M* NaCl, 1 m*M* EDTA, and 2 m*M* Tris. At this stage, intense fluorescence was observed. Subsequent F/T treatment caused a loss in fluorescence due to quenching of that fraction of the complex which had leaked out and had reacted with EDTA. Alternatively, the elution from the column was performed without EDTA. The eluted liposomes, now containing Tb·DPA inside only, were subjected to a F/T cycle. The fluorescence was measured, giving the total Tb complex present. EDTA was added next, and the resulting loss in fluorescence gave the amount of complex which had leaked out. Both variants of this method use only one column eluate to measure total complex and that lost during F/T, thus avoiding errors from nonreproducible column separations. To measure the fluorescence of the Tb·DPA complex, an excitation wavelength of 276 nm and emission wavelength of 545 nm were used. Since the latter happens to be close to double the excitation wavelength, interference by the second harmonic is possible. This can be eliminated by use of a cut-off filter in the emission beam.

IV. RESULTS

This section presents some results of F/T investigations of liposomes obtained in our laboratory by the techniques discussed. The results are illustrative rather than comprehensive and are intended to show the scope and applicability of these methods. In order to place these F/T studies in proper perspective, comparative data on the aggregation of liposomes by other means are included, such as the well-known effects of Ca^{++} ions, and the passage through the transition temperature.

Unless indicated otherwise, the abbreviation F/T refers to a standard freeze-thaw cycle in which samples of 3 mℓ in 13 mm × 100 mm stoppered test tubes were directly immersed into dry ice/acetone in a Dewar flask, held there for 15 min, and thawed in a 25°C water bath.

A. Size and Structural Changes Due to F/T Cycling

Electron micrographs of liposomes before and after F/T treatment were examined and compared with light-scattering measurements taken on the same samples. Table 2 gives

Table 2
SIZE CHANGES OF LIPOSOMES INDUCED BY F/T AND ADDITION OF Ca^{++} — COMPARATIVE DATA FROM ELECTRON MICROSCOPY AND LIGHT SCATTERING (ALL EXPERIMENTS WERE DONE AT 150 m*M* SALT CONCENTRATION)

Injected liposomes

Lipid	Treatment	E/M diam (Å)	Diam incr.	Scatter incr.
PS	None	250	—	—
	F/T	1000	4	2.4
	$(F/T)^3$ [a]	1000	4	1.6
	Ca^{++} [b]	2000	8	7.5
PS/PC (70:30)	None	300	—	—
	F/T	1000	3.3	5
	Ca^{++}	1000	3.3	5

Sonicated liposomes

Lipid	Treatment	E/M diam (Å)	Diam incr.	Scatter incr.
PS	None	400	—	—
	F/T	700	2	1.5
	Ca^{++}	2000	5	5.1
PC	None	300	—	—
	F/T	700	2.3	1.1

[a] Subjected to three successive F/T cycles.
[b] Addition of $CaCl_2$ to a concentration of 5 m*M*.

Table 3
EFFECT OF SALT CONCENTRATION ON INCREASE IN LIGHT SCATTERING INDUCED BY F/T OF INJECTED LIPOSOMES

		Scattering increment factor	
Lipid	Treatment	5 m*M* NaCl	100 m*M* NaCl
PS	F/T	11	2.4
	$(F/T)^3$ [a]	6	1.6
PC	F/T	4	2.5
	$(F/T)^3$ [a]	6	3.8

[a] Subjected to three successive F/T cycles.

average diameters measured from electron micrographs, together with incremental factors of size and of the $(i_{90}/C)_o$ values resulting from F/T and addition of Ca^{++}.
The data show:

1. Liposomes made by injection averaged 250 Å in diameter and were unilamellar. On F/T cycling they were transformed into oligolamellar vesicles of approximately 1000 Å.
2. Liposomes made by brief sonication (10 min) were 350 to 400 Å initially and increased to approximately 750 Å on F/T. This is another manifestation of the relative inertness of sonicated liposomes. A possible reason for this is that sonicated vesicles are well annealed and present a smooth surface, whereas injected ones are irregularly packed and have reactive sites.
3. Repeated F/T cycling of injected liposomes caused no further size increase after the first cycle. This may be caused by the absence of "reactive" SUV after the initial fusion reaction.
4. Scattering increments were roughly proportional to the size increments seen in electron micrographs, as already discussed in Section III. It is not clear why repeated F/T cycles of injected PS vesicles caused a decrease in scattering but no observable size change.

The effect of salt concentration on the F/T response is shown in Table 3. These results indicate:

1. At salt concentrations of 100 m*M* the response of both PS and PC liposomes made by injection was much less than at 5 m*M* salt. As generally observed in polyelectrolytes and proteins, high ionic strength has the effect of screening ionic charges and permitting the structure to become compact. In liposomes evidently the same effect operates. It is somewhat greater for negatively charged PS than for zwitterionic PC, in line with expectation.
2. The effect of repeated F/T cycling was anomalous. The scattering of PS vesicles, at both low and high salt concentration, increased after the first cycle, then declined during further cycles. PC vesicles, in contrast, gradually increased their scattering.

B. Temperature Effects During F/T Treatments

1. Effect of Storage Temperature Level

PC liposomes made by injection were cooled to various temperature levels ranging from

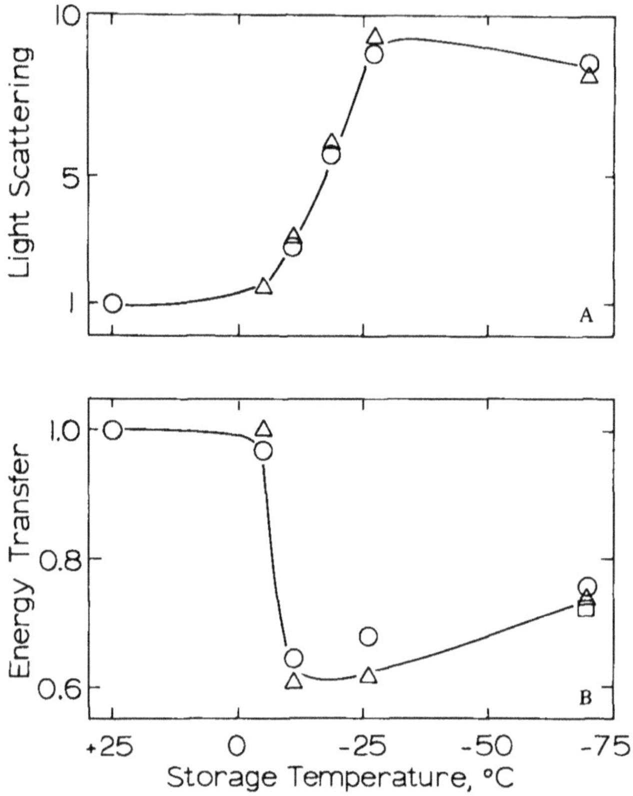

FIGURE 2. Effect of storage temperature for 18 hr, and of cooling and warming rate on structural rearrangement of PC liposomes. (A) Increase in light scattering; (B) Loss in energy transfer. △, Cooled and warmed at 0.3°C/min; ○, Cooled and warmed at 4°C/min; □, Cooled at 750°C/min and warmed at 4°C/min.

0 to −70°C, held there for 18 hr, then warmed to room temperature. The resulting changes were monitored by light scattering and by measuring the decline in energy transfer between dioctadecyloxacarbocyanine and Chl *a*, used as membrane-bound probes. Reduced energy transfer was taken as indication of the loss of these probes during structural rearrangement.[17]

Results are shown in Figure 2. Changes in scattering and energy transfer first appeared at −5°C. Scattering increased gradually with decreasing storage temperature to −27°C and was slightly less for storage at −70°C. Loss in energy transfer followed a similar pattern, but increased more sharply, reaching a maximum at −11°C. The results strongly suggest that the liquid crystal-gel phase transition, which for egg yolk PC occurs between −15 and −7°C,[22] is at least partly responsible for the observed changes. The more gradual change in scattering may mean that aggregation and fusion is a secondary process, subsequent to the initial membrane disruption. If so, then an intermediate storage temperature (e.g., −20°C) could cause loss of a large part of the probes, but only a limited degree of aggregation.

2. *Effect of Slow and Fast Cooling and Warming*

The above experiments were performed at cooling rates varying from 0.3°C/min to 750°C/min, and at warming rates from 0.3°C/min to 4°C/min. As shown in Figure 2, the effects of changing the rates were almost undetectable. The slowest rate was marginally more damaging. This result is in marked contrast to rate effects during F/T of living cells, where the damage is minimal at intermediate freezing rates. Our data show that liposomes, unlike

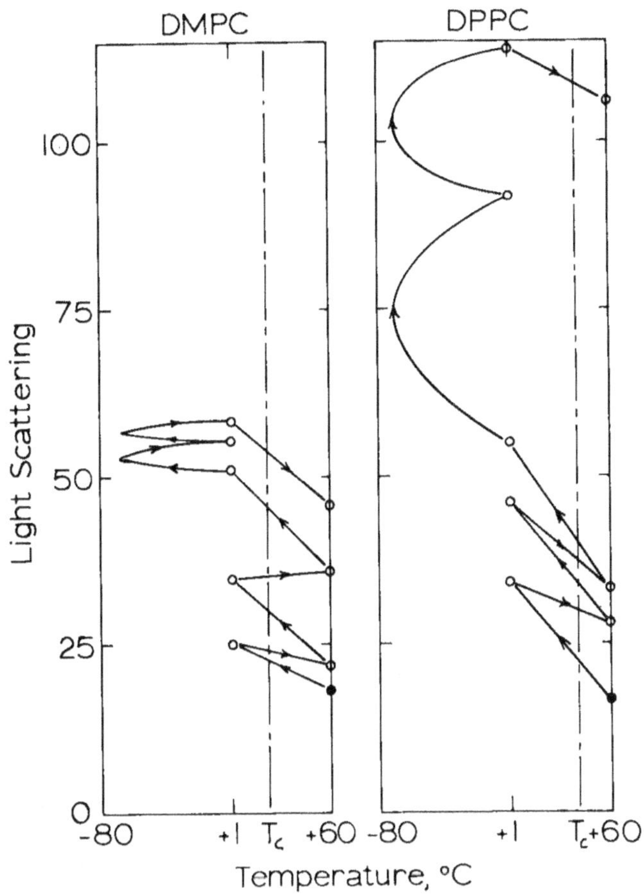

FIGURE 3. Effect of repeated cooling and warming through T_c, and of repeated F/T cycles on aggregation of DMPC and DPPC liposomes, as measured by light scattering.

cells, achieve rapid equilibration even at −70°C. For samples cooled to −70°C by direct immersion in dry ice/acetone, the scattering increase was the same after 15 min as after 18 hr.

3. Separation of F/T Effects from Phase Transition Effects

As is well known, passage of lipid bilayers through their phase transition temperature, T_c, causes destabilization and leads to aggregation of liposomes, as discussed earlier.[20] In order to separate this effect from that of F/T, liposomes made from DMPC and DPPC were tested. These saturated lipids have T_c values of 23 and 41°C, respectively. The liposomes were made by injection at 60°C. They were repeatedly cooled to 1°C and reheated to 60°C, and finally subjected to repeated F/T cycles. Scattering values measured at each stage are plotted in Figure 3. A progressive scattering increase was observed during cycling through T_c. Equilibration apparently was incomplete, however, or would have required further cycling, since subsequent F/T cycles produced additional scattering increases. Differences in scattering at 60° and 1°C evidently are due to the different scattering properties of liposomes above and below T_c, aside from aggregation effects.[30]

The measurements at 60°C show that cycling of DPPC through T_c increased scattering by a factor of 2.05, and subsequent F/T cycling by a further factor of 3.17, giving an overall increase by a factor of 6.5. This agrees well with the scattering increment factors of ap-

Table 4
DEGREE OF COMPLETION OF AGGREGATION REACTION AND FRACTIONAL LOSS OF MEMBRANE PROBES DURING F/T CYCLE OF PS LIPOSOMES

System	Treatment	Relative transfer efficiency (ϕ_{DA})
Mixture of separately labeled liposomes	None	0
	F/T	0.429
Doubly labeled liposomes	None	0.560
	F/T	0.452

Note: Fraction aggregated after F/T $= \frac{0.429}{0.452} = 0.95$. Fractional loss of membrane probes after F/T $= \frac{0.560 - 0.452}{0.560} = 0.19$.

proximately 6.0 obtained on repeated F/T cycling of egg yolk PC and brain PS liposomes. DMPC, in contrast to DPPC, gave incremental scattering factors of only 1.93 for T_c cycling and 1.28 for subsequent F/T cycling, or an overall increment of 2.5.

C. Liposome Aggregation and Fusion

The degree of completeness to which the aggregation and fusion of liposomes proceeds during F/T cycling can be determined by energy transfer between membrane-bound probes. Illustrative data are given in Table 4. DCY was used as donor in one population of liposomes, and Chl *a* as acceptor in another. The liposomes were made by sonication. Energy transfer values shown are relative. The results show that the aggregation reaction was 95% complete after passage through one standard F/T cycle, even though these sonicated liposomes are relatively unreactive. The electron micrographs of such liposomes after F/T show few, if any, SUV of the original size, in agreement with this finding. The data in Table 4 also show that 19% of the membrane probes were lost during the fusion reaction.

D. Incorporation and Loss of Liposome Constituents During F/T

1. Membrane-Bound Probes

The same membrane probes which had been used to measure energy transfer (DCY and Chl *a*) were also selected to test (1) their incorporation into liposomes during F/T if originally present in the outer aqueous solution and (2) their loss from the membrane during F/T into which they had originally been incorporated. Liposomes made by injection from PS were used. The effect of ionic strength was also tested by varying the salt concentration from 5 m*M* to 1.0 *M*.

Results are shown in Figure 4. For Chl *a*, 80% of the external probe was incorporated, and 34% of that originally in the membrane was lost, independently of salt concentration. For the cationic cyanine dye, 75 to 97% of the external dye was incorporated, the lower figure applying to higher salt concentration. The loss of this dye from the membrane could only be estimated since an anomalous fluorescence increase occurred during the F/T cycle, possibly due to formation of a more rigid dye-PS complex. In presence of 1.0 *M* NaCl, at room temperature, 20% of the chlorophyll and 35% of the cyanine were incorporated from the external solution, and slight losses of both these probes from the membranes occurred. F/T treatment thus caused extensive equilibration of amphiphilic probes between membrane and aqueous solution, in either direction. High salt concentrations somewhat repressed the exchange of the cationic cyanine dye during F/T, but promoted its exchange at room temperature.

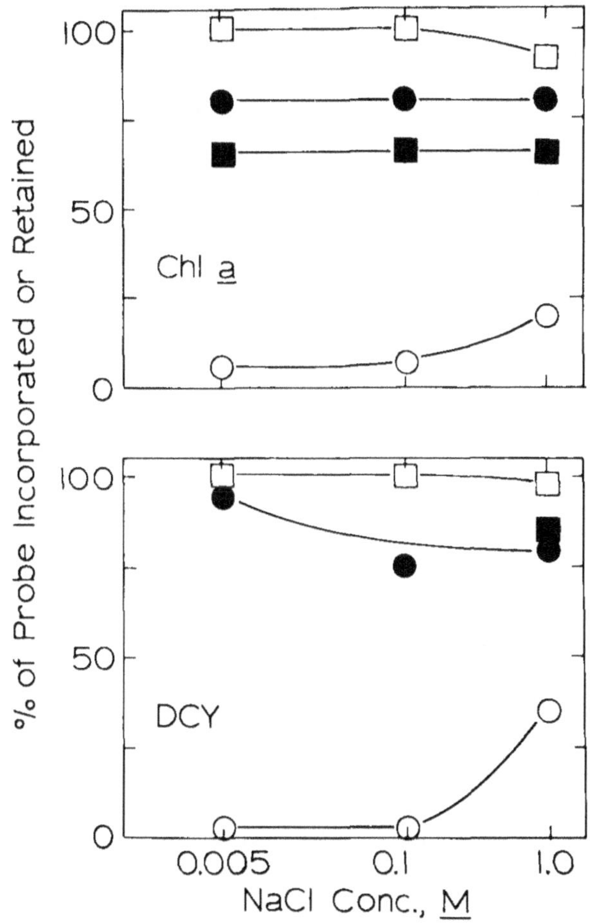

FIGURE 4. Uptake and loss of the membrane-bound probes Chl *a* and DCY at room temperature and after F/T, as function of NaCl concentration. Circles: probe added to external solution after formation of liposomes. ○, at room temp.; ●, after F/T. Squares: probe incorporated into liposomes during formation by injection, □, at room temp., ■, after F/T.

2. Water-Soluble Solutes

Uptake from the external solution into the inner aqueous liposome center was measured with CF as the water-soluble probe. PS liposomes were made by injection, after which CF was added to a concentration of 5 m*M*. The sample was subjected to a F/T cycle and then passed through a Sephadex® G-75 column to remove CF from the external solution. The fluorescence of the eluted liposome sample was read, giving the amount of CF trapped. The data in Table 5 show that F/T results in capture of external CF by liposomes. This method is useful for comparing trapping efficiencies of different systems.

The loss of solutes from the inner aqueous compartment was determined by measuring the loss of fluorescence of trapped Tb·DPA complex as it came in contact with EDTA present in the external solution. Liposomes were made from PC and PS by sonication. As shown in Table 5, a standard F/T cycle caused a loss of 22% of the complex from PC vesicles, but a loss of 70% from PS vesicles. This dramatically shows the much greater disruption of PS liposomes by F/T at low salt concentration, compared to PC vesicles. It also shows that either the liposomes open up during the F/T process, or temporarily become highly

Table 5
CAPTURE AND LOSS OF AQUEOUS SOLUTES DURING F/T

Trapping of External CF by Liposomes

System	CF fluorescence	μmol CF trapped per mg of liposomes
(1) 5 mg liposomes/mℓ + 2 μmol CF/mℓ in external solution	680	—
(2) System (1), after passage through Sephadex® column	0.15	8.8×10^{-5} (= leakage or mechanical carry-over)
(3) System (1) after F/T, followed by passage through Sephadex® column	20.6	$\frac{20.6}{680} \times \frac{2}{5} = 1.2 \times 10^{-2}$

Loss of Tb·DPA Complex from Inner Aqueous Compartment

System	Treatment	Tb·DPA fluorescence as % of original value
PC	None	100
	F/T	78
PS	None	100
	F/T	30
PS + glycerol, 10%	None	100
	F/T	76

Note: Lipid: PS/PC (80:20).

permeable to small molecules and ions. Additional data in Table 5 show that the loss from the inner compartment during F/T was much less in presence of 10% glycerol in the external solution.

E. Effects of Additives

1. Cholesterol

The effect of cholesterol on the F/T-response of liposomes is of considerable interest, since cholesterol has diverse effects on lipid bilayers, as reported in numerous studies.[22] The effect of cholesterol, when present in the membrane of PS liposomes at concentrations of up to 50% (v/v), was measured by light scattering. Incremental factors in presence of 5 and 100 m*M* NaCl are shown in Table 6. Cholesterol clearly has a destabilizing effect which is much more pronounced at low salt concentration. This result agrees qualitatively with the increased glucose release from DPPC liposomes on F/T treatment with increasing cholesterol content.[13,41]

This destabilizing effect could not have been predicted since cholesterol has both stabilizing and destabilizing effects on bilayers, depending on temperature and composition.[16] The observed destabilization may be due to the fact that cholesterol does not uniformly distribute itself in bilayers but forms complexes with phospholipids, thus creating discontinuities which act as nuclei for structural changes. Cholesterol was found to stabilize liposomes against Ca^{++}-induced fusion, in contrast to its effect during F/T.[18]

2. Glycoprotein

The effect of serum glycoprotein (Cohn Fraction VI) in the membrane, at 25% by weight, during F/T was tested by light scattering. At 5 m*M* salt, it considerably reduced the scattering increment, as shown in Table 6. Energy transfer measurements, discussed in the next section, gave concordant results.

Table 6
EFFECT OF CHOLESTEROL AND GLYCOPROTEIN ON F/T RESPONSE OF PS LIPOSOMES — LIGHT-SCATTERING INCREMENT FACTORS AT LOW AND HIGH SALT CONCENTRATION

	No additive	% Chol (v/v) 10	30	50	Glycoprotein (25% by weight)
In 5 m*M* NaCl	6.25	7.4		19.3	1.7
In 100 m*M* NaCl	1.9	2.1	4.5	3.8	1.0

3. Nonionic External Solutes

Glycerol, ethylene glycol, dimethyl sulfoxide, and sucrose are among the substances used as cryoprotective agents in the freezing of cells and tissues for long-term storage. Glycerol has been shown to prevent hemolysis of erythrocytes on freezing. Such compounds, used in concentrations of 10 to 20% by weight, are known to reduce the amount of ice formed. In whole cells, therefore, they reduce both membrane damage by ice formation, and by dehydration.[10]

The effect of glycerol and DMSO on the perturbation of injected PC liposomes during F/T was measured by light scattering and energy transfer.[17] The approximately sixfold increase in light scattering resulting from F/T became progressively smaller for increasing concentrations of glycerol or DMSO, up to 50%, in the external solution. This result is partly misleading, however, since such agents can also dissolve phospholipids, i.e., raise their critical micelle concentration.

A more accurate picture is provided by energy transfer measurements. Using PC liposomes labeled with DCY and Chl *a* as donor and acceptor probes within the bilayer, it was found that glycerol and DMSO at room temperature had no effect below 10%, but caused extensive loss of energy transfer at higher concentrations. On F/T treatment, both these agents at a concentration of 10% prevented any loss of energy transfer which, in their absence, was 30 to 40%. This agrees, qualitatively, with the reduced loss of trapped aqueous solutes during F/T seen in presence of glycerol.

Sucrose has a similar protective effect. At concentrations of 100 m*M* or more, it reduces the trapping volume of LUV liposomes made by rapid freezing, thawing, and sonication.[40] This may be interpreted as a diminished degree of fusion and rearrangement of the SUV originally present. Glycerol was found to have no protective effect on Ca^{++}-induced aggregation and fusion, as measured by light scattering. Since glycerol and similar agents prevent ice formation, this result is good evidence that ice formation is at least one of the factors which induce liposome fusion upon freezing.

Liposomes containing 25% by weight serum glycoprotein in the membrane showed no loss of energy transfer during F/T cycling, thus confirming the result obtained by light scattering. Such protein-containing liposomes were also resistant to the destructive action, at room temperature, of glycerol and DMSO at concentration over 10%.

ACKNOWLEDGMENTS

This work was supported by a grant of the Charles and Johanna Busch Memorial Fund. Thanks are due to Dr. Suzanne M. Gibson for expert assistance in this research.

REFERENCES

1. **Ryman, B. E. and Tyrrell, D. A.,** Liposomes — methodology and applications, in *Lysosomes in Applied Biology and Therapeutics*, Vol. 6, Dingle, J. T., Jacques, P. J., and Shaw, I. H., Eds., North-Holland, New York, 1979, chap. 19.
2. **Gregoriadis, G.,** Liposomes, in *Drug Carriers in Biology and Medicine*, Gregoriadis, G., Ed., Academic Press, New York, 1979, chap. 14.
3. **Juliano, R. L. and Layton, D.,** Liposomes as a drug delivery system, in *Drug Delivery Systems*, Juliano, R. L., Ed., Oxford University Press, London, 1980, chap. 6.
4. **Laane, C., Ford, W. E., Otvos, J. W., and Calvin, M.,** Photosensitized electron transport across lipid vesicle walls, *Proc. Natl. Acad. Sci. U.S.A.*, 78, 2017, 1981.
5. **Fendler, J. H. and Fendler, E. J.,** *Catalysis in Micellar Systems*, Academic Press, New York, 1975.
6. **Heber, U., Volger, H., Overbeck, V., and Santarius, K. A.,** Membrane damage and protection during freezing, in *Proteins at Low Temperatures*, Fennema, O., Ed., American Chemical Society, Washington, D.C., 1979, chap. 8.
7. **Mazur, P.,** Causes of injury in frozen and thawed cells, *Fed. Proc.*, 24, S175, 1965.
8. **Pribor, D. B.,** Biological interactions between cell membranes and glycerol or DMSO, *Cryobiology*, 12, 309, 1975.
9. **Miller, R. H. and Mazur, P.,** Survival of frozen-thawed human red cells as a function of cooling and warming velocities, *Cryobiology*, 13, 404, 1976.
10. **Farrant, J.,** Water transport and cell survival in cryobiological procedures, *Philos. Trans. R. Soc. London*, B. 278, 191, 1977.
11. **Rendi, R.,** Water extrusion in isolated subcellular fractions. VI. Osmotic properties of swollen phospholipid suspensions, *Biochim. Biophys. Acta*, 135, 33, 1967.
12. **Siminovitch, D. and Chapman, D.,** Liposome bilayer model systems of freezing living cells, *FEBS Lett.*, 16, 207, 1971.
13. **Morris, G. J. and McGrath, J. J.,** The response of multilamellar liposomes to freezing and thawing, *Cryobiology*, 18, 390, 1981.
14. **Yu, J. and Branton, D.,** Reconstitution of intramembrane particles in recombinants of erythrocyte protein Band 3 and lipid, *Proc. Natl. Acad. Sci. U.S.A.*, 73, 3891, 1976.
15. **Hui, S. W., Stewart, T. P., Boni, L. T., and Yeagle, P. L.,** Membrane fusion through point defects in bilayers, *Science*, 212, 921, 1981.
16. **Cullis, P. R. and deKruijff, B.,** The polymorphic phase behavior of phosphatidylethanolamines of natural and synthetic origin, *Biochim. Biophys. Acta*, 513, 31, 1978.
17. **Strauss, G. and Ingenito, E. P.,** Stabilization of liposome bilayers to freezing and thawing: effects of cryoprotective agents and membrane proteins, *Cryobiology*, 17, 508, 1980.
18. **Papahadjopoulos, D., Poste, G., Schaeffer, B. E., and Vail, W. J.,** Membrane fusion and molecular segregation in phospholipid vesicles, *Biochim. Biophys. Acta*, 352, 10, 1974.
19. **Prestegard, J. H. and Fellmeth, B.,** Fusion of dimyristoyl lecithin vesicles as studied by proton magnetic resonance spectroscopy, *Biochemistry*, 13, 1122, 1974.
20. **Martin, F. J. and MacDonald, R. C.,** Phospholipid exchange between bilayer membrane vesicles, *Biochemistry*, 15, 321, 1976.
21. **Verkleij, A. J., vanEchteld, C. J. A., Gerritsen, W. J., Cullis, P. R., and deKruijff, B.,** The lipidic particle as an intermediate structure in membrane fusion processes and bilayer to hexagonal H_{II} transitions, *Biochim. Biophys. Acta*, 600, 620, 1980.
22. **Szoka, F., Jr. and Papahadjopoulos, D.,** Comparative properties and methods of preparation of lipid vesicles (liposomes), *Ann. Rev. Biophys. Bioeng.*, 9, 467, 1930.
23. **Batzri, S. and Korn, E. D.,** Single bilayer liposomes prepared without sonication, *Biochim. Biophys. Acta*, 298, 1015, 1973.
24. **Kremer, J. M. H., v.d. Esker, M. W. J., Pathmamanoharan, C., and Wiersema, P. H.,** Vesicles of variable diameter prepared by a modified injection method, *Biochemistry*, 16, 3932, 1977.
25. **Huang, C.,** Studies on phosphatidyl vesicles. Formation and physical characteristics, *Biochemistry*, 8, 344, 1969.
26. **Meryman, H. T.,** Physical limitations of the rapid freezing method, *Proc. R. Soc. London B*, 147, 452, 1957.
27. **Bangham, A. D., Hill, M. W., and Miller, N. G. A.,** Preparation and use of liposomes as models of biological membranes, *Meth. Membr. Biol.*, 1, 1, 1974.
28. **Olson, F., Hunt, C. A., Szoka, F. C., Vail, W. J., and Papahadjopoulos, D.,** Preparation of liposomes of defined size distribution by extrusion through polycarbonate membranes, *Biochim. Biophys. Acta*, 557, 9, 1979.
29. **Kerker, M.,** Light scattering, in *Chemistry and Physics of Interfaces II*, Ross, S., Ed., American Chemical Society, Washington, D.C., 1971, 170.

30. **Chong, C. S. and Colbow, K.,** Light scattering and turbidity measurements on lipid vesicles, *Biochim. Biophys. Acta*, 436, 260, 1976.
31. **Zimm, B. H.,** Apparatus and methods for measurement and interpretation of the angular variation of light scattering, *J. Chem. Phys.*, 16, 1099, 1948.
32. **Goll, J. H. and Stock, G. B.,** Determination by photon correlation spectroscopy of particle size distributions in lipid vesicle suspensions, *Biophys. J.*, 191, 265, 1977.
33. **Chen, F. C., Chrezeszczyk, A., and Chu, B.,** Quasielastic laser light scattering of phosphatidylcholine vesicles, *J. Chem. Phys.*, 64, 3403, 1976.
34. **Gibson, G. A. and Loew, L. M.,** Phospholipid vesicle fusion monitored by fluorescence energy transfer, *Biochem. Biophys. Res. Commun.*, 88, 135, 1979.
35. **Vanderwerf, P. and Ullman, E. F.,** Monitoring of phospholipid fusion by fluorescence energy transfer between membrane-bound dye labels, *Biochim. Biophys. Acta*, 596, 302, 1980.
36. **Mehreteab, A. and Strauss, G.,** Energy transfer and energy losses in bilayer membrane vesicles (liposomes), *Photochem. Photobiol.*, 28, 369, 1978.
37. **Weinstein, J. N., Yoshikami, S., Henkart, P., Blumenthal, R., and Hagins, W. A.,** Liposome-cell interaction: transfer and intracellular release of a trapped fluorescent marker, *Science*, 195, 489, 1977.
38. **Blumenthal, R., Weinstein, J. N., Sharrow, S. O., and Henkart, P.,** Liposome-lymphocyte interaction, *Proc. Natl. Acad. Sci. U.S.A.*, 74, 5603, 1977.
39. **Wilschut, J. and Papahadjopoulos, D.,** Ca^{++}-induced fusion of phospholipid vesicles monitored by mixing of aqueous contents, *Nature (London)*, 281, 690, 1979.
40. **Pick, U.,** Liposomes with a large trapping capacity prepared by freezing and thawing of sonicated phospholipid mixtures, *Arch. Biochem. Biophys.*, 212, 186, 1981.
41. **Morris, G. J.,** The response of liposomes to various rates of cooling to −196°C: effect of phospholipid: cholesterol ratio, *Cryobiology*, 19, 215, 1982.

Chapter 16

FREEZING OF LIPOSOMES

Patrick Machy and Lee D. Leserman

TABLE OF CONTENTS

I. INTRODUCTION

There are excellent reasons to find conditions under which liposomes can be stored for long periods without aggregation, fusion, or loss of encapsulated contents. For research purposes, comparison of results from laboratory to laboratory would be ameliorated by exchange of liposome preparations: this requires stable conditions for shipment. In our laboratory small preparations of liposomes are frequently made which differ only slightly with respect to lipid composition, coupled ligand, or entrapped aqueous contents. It would be helpful if these could be stored so that the preparation can be used over longer periods of time. Clinical studies using liposomes will require very large preparations of liposomes to be used over extended trial periods. It will be necessary to store liposomes during this time.

Though in our experience, cholesterol-containing liposomes sterilized by membrane-filtration can be stored indefinitely at 4°C, low temperature can by itself provoke the aggregation[1] and fusion[2] of liposomes composed of dipalmitoyl phosphatidylcholine (DPPC). This effect was not seen at temperatures above the transition temperature of the lipid, but passage of liposomes through this temperature will by itself provoke significant leakage of contents.[3]

Conditions now are well described for the safe freezing of most mammalian cells. Because of the wide availability, at least in research laboratories, of suitable conditions for the long-term maintenance of frozen reagents, we have evaluated freezing of liposomes as a means of long-term storage. Freezing of liposomes has been reported by Strauss and Ingenito,[4] who found significant destabilization of liposomes when frozen in the absence of cryoprotective agents. This destabilization could be overcome by the addition of amphipathic serum proteins, presumably by altering phospholipid hydrophobic interactions. Because stable entrapment of low-molecular-weight molecules usually requires the addition of cholesterol,[5] we have concentrated on the freezing of cholesterol-containing liposomes. We have also examined the effect of liposome size on freezing, reasoning that since freezing damage is related to osmatic dehydration,[4] then osmatically insensitive small liposomes[6] would be different in their behavior than osmotically sensitive large liposomes. We report here that under appropriate conditions, liposomes can be frozen with no change in size, encapsulated contents, or behavior in bioassays.

II. MATERIALS AND METHODS

A. Preparation of Liposomes

Small unilamellar liposomes (SUV) containing 20 m*M* methotrexate (MTX) (Division of Cancer Treatment, National Cancer Institute) and/or 40 m*M* recrystallized and LH20 column purified[7] carboxyfluorescein[8] (CF) (Eastman) in 100 m*M* $NaHCO_3$, pH 7.5 and composed of 64 mol % DPPC (Sigma), 35 mol % cholesterol (Chol) (Applied Sciences Laboratories) and 1 mol % dipalmitoyl phosphatidylethanolamine (DPPE) (Calbiochem) modified with *N*-hydroxysuccinimidyl 3-(2-pyridyldithio)-propionate[9] (SPDP) (Pharmacia) (40 μmol total lipid) were made by probe sonication at 50°C, as described in detail in Chapter 2 of Volume III.

For formation of SUV, MLV (multilamellar vesicles), and LUV (large unilamellar vesicles) of the same lipid composition, 40 μmol total lipid [54 mol % dimyristoyl phosphatidylcholine (DMPC) (Sigma), 10 mol % dipalmitoyl phosphatidylserine (DPPS) (available from Avanti Polar Lipids, Birmingham, Ala.); 35 mol % Chol and 1 mol % SPDP modified DPPE (DPPE-DTP)] was used as detailed below.

SUV — To the above dried lipids was added 3 mℓ of a mixture of 20 m*M* MTX and 40 m*M* CF. The mixture was sonicated with a probe sonicator at 40°C for 30 min.

MLV — Multilamellar vesicles were made by adding 3 mℓ of 40 m*M* CF solution

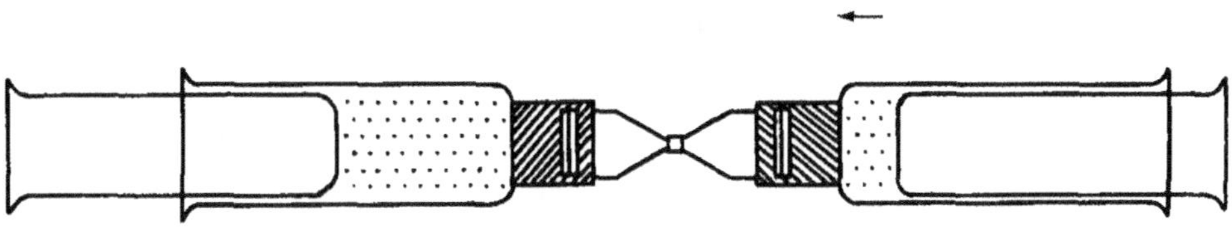

FIGURE 1. Schematic representation of the formation of a stable emulsion using syringes; 1 mℓ of CF and MTX was injected into 3 mℓ of organic solvent containing lipids via 2 Luer-lock syringes and a double hubbed needle. The emulsion was formed by the passage back and forth of the two pistons and was stable for many hours. (Magnification × 20.)

prewarmed to 45°C to the dried lipid and vortexing for 5 min. The resulting MLV were then sized by sequential passage through 0.4-μm and 0.2-μm Unipore polycarbonate filters (Biorad).[10]

LUV — Large unilamellar vesicles were made according to the reverse-phase-evaporation (REV) technique of Szoka and Papahadjopoulos.[11] Lipid was dissolved in 1 mℓ chloroform and 2 mℓ diisopropyl ether (Merck). A mixture of 1 mℓ of CF and MTX was then added. The emulsion was made by a syringe system (see below). The lipid-aqueous emulsion was transferred to a 15 mℓ rotovap tube, warmed to 45°C, and the organic phase was removed at 360 mmHg pressure. Liposomes were then filtered through 0.4 μm and 0.2 μm Unipore filters as for MLV.

B. Formation of a Stable Emulsion Without Sonication

Concern has been expressed for the stability of certain molecules, notably DNA, after exposure to the bath sonication required to make an emulsion for reverse phase evaporation liposomes.[12] To eliminate this sonication stress we have adapted the classical immunologic method for making Freund's adjuvant emulsions.[13] The aqueous and organic phase are placed in separate glass syringes fitted with Luer locks. The syringes are joined by a double hubbed needle (Figure 1) or by a metal three-way stopcock. The aqueous phase is injected into the organic phase, then the mixture is passed rapidly 10 to 20 times from syringe to syringe. Some leakage of the material around the barrel of the syringe occurs; this can be eliminated by the use of gas-tight syringes. The emulsion thus achieved is stable during the period required for the evaporation of the organic phase.

C. Coupling Reaction Between Liposomes and Monoclonal Antibody or Protein A

SUV composed of DPPC, Chol, DPPE-DTP were incubated with 39G, an IgG2a, kappa, anti-H-2 I-E^k monoclonal antibody,[14] iodinated by the chloramine T method,[15] and modified by 5 mol SPDP per mole antibody, at a final concentration of 180 μg/mℓ.

SUV and LUV composed of DMPC, DPPS, Chol, and DPPE-DTP were incubated with iodinated protein A modified by 10 mol SPDP per mole protein A at a final concentration of 200 μg/mℓ. The final concentration of total lipid was 5.1 m*M*. The coupling reaction was performed in liposome buffer (L-buffer; NaCl, 145 m*M*; Hepes, 10 m*M*; pH 8) for 24 hr. Fifty percent of the antibody became fixed to the DPPC, Chol, DPPE-DTP liposomes. With protein A the fixation was 17% to SUV and 11% to LUV, respectively.

D. Freezing and Thawing Liposomes

Liposomes were dialyzed against L-buffer (pH 7.45) overnight and sterilized by filtration through 0.45-μm Millipore filters. Aliquots of liposomes were placed into standard cell-freezing vials after a 1 to 1 dilution with RPMI 1640 medium (Gibco) supplemented with

7% heat-inactivated fetal calf serum (FCS) (Gibco), penicillin, and streptomycin, or with the same medium containing 5, 10, or 20% (vol/vol) dimethylsulfoxide (DMSO) (Prolabo). The addition of DMSO dilutions into liposome solutions was made at 4°C dropwisc. Liposomes were then placed in a cotton lined box in a freezer at − 80°C; the rate of temperature decrease was approximately 3°C/min. After freezing overnight the liposomes were transferred to liquid nitrogen (− 180°C) for long-term storage. Thawing from − 80°C or liquid nitrogen was made by immersion in a water bath at 37°C until the last ice crystals melted.

E. Leakage Studies

CF leakage was estimated by measurements of the fluorescence with an Aminco SPF spectrofluorimeter (excitation: 488 nm; emission 520 nm) before and after releasing entrapped CF with Triton® X-100 detergent (final concentration: 0.5%). Leakage was assessed by dialysis of liposomes against a 100-fold excess of L-buffer (pH 7.45) overnight and measuring CF in the dialysate and CF remaining liposome associated.

F. Size Distribution of Liposomes

The size of liposomes before and after freezing was measured by using a 1 cm × 34 cm Sephacryl® S-1000 column (Pharmacia).[16] The column was presaturated with 20 μmol lipids (64 mol % DMPC; 35 mol % Chol, 1 mol % DPPE-DTP) in 1 mℓ L-buffer (pH 7.45) and with 1 mℓ FCS to avoid nonspecific adsorption of lipids and proteins during chromatography. LUV (mean diameter about 0.4 μm) containing CF and free CF were used to calibrate the column. The void volume (V_0) was 11.5 mℓ; the total volume (V_t) was 29.5 mℓ. The column was run at a flow rate of 3 mℓ/hr. Liposomes were visualized by fluorescence and by measurement of ^{125}I-radiolabeled protein covalently coupled to liposomes.

G. *Staphylococcus aureus* Precipitation of Antibody-Bearing Liposomes

Paraformaldehyde fixed *S. Aureus* bacteria (Strain Cowan I) were prepared as described.[17] The bacteria were washed three times before use with L-buffer (pH 7.45), then resuspended to 10% (vol/vol) in the same buffer; 10 μℓ of antibody-bearing liposomes were diluted to 1 mℓ with 0.2% bovine serum albumin in L-buffer. To some tubes was added 100 μℓ of the *S. aureus* suspension, to others was added 100 μℓ of buffer. After 30 min incubation at 4°C, tubes were spun for 10 min at 2300 × *g*. The bacterial pellets were counted for ^{125}I and measured for fluorescence after addition of 10 μℓ of 20% Triton® X-100 and 900 μℓ of buffer and centrifugation for 10 min. Supernatants were counted and measured for fluorescence before and after Triton®. The percentage of liposomes precipitated was calculated as described.[18]

H. Cell Cultures

Spleen cells from CBA/J (CSEAL-CNRS, Orléans, France) mice were treated with 0.83% NH_4Cl in order to lyse red blood cells. B cells were obtained from spleen cells depleted of T cells as described in detail;[19] 30 × 10^6 cells were incubated at 37°C in 7% CO_2 in tissue culture flasks with 20 mℓ of RPMI 1640 medium supplemented with 5% FCS, 2 m*M* glutamine, 5 × 10^{-5} *M* 2-mercaptoethanol, penicillin, and streptomycin.

For mitogen-induced proliferation, purified B cells were incubated with 20 μg/mℓ of lipopolysaccharide (Difco, LPS from *E. coli* n° 055: B) as previously described.[19] L cells were cultured in RPMI medium supplemented with 5% FCS.

I. Measurement of Cell-Associated CF

After 40 hr of incubation B lymphoblasts were resuspended in supplemented RPMI medium without mitogen at a concentration of 8 × 10^6 cells per milliliter. L cells were incubated with 1.5 m*M* EDTA in order to resuspend the adherent cells, washed, and resuspended as

Table 1
LEAKAGE COMPARISON BETWEEN MLV, LUV, AND SUV[a] BEFORE AND AFTER FREEZING OF LIPOSOMES WITH AND WITHOUT DMSO

	Leakage before freezing (%)			Leakage after freezing (%)		
	SUV	LUV	MLV	SUV[b]	LUV	MLV
−DMSO[c]	0	0	0	11 ± 4	64 ± 6	55 ± 2
+DMSO	0	28 ± 5	24 ± 1	1 ± 1	27 ± 8	22 ± 1

Note: Results are shown as mean ± SEM of three experiments.

[a] Liposomes were composed of 40 μmol-total lipid (54 mol % DMPC; 10 mol % DPPS; 35 mol % Chol; 1 mol % DPPE-DTP). Liposomes were coupled to protein A except for MLV. Without protein coupling the same results were obtained.

[b] When SUV were composed of 64 mol % DPPC; 35 mol % Chol, and 1 mol % DPPE-DTP, no significant leakage was observed.

[c] Liposomes were mixed with an equal volume of medium containing 7% FCS plus or minus 20% DMSO.

B cells; 50-μℓ aliquots were placed in conical wells of 96 well microtiter plates. B cells were incubated directly with 39G antibody-bearing liposomes. L cells were incubated with protein A-bearing liposomes after incubation of the cells with H100-30/33, an IgG2b, kappa, anti-H-2K^k monoclonal antibody[20] at 20 μg/mℓ for 30 min at 4°C and one wash by centrifugation with 100 μℓ of L-buffer (pH 7.45). Cells were incubated with liposomes at 37°C for 3 hr and washed four times by centrifugation in L-buffer. The fluorescence of cell-associated CF was measured as described previously.[19]

J. MTX-Mediated Inhibition of d(^{3}H)Urd Incorporation

To flat-bottomed wells of microtiter plates containing 10^5 L cells was added anti-H-2K^k antibody (H100-30/33) at a final concentration of 20 μg/mℓ. After 30 min incubation at 4°C cells were washed once by centrifugation in 100 μℓ RPMI medium. The cells were then resuspended in supplemented medium. Dilutions of free MTX or protein A-bearing liposomes containing 20 m*M* MTX were added to the cultures. After 3 hr of incubation at 37°C, 0.5 μCi of d(^{3}H)Urd was added to the cultures. After an additional 16 hr at 37°C cells were collected and the concentration of incorporated d(^{3}H)Urd measured as described elsewhere.[21]

III. RESULTS

A. Leakage of Liposome Contents as a Consequence of Freezing

SUV, MLV, and LUV of the same lipid composition (54 mol % DMPC, 10 mol % DPPS, 35 mol % Chol, 1 mol % DPPE-DTP) were frozen with or without the cryoprotective agent DMSO. As shown in Table 1, before freezing there was no leakage of liposome contents in the absence of DMSO. Liposomes frozen in the absence of DMSO leaked a small amount (SUV) or most (LUV and MLV) of their contents. Table 1 and Figures 2 and 3 show that the addition of DMSO caused a concentration-dependent increase in liposome leakage, but also a concentration-dependent reduction in the additional leakage caused by freezing. At a final concentration of 10% DMSO there was no additional leakage caused by freezing.

The leakage induced by either DMSO or freezing was an all or none event; liposomes leaked either all of their contents or not at all. This was confirmed by rechromatography of liposomes after freezing. The ratio of fluorescence of liposomes before and after disruption by detergent, which is related to the CF concentration in liposomes (see Chapter 10, Volume

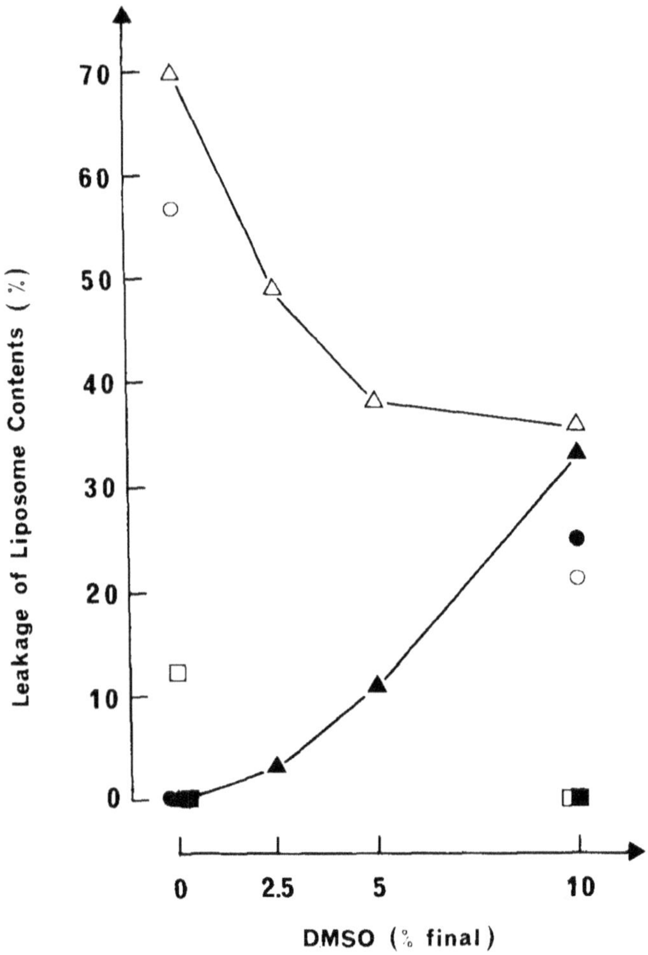

FIGURE 2. Effect of DMSO on the leakage of liposomes before and after freezing. Solutions of MLV, LUV, and SUV were diluted by the addition of RPMI 1640 medium containing 7% FCS and different proportions of DMSO. The dilution was made by dropwise addition of medium into the liposome solutions. One part of the liposome preparation was dialyzed overnight against a 100-fold excess of L-buffer (pH 7.45). The other part was frozen at −80°C, thawed, and dialyzed for measurements of fluorescence before and after addition of Triton® X-100 (0.5% final). (▲) LUV; (●) MLV; (■) SUV of the same lipid composition, before freezing. (△) LUV; (○) MLV; (□) SUV after freezing and thawing. The results were substantially similar if FCS was omitted from the freezing solution (data not shown).

III) had not changed (data not shown). Thus "surviving" liposomes retained the initial aqueous contents.

Liposomes (SUV) made of 64 mol % DPPC, 35 mol % Chol, and 1 mol % DPPE-DTP did not leak at all, either as a consequence of the addition of 10% DMSO, or of freezing in its absence (Table 2). Similarly, the process of freezing did not affect the quantity of stability of monoclonal antibody coupled to the liposomes prior to freezing. Table 2 shows that the same proportions of ^{125}I-labeled antibody bound to the surface of liposomes and CF in liposomes, respectively, were precipitated by *S. aureus*, which express an immunoglobulin binding protein (protein A) on their surface.

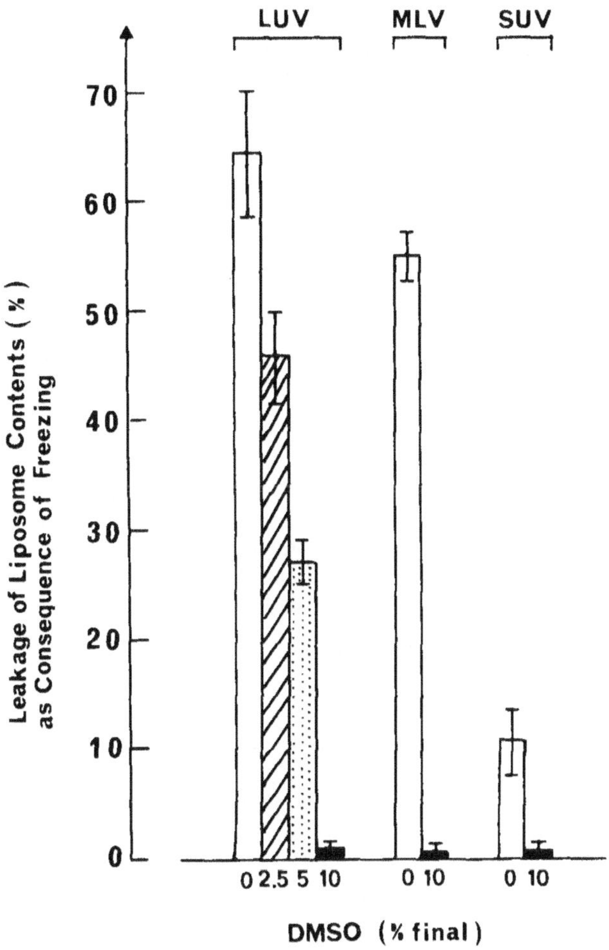

FIGURE 3. Comparison of leakage of LUV, MLV, and SUV contents as consequence of freezing. Bars represent the mean ± SD of three experiments.

B. Size Distribution of Liposomes Before and After Freezing

The freezing of liposomes did not affect their size distribution, that is, it caused neither aggregation nor fusion of liposomes. Figure 4 shows the results of passage of SUV, MLV, and LUV over a column of Sephacryl® S-1000 superfine, before and after freezing of the liposome preparations, as confirmed for both liposome-encapsulated CF and liposome-bound radiolabeled protein. The SUV made with 10% DPPS had a larger diameter (800 Å as compared to 500 to 600 Å) than those lacking this charged phospholipid.

C. Specific Binding of Liposomes to Target Cells

We have compared the binding to mouse B cell blasts of frozen and thawed or unfrozen liposomes composed of DPPC : Chol : DPPE-DTP and bearing on their surface antibody with specificity for a determinant expressed by the cells. As shown in Table 3 the binding was specific, in that it was inhibited by an excess in free solution of the antibody bound to the liposomes, but not an irrelevant antibody. Freezing did not alter the magnitude or specificity of liposome binding.

Similarly, Table 4 indicates that DMPC : DPPS : Chol : DPPE-DTP SUV, or LUV bearing covalently coupled protein A showed cell-specific binding which was unchanged by freezing

Table 2
LEAKAGE AND *S. AUREUS* PRECIPITATION COMPARISON OF LIPOSOMES[a] COVALENTLY COUPLED TO MONOCLONAL ANTIBODY 39G BEFORE AND AFTER FREEZING OF LIPOSOMES WITH AND WITHOUT DMSO

	Before freezing			After freezing		
	Leakage[b] (%)	*S. aureus* precipitation[c] (%) ^{125}I	*S. aureus* precipitation[c] (%) CF	Leakage (%)	*S. aureus* precipitation (%) ^{125}I	*S. aureus* precipitation (%) CF
−DMSO	0	76	60	0	77	61
+DMSO[d]	0	80	61	2	82	60

[a] SUV were composed of 40 μmol total lipids (64 mol % DPPC; 35 mol % Chol; 1 mol % DPPE-DTP).
[b] The same values were obtained when liposomes were not coupled to the monoclonal antibody.
[c] When liposomes were not coupled to the antibody only 3% of CF was precipitated.
[d] DMSO was added at a concentration of 10% final.

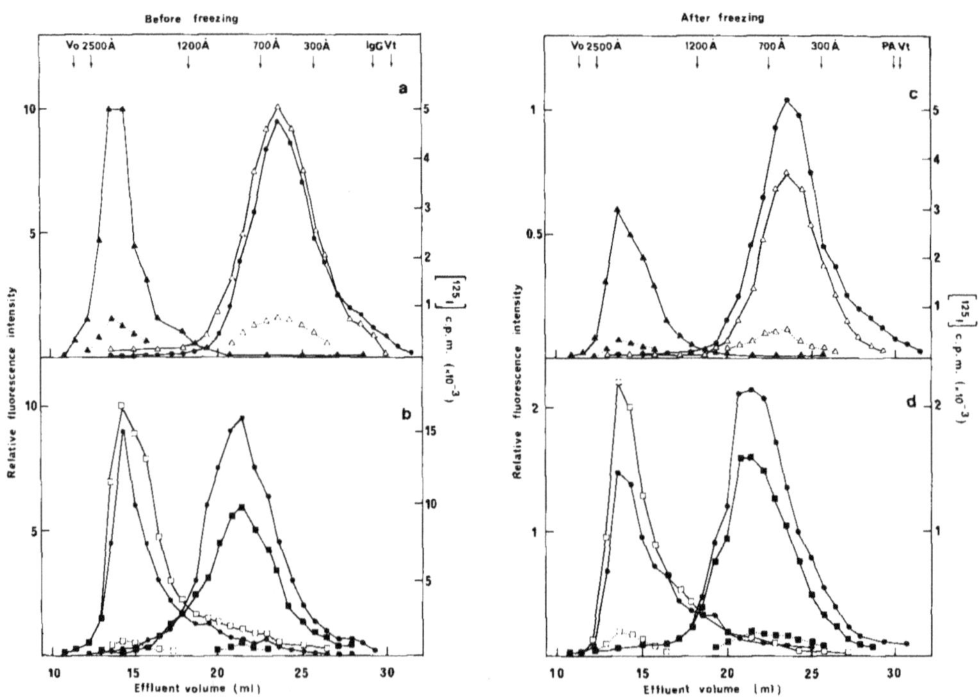

FIGURE 4. Chromatography of liposomes on a Sephacryl® S-100 column. Aliquots of liposome preparations were passed over a Sephacryl® S-1000 column as described in Section II. Sample size was 0.7 mℓ in L-buffer (pH 7.45). Aliquots of the effluents were counted in a gamma counter when liposomes were coupled to radiolabeled antibody or protein A and measured for fluorescence before and after addition of detergent. The column was calibrated according to Nozaki et al.[16] (a and b) before freezing; (c and d) after freezing and dialysis. (△) SUV composed of DPPC, Chol, DPPE-DTP; (▲) MLV; (□) LUV and (■) SUV composed of DMPC, DPPS, Chol, DPPE-DTP. (----) before and (——) after lysis of liposomes. (●) ^{125}I protein coupled to liposomes (a and c, anti-H-2 I-E^k monoclonal antibody 39G coupled to SUV composed of DPPC, Chol, DPPE-DTP; b and d, protein A coupled to LUV and SUV composed of DMPC, DPPS, Chol, DPPE-DTP). Column markers: IgG, immunoglobulin G; P.A., protein A; V_t, total volume; V_o, void volume.

Table 3
SPECIFIC BINDING OF LIPOSOMES[a] BEARING ANTI-H-2 I-E^k MONOCLONAL ANTIBODY 39G TO B CELL BLASTS BEFORE AND AFTER FREEZING

	Cell-associated CF (pmol)[b]		
		After freezing[c]	
Inhibitor added	Before freezing	−DMSO	+DMSO
None	12.2 ± 2	12.8 ± 2	17 ± 2
39G[d]	0.3 ± 0.1	0.3 ± 0.2	0.3 ± 0.1
H100-5/28	9.6 ± 0.5	8.8 ± 1	12 ± 1

Note: Results are shown as mean ± SEM of three experiments.

[a] SUV were composed of 64 mol % DPPC; 35 mol % Chol; 1 mol % DPPE-DTP.
[b] 4 × 10^5 cells were incubated with antibody-bearing liposomes containing a total of 306-318 pmol of CF.
[c] Liposomes were not dialyzed because no leakage was observed. They were tested as such directly after thawing.
[d] Before addition of liposomes, cells were incubated with relevant (39G) or irrelevant (H100-5/28, anti-H-2K^k)[20] antibodies free in solution at a final concentration of 20 μg/mℓ for 10 to 15 min.

of the liposomes. The binding of the liposomes depended on prior incubation of the cells with an antibody capable of binding protein A, and the liposome-bound protein A could be competitively inhibited by free protein A.

D. Drug Transfer from Liposomes to Cells

In addition to unchanged specific binding we have verified that drug delivery is unaffected by freezing of liposomes. After thawing DMPC DPPS : Chol : DPPE-DTP SUV, and LUV were dialyzed to remove leaked contents and DMSO, then added to cultures of L cells, and compared to free MTX and unfrozen liposomes of the same composition. MTX-mediated inhibition of d(^{3}H)Urd incorporation was identical for frozen or unfrozen liposomes (Figure 5). Interestingly, LUV, despite their higher level of binding to cells (Table 4), were not better than SUV in terms of drug delivery.

IV. DISCUSSION

In order to relate the effects of freezing on liposomes of identical lipid composition, liposomes were made of 54 mol % DMPC, 10 mol % DPPS, 35 mol % Chol, and 1 mol % DPPE-DTP. This change from our standard lipid composition of 64 mol % DPPC, 35% Chol, and 1% DPPE-DTP (see Chapter 2, Volume III) was made necessary by our inability to make consistently good REV liposomes in the absence of PS. The addition of PS somewhat reduces the efficiency of protein coupling and increases the size of small sonicated liposomes (Figure 4), but does not increase either the leakage of liposomes or their ability to specifically bind and to deliver their contents into target cells.

The resistance of liposomes to freezing was correlated inversely with size. Small liposomes composed of DPPC : Chol : DPPE-DTP could be frozen as shown here in medium free of cryoprotective agents, without any detectable leakage, aggregation, or change in ability to specifically target drug. Although we used medium containing FCS, this could be omitted, and SUV could be frozen in simple buffer solution without any loss of contents (data not

Table 4
SPECIFIC BINDING OF LUV- AND SUV-[a] BEARING PROTEIN A BEFORE AND AFTER FREEZING TO L CELLS PREINCUBATED WITH ANTI-H-2K^k MONOCLONAL ANTIBODY (H100-30/33)

		Cell-associated CF (pmol)[b]				
		SUV			LUV	
			After freezing[e]			
Antibody[c]	Free protein A[d]	Before freezing	−DMSO	+DMSO	Before freezing	After freezing[e] (+DMSO)
anti-H-2K^k	−	12.2 ± 2	12.4 ± 1.5	12.0 ± 2	32.0 ± 3	32.0 ± 2
(H100-30/33)	+	0.16 ± 0.07	0.2 ± 0.05	0.2 ± 0.1	0.3 ± 0.1	0.3 ± 0.1
None	−	0.2 ± 0.1	0.2 ± 0.09	0.1 ± 0.05	0.4 ± 0.1	0.4 ± 0.1

Note: Results are shown as mean ± SEM of three experiments.

[a] LUV and SUV were composed of 54 mol % DMPC; 10 mol % DPPS; 35 mol % Chol; 1 mol % DPPE-DTP.
[b] 4 × 10^5 cells were incubated with protein A-bearing liposomes containing a total of 212 pmol of CF.
[c] Cells were preincubated with 20 μg/mℓ final of antibody and washed before addition of protein A-bearing liposomes.
[d] Before addition of liposomes, cells were incubated for 10 min with 50 μg/mℓ final protein A.
[e] After thawing liposomes were dialyzed overnight before testing.

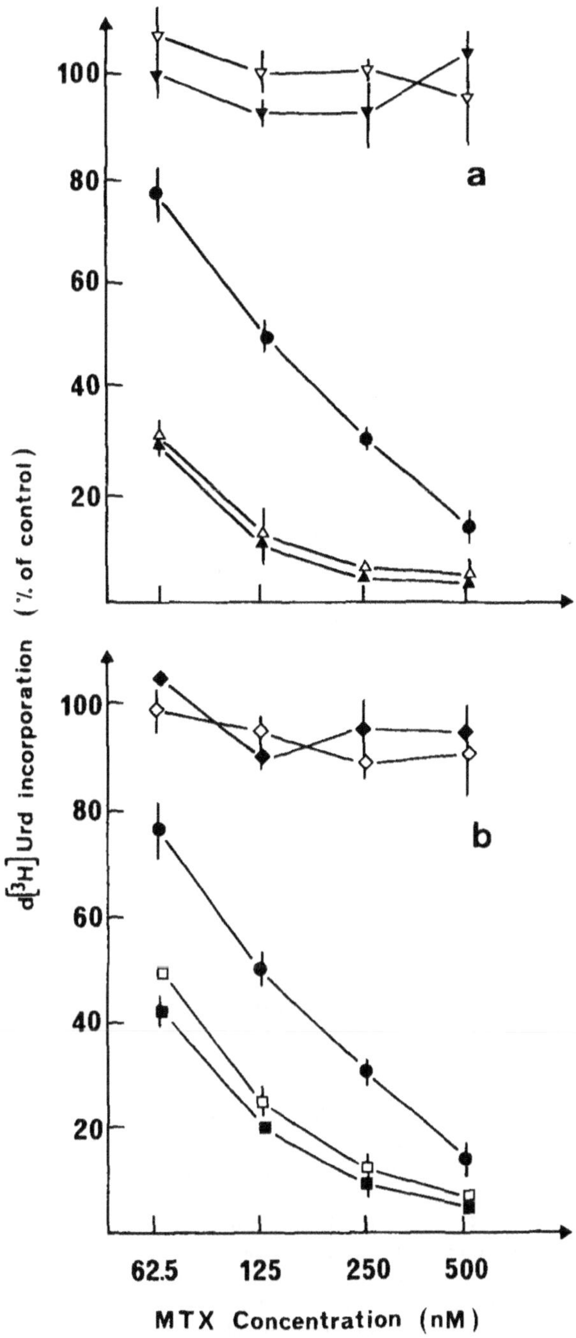

FIGURE 5. MTX-mediated inhibition of d(^{3}H)Urd incorporation. L cells (10^5) were preincubated with anti-H-2K^k monoclonal antibody (H 100-30/33) before addition of LUV or SUV of the same lipid composition (DMPC, *DPPS, Chol, DPPE-DTP) coupled to protein A as described in Section II.* (●) *free MTX in solution. (a)* SUV-containing MTX (△) before and (▲) after freezing (± DMSO) and dialysis; (▽) before and (▼) after freezing (± DMSO) and dialysis when incubated without antibody-bearing cells. The same results were obtained with nondialyzed protein A-bearing SUV composed of DPPC, Chol, DPPE-DTP (data not shown). (b) LUV-containing MTX (□) before and (■) after freezing (+ DMSO) and dialysis; (◇) before and (◆) after freezing (+ DMSO) and dialysis without antibody-bearing cells. Control d(^{3}H)Urd incorporation: 120,000 ± 10,000 cpm. Bars represent *mean ± SD of four experiments.*

shown). Sonicated liposomes of slightly larger size (made with the charged phospholipid DPPS) were somewhat less resistant; approximately 10% of contents leaked upon freezing. This leakage could be entirely eliminated if liposomes were frozen in 10% DMSO, which did not by itself induce any leakage.

The behavior of MLV and LUV was different than that of SUV. Most MLV and LUV lost their contents when frozen in the absence of DMSO. The leakage caused by freezing could be eliminated by the addition of 10% DMSO, but this agent was by itself responsible for a significant amount of leakage from liposomes.

After leaked contents and DMSO were eliminated by dialysis or column chromatography, the remaining liposomes contained the initial CF concentration, as measured by the quenching value, which is a sensitive index of CF concentration (see Chapter 10, Volume III). Thus, liposomes do not "leak", at least with respect to these low-molecular-weight encapsulated materials. Rather they release all of their contents, or none at all.

Thus we can freeze, presumably indefinitely, and recover for use the majority of LUV and MLV with no detectable change, save the loss of a fraction of the liposomes, and SUV without either leakage or change in size or specificity. The ability to freeze successfully was correlated with those properties associated with stability in the nonfrozen state, namely, small size and the presence of cholesterol.

ACKNOWLEDGMENTS

Jacques Barbet suggested the syringe technique for making a stable lipid emulsion. This work was supported by Institutional grants from the Institut National de la Santé et de la Recherche Médicale and the Centre National de la Recherche Scientifique, by a contract under the program "Institut National de la Santé et de la Recherche Scientifique-National Cancer Institue Recherches Cliniques en Cancérologie", and by a grant from the Association pour le Développement de la Recherche Contre le Cancer.

REFERENCES

1. **Wong, M. and Thompson, T. E.,** Aggregation of dipalmitoyl phosphatidylcholine vesicles, *Biochemistry*, 21, 4133, 1982.
2. **Pick, U.,** Liposomes with a large trapping capacity prepared by freezing and thawing of sonicated phospholipid mixtures, *Arch. Biochem. Biophys.*, 212, 186, 1981.
3. **Yatvin, M. B., Weinstein, J. N., Dennis, W. H., and Blumenthal, R.,** Design of liposomes for enhanced local release of drugs by hyperthermia, *Science*, 202, 1290, 1978.
4. **Strauss, G. and Ingenito, E. P.,** Stabilization of liposome bilayers to freezing and thawing: effects of cryoprotective agents and membrane proteins, *Cryobiology*, 17, 508, 1980.
5. **Kirby, C., Clarke, J., and Gregoriadis, G.,** Effect of the cholesterol content of small unilamellar liposomes on their stability in vivo and in vitro, *Biochem. J.*, 186, 591, 1980.
6. **Wilschut, J.,** Preparation and properties of phospholipid vesicles, in *Méthodologie des Liposomes*, Leserman, L. D. and Barbet, J., Eds., Editions INSERM, Paris, 1982, 9.
7. **Ralston, E., Hjemeland, L. M., Klausner, R. D., Weinstein, J. N., and Blumenthal, R.,** Carboxyfluorescein as a probe for liposome-cell interactions. Effect of impurities, and purification of the dye, *Biochim. Biophys. Acta*, 649, 133, 1981.
8. **Weinstein, J. N., Yoshikami, S., Henkart, P., Blumenthal, R., and Hagins, W. A.,** Liposome-cell interaction: transfer and intracellular release of a trapped fluorescent marker, *Science*, 195, 489, 1977.
9. **Leserman, L. D., Barbet, J., Kourilsky, F., and Weinstein, J. N.,** Targeting to cells of fluorescent liposomes covalently coupled with monoclonal antibody or protein A, *Nature (London)*, 288, 602, 1980.
10. **Olson, F., Hunt, C. A., Szoka, F. C., Vail, W. J., and Papahadjopoulos, D.,** Preparation of liposomes of defined size distribution by extrusion through polycarbonate membranes, *Biochim. Biophys. Acta*, 557, 9, 1979.

11. **Szoka, F. and Papahadjopoulos, D.,** Procedure for preparation of liposomes with large internal aqueous space and high capture by reverse-phase evaporation, *Proc. Natl. Acad. Sci. U.S.A.*, 75, 4194, 1978.
12. **Nicolau, C.,** Transfert de DNA dans les cellules eukaryotes médié par les liposomes: introduction et expression de genes dans la cellule hôte, in *Méthodologie des Liposomes*, Leserman, L. D. and Barbet, J., Eds., Editions INSERM, Paris, 1982, 49.
13. **Mishell, B. B. and Shiigi, S. M., Eds.,** *Selected Methods in Cellular Immunology*, W. H. Freeman, San Francisco, 1980, 467.
14. **Pierres, M., Devaux, C., Dosseto, M., and Marchetto, S.,** Clonal analysis of B- and T-cell responses to Ia antigens. I. Topology of epitope regions on I-A^k and I-E^k molecules analyzed with 35 monoclonal alloantibodies, *Immunogenetics*, 14, 481, 1981.
15. **Hunter, W. R. and Greenwood, F. C.,** Preparation of iodine-131 labelled human growth hormone of high specific activity, *Nature (London)*, 194, 495, 1962.
16. **Nozaki, Y., Lasic, D. D., Tanford, C., and Reynolds, J. A.,** Size analysis of phospholipid vesicle preparations, *Science*, 217, 366, 1982.
17. **Goding, J. W.,** Use of staphylococcal protein A as an immunological reagent, *J. Immunol. Methods*, 20, 241, 1978.
18. **Barbet, J., Machy, P., and Leserman, L. D.,** Monoclonal antibody covalently coupled to liposomes: specific targeting to cells, *J. Supramol. Struct. Cell. Biochem.*, 16, 243, 1981.
19. **Machy, P., Barbet, J., and Leserman, L. D.,** Differential endocytosis of T and B lymphocyte surface molecules evaluated with antibody-bearing fluorescent liposomes, *Proc. Natl. Acad. Sci. U.S.A.*, 79, 4148, 1982.
20. **Lemke, H., Hämmerling, G. J., and Hämmerling, U.,** Fine specificity analysis with monoclonal antibodies of antigens controlled by the major histocompatibility complex and by the Qa/TL region in mice, *Immunol. Rev.*, 47, 175, 1979.
21. **Leserman, L. D., Weinstein, J. N., Blumenthal, R., and Terry, W. D.,** Receptor-mediated endocytosis of antibody-opsonized liposomes by tumor cells, *Proc. Natl. Acad. Sci. U.S.A.*, 77, 4089, 1980.

Chapter 17

STABILITY TESTING OF LIPOSOMES DURING STORAGE

Sven Frøkjaer, Ejnar L. Hjorth, and Ole Wørts

TABLE OF CONTENTS

I. INTRODUCTION

From a pharmaceutical point of view it is important to obtain information about the stability of liposomes during storage in order to be able to evaluate the potential use of liposomes as a carrier system over a certain period of time. Stability of a pharmaceutical product is defined as the capability of a formulation to remain within certain limits regarding its chemical, physical, and microbiological stability.[1] The initial step in designing a stability testing program is to define the relevant product parameters to be investigated.[2-4] The next and very important step is to select the analytical methods to be employed because evaluation of stability obviously depends on the specificity and sensitivity of the methods used.[5]

II. STABILITY TESTING PROGRAM

All stability testing programs consist of two sections: (1) product characterization after preparation and (2) product stability during storage. Before implementing a stability testing program the first step is to select the relevant product parameters to be evaluated looking at liposomes from a chemical, physical, and microbiological point of view. Liposomes are vesicles in the size range from 25 nm to 100 μm formed in general from amphiphilic substances like phospholipids dispersed in excess amount of water.[6,7] From a physicochemical point of view liposomes are heterogeneous systems with potential chemical and physical stability problems. Chemically, phospholipids are susceptible to hydrolysis.[8] Furthermore, phospholipids containing unsaturated fatty acids as well as cholesterol which is often used in liposome compositions can undergo oxidation.[9,10] Physically, liposomes are vesicles with a certain size distribution determined by the method used for preparation of liposomes. During storage changes in size can occur due to aggregation or fusion of vesicles.[11-13] Another physical stability problem is retention of entrapped substances due to the physicochemical changes in liposome preparations taking place during storage, which can lead to modified permeability properties of the liposome membrane.

Microbiologically, liposomes represent a major problem, when it is necessary to use sterile preparations, because it is not possible to use heat or radiation methods for sterilization of liposomes without accepting an accelerated degradation. The only mild method available is sterilization by filtration and since it is necessary to use filters with a pore size of 0.2 μm to ensure sterility, it puts limitations on the size range of liposomes available for different applications. When filtering liposomes at a temperature above phase transition temperature, phospolipid vesicles larger than 0.2 μm can be filtered because of the increased flexibility of the phospholipid membrane in its liquid crystalline phase.[14,15]

Another aspect to be taken into consideration before starting stability testing is to evaluate the potential influence of different formulation factors such as pH, ionic strength, buffer system, and solvent system.[4] The influence of one of these factors, pH, on the degradation of distearoyl phosphatidylcholine (DSPC) has been investigated. The initial degradation by hydrolysis of DSPC is the formation of lyso-DSPC, which takes place at both acid and alkaline pH. The pH-rate profile for the observed pseudo-first-order rate constant, k_{obs}, for the degradation of DSPC in aqueous solution at 70°C is shown in Figure 1. A stability optimum is observed at pH = 6.5 (ionic strength = 0.5).

Yet another aspect to consider is the environmental conditions which may influence stability of liposomes. Different physical factors like temperature, light, oxygen, and heavy metal ions may initiate chemical and physical reactions such as hydrolysis, oxidation, photochemical decomposition, and changes in size distribution of vesicles. One environmental factor that significantly affects stability is temperature. The effect of temperature on the rate of hydrolysis can be expressed by the Arrhenius equation:

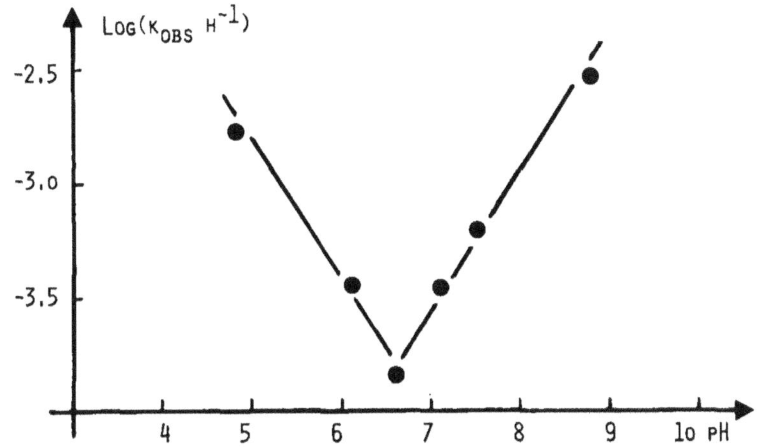

FIGURE 1. pH-rate profile for degradation of DSPC in aqueous solution at 70°C. (From Frøkjaer, S., Hjorth, E. L., and Wørts, O., *Optimization of Drug Delivery*, Alfred Benzon Symp. 17, Bundgaard, H., Bagger Hansen, A., and Kofod, H., Eds., Munksgaard, Copenhagen, 1982, 384. With permission.)

$$k = s \cdot e^{-E_A/RT} \quad (1)$$

or

$$\ln k = \frac{-E_A}{RT} + \ln s \quad (2)$$

where k is the rate constant, s is the frequency factor, E_A is the activation energy, R is the gas constant, and T is the absolute temperature.

By plotting ln k_{obs} as a function of the reciprocal value of T an Arrhenius plot of the pseudo-first-order degradation of DSPC at pH = 6.5 is shown in Figure 2. A break in the Arrhenius plot is observed around the phase transition temperature (T_c = 52°C) of DSPC showing that the degradation of DSPC proceeds relatively more rapidly in the gel phase than in the liquid crystalline phase above T_c. Another consequence of the break in the Arrhenius plot is that temperature-accelerated stability studies can only be performed up to a maximum temperature below the phase transition temperature, which means initial prediction of stability involves prolonged stability studies to sort out the influence of the different parameters on the stability of liposome preparations.

Oxidation is a chemical reaction of potential importance for the stability of liposome preparations. Oxidations are chain reactions which proceed under the influence of molecular oxygen. The oxidation of unsaturated phospholipids or cholesterol is initiated by formation of free radicals from the organic compounds by the action of heat, light, or heavy metal ions. The oxidative degradation can be monitored by a nonspecific method, oxidation index as described by Klein.[9]

In our experiments we have tried to prevent oxidation from occurring throughout the period of storage by using water with a low level of both oxygen and heavy metal ions and by having the liposome preparations under argon protected from light during preparation and storage. Argon is used because, in contrast to the more commonly used nitrogen, it is heavier than air. After having considered all product parameters, formulation factors, and environmental factors the stability testing program can be initiated.

On the basis of the previously mentioned considerations we have selected a number of product parameters to be analyzed in order to be able to characterize our liposome preparation. These are total phosphorus (method A), phospholipid composition (method B), lyso-PC

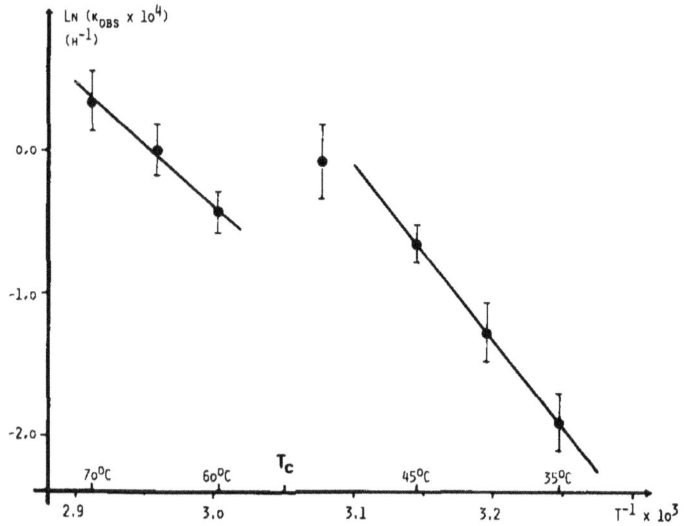

FIGURE 2. Arrhenius plot of pseudo-first-order rate constant for degradation of DSPC at pH 6.5. T_c: phase transition temperature for DSPC. (From Frøkjaer, S., Hjorth, E. L., and Wørts, O., *Optimization of Drug Delivery*, Alfred Benzon Symp. 17, Bundgaard, H., Bagger Hansen, A., and Kofod, H., Eds., Munksgaard, Copenhagen, 1982, 384. With permission.)

(method C), cholesterol (method D), vesicles size (method E), and release of entrapped material. Product stability is evaluated by analyzing the following product parameters: degradation products (method B), lyso-PC (method C), vesicle size (method E), and release of entrapped material.

III. ANALYTICAL METHODS

A. Total Phosphorus

Total phosphorus is determined by a modified procedure of McClare[16] where the ascorbic acid concentration is increased to well above the concentration needed for maximum color development.[17]

Reagent — 1.0 g ammonium molybdate and 8.0 mℓ of perchloric acid, 70% in 60 mℓ of water (solution A). 1.0 g ascorbic acid is dissolved in 40 mℓ of water and added to solution A whereupon the solution becomes pale yellow. The reagent is prepared daily.

Standard phosphate solution — Na_2HPO_4 dried to constant weight at 110°C is used as inorganic standard. 35.149 mg Na_2HPO_4 in 100 mℓ water; this solution gives a working solution containing 76.74 μg P per milliliter.

Procedure — The sample containing 0.5 to 10 μg P is pipetted into a Pyrex (no 24/1/6) stoppered tube (15 mℓ). The tubes are only used for phosphate analysis. The sample is evaporated to complete dryness at 100°C. Then, 0.5 mℓ perchloric acid (70%) is added and the sample digested by refluxing for 30 min in a thermostated preheated block (45 holes metal block) at 210 ± 5°C. The dimensions of the tubes compared to the holes in the block ensure that perchloric acid refluxes without reaching the neck of the tubes. The digestion takes place in a fume chamber. After cooling, 9.5 mℓ of the reagent is added to the digested sample. A stopper is firmly placed in the tube and, after mixing, the tube is heated to 50°C in a thermostat for 1 hr. The blue color which is developed is stable for several hours. The absorbance of the cooled solution is measured against water at 825 nm (Pye Unicam SP 800 spectrophotometer). For calibration, 7 μg P standards and blanks are carried simultaneously through the procedure. The standard curve is linear for up to 10 μg phosphorus. Standard deviation is ± 1% at a sensitivity of about 7 μg phosphorus.

B. Phospholipids

Working principle — Phospholipids and degradation products are determined after being extracted from liposomes. The lipid mixture is separated in a two-dimensional thin-layer chromatographic (TLC) system. Spots corresponding to single constituents are removed and assayed for total phosphorus. The chromatographic method is described by Poorthuis et al.[18] with modifications as given by Spanner[19] in order to obtain a better separation of PS and lyso-PC by using an acidic medium in the second dimension.

Materials — The chromatographic chamber is an all-glass tank with ground glass cover (internal dimensions approximately 7 cm × 23 cm × 22 cm, d × w × h). The tank is lined by filter paper in order to aid the saturation of the chamber with solvent vapor. 1 to 2 hr before use 100 mℓ of solvent is added to each chamber and the lid replaced. The I_2-staining chamber is a tank as described above (without filter paper). 1 hr before use, approximately 15 g of crystalline iodine is placed at the bottom of the tank and the lid is replaced. Chromatographic plates are TLC plates silica gel 60 (without fluorescent indicator), precoated, layer thickness 0.25 mm and area of 20 cm × 20 cm for two-dimensional and 5 cm × 20 cm for one-dimensional TLC. Solvent systems used are for first dimension, $CHCl_3$:CH_3OH:H_2O:NH_3 (70:30:3:2, v/v/v/v) and for second dimension, $CHCl_3$:CH_3OH:CH_3COOH:H_2O (66:30:9:1,5, v/v/v/v).

Procedure — Liposome preparation corresponding to approximately 5 mg of phospholipid is adjusted to pH = 1.5 by addition of 1 *N* hydrochloric acid. Distilled water is added to a final volume of 1 mℓ. Lipids are then extracted by vortexing with 1 mℓ of chloroform:methanol, 2:1. After centrifugation the organic phase is withdrawn. Using a Hamilton syringe, a volume of lipid extract corresponding to 0.5 to 1 mg of phospholipid is applied to the TLC-plate as a single, narrow spot in the lower left corner 2 cm from each border. The plate is eluted 10 cm in each direction in the solvent systems specified above. Between the first and second elution, the plate must be thoroughly dried in a stream of air for 1 hr. After elution, the plate is dried in a stream of air and placed in the I_2-chamber. After 5 to 15 min, the lipids will be visible as yellowish-brown spots on a yellowish-white background. The spots are marked with a pencil, and silica gel in each spot is scraped off with a razor blade and transferred quantitatively to 10-mℓ test tubes. In principle, total phosphorous can be determined directly on the scraped off silica gel. However, we have noticed that different batches of plates may contain variable amounts of phosphate which will compromise the determination of phosphorous in spots with low contents of phospholipids. It is therefore advisable to extract lipids from the gel prior to the determination of phosphorous. The scraped off silica gel is extracted in turn with the following media: (1) 3 mℓ $CHCl_3$:CH_3OH:H_2O, (65:25:4), (2) 2 mℓ $CHCl_3$:CH_3OH:H_2O, (65:25:4), (3) 2 mℓ CH_3OH, (4) 2 mℓ CH_3OH:CH_3COOH:H_2O, (94:1:5). After each extraction, the silica gel is isolated by centrifugation. The pooled extracts are transferred to a 10 mℓ measuring flask and adjusted to volume by addition of $CHCl_3$:CH_3OH:H_2O (65:25:4). Depending upon phospholipid content in the various spots, an amount corresponding to $\leq$ 7 μg P is transferred to reagent tubes, and assayed for phosphorous as described in method A.

C. Lyso-PC

As shown in Figure 3, the PC molecule contains a total of four ester bonds where hydrolysis can occur. Consequently, hydrolysis may result in numerous degradation products which can be monitored simultaneously only with difficulty. However, we and others[20] have found that one of the main degradation products in the initial step of the chemical degradation is the formation of lyso-PC through hydrolysis of the ester bond at C_2 in the glycerol moiety.

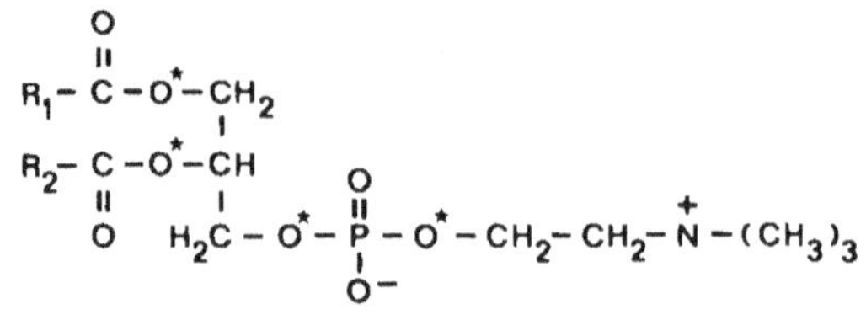

FIGURE 3. Structure of PC. R_1 and R_2 are hydrocarbon moieties of fatty acids.

The formation of lyso-compounds has therefore been chosen as a measure for the chemical stability of phospholipids.

Furthermore, the presence of lyso-PC in a lipid bilayer is known to induce changes in the physicochemical properties of the bilayer, for instance enhancement of the permeability of the lipid membrane.[21] It is therefore very important that the amount of lyso-PC formed during storage is kept at a sufficiently low level.

Working principle — Phospholipids are extracted from liposomes, and the lipid mixture is separated into PC and lyso-PC in a TLC system. Spots corresponding to phospholipid constituents are scraped off and assayed for total phosphorus content. All determinations are carried out in duplicate.

Materials — Materials are as described for method B except that only first dimension solvent system is used.

Procedure — Lipids are extracted from the liposome preparation as described in method B. Using a Hamilton syringe, 2 × 50 μℓ of lipid extracts (0.5 to 1 mg phospholipid) are applied as single, narrow spots on a TLC-plate, 2 cm from the bottom and 1.5 cm from the right and left borders, respectively. The plate is then placed in the chromatographic chamber. In order to obtain satisfactory separation, it is essential that the solvent front is allowed to elute as far as possible (approximately 12 cm). After elution, the plate is dried in a stream of air and placed in the iodine chamber. After 5 to 15 min, lipids will be visible as yellowish-brown spots on a yellowish-white background. Individual spots are marked, scraped off, extracted according to method B and assayed for phosphorus content as described in method A. Detection limits for lyso-DSPC and lyso-egg lecithin are ≃ 5 μg and ≃ 0.5 μg, respectively.

D. Cholesterol

Working principle — The method is based upon a commercial test combination which uses the so-called catalase method as described by Röschlau et al.[22] Cholesterol is determined as the sum of esterified and nonesterified cholesterol. In order to obtain better linearity of the standard curve, we extract cholesterol from liposomes and standards prior to analysis as described by Johnson.[23] All determinations are carried out in duplicate, and cholesterol content is calculated from a standard curve obtained from similarly treated cholesterol standards.

Materials — Test-combination cholesterol: Boehringer Mannheim GmbH, Cat no. 124079 (50 tests) or 124087 (180 tests). Preciset® cholesterol standards: Boehringer Mannheim GmbH, Cat no. 125512.

Procedure — The following description refers to the instruction sheet which accompanies the test combination. To 100 μℓ of liposome (0.5 to 3.0 mg Chol per milliliter) dispersion or standards is added 200 μℓ of a suitable buffer (e.g., 10 m*M* tris, 140 m*M* NaCl, pH = 7.4), 400 μℓ of methanol, and 800 μℓ of chloroform. Extraction of cholesterol is carried out for 15 min on an end-over-end mixer at maximum speed, after which the two phases are separated by centrifugation. 300 μℓ of the organic phase is evaporated to dryness, either in a rotating evaporator or overnight at room temperature. Each of the dried down residues is then redissolved in 200 μℓ of isopropanol. To each isopropanol solution is added 5.00 mℓ of "solution 5" which is prepared as specified in the instructions (Boehringer Mannheim GmbH, Cat no. 124079 or 124087). The final solution is "SAMPLE BLANK", and from this point the assay is carried out according to the instruction sheet. The standard curve is linear for up to 3.0 mg Chol per milliliter.

E. Vesicle Size

The average size and size distribution of liposomes are important parameters in respect to the biological fate of liposomes as shown by several authors.[24-27] A valid morphological characterization of liposomes can be obtained by freeze-fracture electron microscopy,[28] but as a routine method for stability testing under storage a detailed analysis by electron microscopy is regarded as too laborious. We have chosen dynamic light scattering (quasielastic laser light scattering) as a standard technique for monitoring changes in the size of liposomes, although this technique has some limitations when used on polydisperse samples of liposomes, where the autocorrelation function is a sum of exponentials rather than the single exponential obtained from a monodisperse system.[29-32]

Working principle — Dynamic light scattering is a technique for measuring the diffusion coefficient of macromolecules or colloidal particles by the analysis of the time dependence of intensity fluctuations in scattered light due to Brownian motion of the structures in solution. A detailed theoretical background is given in various excellent advanced books.[33,34] The time behavior of the fluctuations is described quantitatively by the intensity autocorrelation function

$$g^{(2)}(\tau) = \frac{\langle I(t)\ I(t+\tau)\rangle}{\langle I\rangle^2} \tag{3}$$

where I is the intensity of scattered light, and τ is the correlation delay time. For noninteracting systems of monodisperse structures which are small compared to the incident light wavelength, the field autocorrelation function is given by

$$/g^{(1)}(\tau)/ = \exp(-\Gamma\tau) \tag{4}$$

$\Gamma = DK^2$ where D in case of spheres is the translational diffusion coefficient, and K is the magnitude of the scattering vector.

$$K = \frac{4\pi}{\lambda} \cdot n \cdot \sin\frac{\theta}{2} \tag{5}$$

λ is the light wavelength, Θ the scattering angle, and n is the refractive index of the solution. For polydisperse systems $/g^{(1)}(\tau)/$ becomes a superposition of single exponentials

$$/g^{(1)}(\tau)/ = \int_0^\infty G(\Gamma) \exp(-\Gamma\tau)d\Gamma \tag{6}$$

where G (Γ) is a normalized distribution function. Provided the light has Gaussian statistics the intensity autocorrelation function which is measured by a correlator is related to the field autocorrelation function $g^{(1)}(\tau)$ by

$$g^{(2)}(\tau) = 1 + /g^{(1)}(\tau)/^2 \tag{7}$$

A commonly used method of analyzing dynamic light scattering data is the method of cumulants developed by Koppel.[29] This method expands the right-hand side of Equation 6 about the mean $\overline{\Gamma}$, in a power series:

$$\ln/g^{(1)}(\tau)/ = -\bar{\Gamma}\tau + \frac{1}{2!}\mu_2 \cdot \tau^2 - \frac{1}{3!} \cdot \mu_3 \cdot \tau^3 + \qquad (8)$$

The cumulants μ_2, μ_3, etc., are related to the moments of the distribution, but in the data analysis it is difficult to go reliably beyond μ_2 which is a measure of polydispersity of the system indicating the width of the distribution. Data analysis is then a quite straightforward procedure involving a least-squares polynomial fit (quadratic) of the measured $\ln/g^{(2)}(\tau) - 1/ = 2\ln/g^{(1)}(\tau)/$ as a function of τ. The number of points for this fitting procedure corresponds to the number of channels in the autocorrelator used. Once an estimate of Γ has been obtained the hydrodynamic radius r_h for spheres like liposomes can be obtained by the Stokes-Einstein relation

$$D = k_B\, T/(6\pi\eta r_h) \qquad (9)$$

where k_B is Boltzmann's constant, T the absolute temperature, and η is the solvent viscosity. In a polydisperse system the measured diffusion coefficient is the so-called Z-average diffusion coefficient

$$D_z = \frac{\Sigma N_i\, M_i^2\, D_i}{\Sigma N_i\, M_i^2} \qquad (10)$$

where N_i is the number of structures of species i having a molecular weight M_i. From Equation 10 it can be seen that D_z, and therefore the hydrodynamic radius, is very sensitive to a contamination of larger structures and/or a broad size distribution. From a drug carrier point of view this size-weighted radius may very well be more relevant as a meaningful pharmaceutical parameter than a number average radius. This is because the amount of entrapped drug is either surface or volume-weighted, depending on the actual location of the drug in the carrier.

Apparatus — A Malvern Photon Correlation Spectrometer equipped with a 128 channel Malvern Autocorrelator K 7025 and a Hughes 10 mW He-Ne laser is used to measure the size of liposomes by dynamic light scattering. Figure 4 shows a block diagram of the equipment. The incident beam from the He-Ne laser (λ = 6328 Å) is focused into the sample cell. The sample cells, 1-mℓ glass ampules for single use, are placed in a vat filled with a clean index matching fluid (toluene) which is electrically thermostated (± 0.05°C) and mounted on a goniometer table. The light scattered from the sample is monitored using a low dark count photomultiplier (EMI 9863) mounted on the goniometer which has a vernier scale capable of measuring the scattering angle within ±0.2°. The scattering angle can be varied from 25 to 145°, but is normally set at 90°. The amplified output of the photomultiplier is fed directly into the autocorrelator interfaced with a Commerdore 2002, 32 K bytes computer for on-line data processing. The Malvern Autocorrelator is a multibit correlator with a variable delay time (τ) between 0.1 μsec and 1 sec. For liposomes with a Z-average diameter of about 100 nm a delay time setting at 5 to 10 μsec is used for routine experiments on the 128-channel correlator.

Procedure — In order to remove particulate matter which may interfere with the light-scattering measurements, all solutions used for sample preparation are clarified by filtration through a Millipore filter, Millex-GS 0.22 μm. The first 5 to 10 mℓ of the filtered solution are discarded. Throughout the preparation of samples great care should be taken to ensure the cleanliness of the equipment used and to reduce particle contamination by working in a clean bench. By these precautions it is possible to obtain solutions with negligible background scatter and without an autocorrelation curve when measured under the same setting conditions as used for measuring liposomes. Depending on the particle size and lipid concentration of

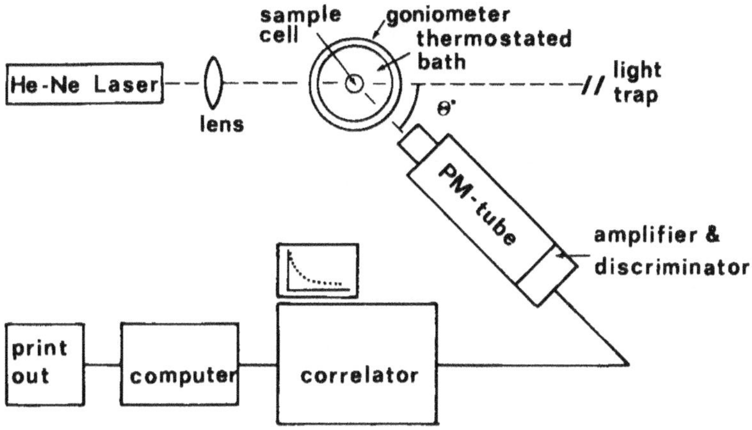

FIGURE 4. Block diagram of the light-scattering equipment.

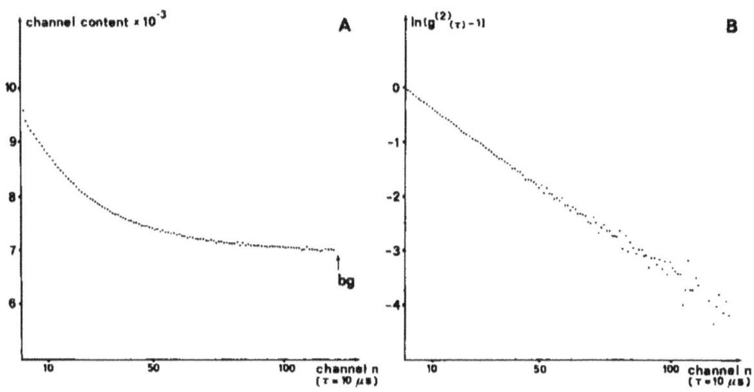

FIGURE 5. Graphic representation of data from a dynamic light-scattering experiment on liposomes (DSPC:Chol, molar ratio 2:1). Experimental conditions: phospholipid concentration = 0.2 μmol/mℓ, Temp.: 25°C, scattering angle = 90°, Delay time per channel = 10 μsec, total experimental duration: 10 sec. Results: Z-average diameter: 97 nm, polydispersity : 0.1. (A) The autocorrelation function including a background count (bg) with uncorrelated data, which must be subtracted before data analysis. (B) The normalized plot of the autocorrelation junction corresponding to A.

the actual liposome preparation it is normally necessary to dilute the preparation in order to obtain a suitable total scatter intensity from the sample. All measurements are made at least three times at 25°C and at a scattering angle of 90°. The SD of the measurements is about ±4% of the measured liposome diameter. An example of an autocorrelation function for a liposome preparation is given in Figure 5A and the corresponding normalized plot of ln/ $g^{(2)}(\tau)$ − 1/ as a function of τ (channel number) is given in Figure 5B. By evaluating data from a dynamic light-scattering experiment one should be aware that almost any correlation function can yield some form of result which does not necessarily have much relevance to the supposed measurement, for instance because of particle-particle interaction, rotational contribution, or foreign particles. It is therefore essential to ensure initially that the method in fact measures the parameter of interest by using a suitable alternative method which, in the case of assaying size of liposomes, could be electron microscopy or any other method for the quantitative analysis of the particle size distribution, for instance sedimentation field flow fractionation.[35]

IV. RELEASE OF ENTRAPPED MATERIAL

The ability of liposomes to retain the entrapped drug molecules within the structure in order to obtain a reasonable shelf life (i.e., minimum 1 year) is of course of great importance from a pharmaceutical point of view. However, it has not been the intention of this chapter to cover the various analytical procedures for monitoring drug release because adequate methods are very much dependent on the specific drug substance and its actual localization within the liposome structure. Although only a very few studies have been made on long-term storage of liposomes, it has been shown that retention of low molecular substances (e.g., sucrose)[36] as well as macromolecules (e.g., bovine serum albumin)[8] is possible for periods of time long enough as to fulfill the pharmaceutical demands.

REFERENCES

1. **Lintner, C. J.,** Stability of pharmaceutical products, in *Remington's Pharmaceutical Sciences,* 15th ed., Mack Publishing, Easton, Pa., 1975, chap. 81.
2. **Mollica, J. A., Ahuja, S., and Cohen, J.,** Stability of pharmaceuticals, *J. Pharm. Sci.,* 67, 443, 1978.
3. **Connors, K. A., Amidon, G. L., and Kennon, L.,** *Chemical Stability of Pharmaceuticals,* John Wiley & Sons, New York, 1979.
4. **Florence, A. T. and Attwood, D.,** *Physicochemical Principles of Pharmacy,* Macmillan Press, London, 1981, chap. 11.
5. **Grimm, W. and Schepky, G.,** Chemische untersuchungsmetoden, in *Stabilitätsprüfung in der Pharmazie,* Editio Cantor, Aulendorf, West Germany, 1980, chap. 2.
6. **Bangham, A. D.,** Development of the liposome concept, in *Liposomes in Biological Systems,* Gregoriadis, G. and Allison, A. C., Eds., John Wiley & Sons, Chichester, U.K., 1980, chap. 1.
7. **Tanford, Ch.,** Biological lipids, in *The Hydrophobic Effect,* 2nd ed., John Wiley & Sons, New York, 1980, chap. 11.
8. **Frøkjaer, S., Hjorth, E. L., and Wørts, O.,** Stability and storage of liposomes, in *Optimization of Drug Delivery,* Alfred Benzon Symp. 17, Bundgaard, H., Bagger Hansen, A., and Kofod, H., Eds., Munksgaard, Copenhagen, 1982, 384.
9. **Klein, R. A.,** The detection of oxidation in liposome preparations, *Biochim. Biophys. Acta,* 210, 486, 1970.
10. **Weiner, N. D., Bruning, W. C., and Felmeister, A.,** Influence of lecithins on autoxidation of cholesterol from aqueous dispersions at 85°C, *J. Pharm. Sci.,* 62, 1202, 1973.
11. **Prestegard, J. H. and Fellmeth, B.,** Fusion of dimyr istoyllecithin vesicles as studied by proton magnetic resonance spectroscopy, *Biochemistry,* 13, 1122, 1974.
12. **Larrabee, A.,** Time-dependent changes in the size distribution of distearoylphosphatidyl choline vesicles, *Biochemistry,* 18, 3321, 1979.
13. **Schullery, S. E., Schmidt, C. F., Felgner, P., Tillack, T. W., and Thompson, T. E.,** Fusion of dipalmitoyl phosphatidylcholine vesicles, *Biochemistry,*19, 3919, 1980.
14. **Olson, F., Hunt, C. A., Szoka, F. C., Vail, W. J., and Papahadjopoulos, D.,** Preparation of liposomes of defined size distribution by extrusion through polycarbonate membranes, *Biochim. Biophys. Acta,* 557, 9, 1979.
15. **Szoka, F. and Papahadjopoulos, D.,** Procedure for preparation of liposomes with large internal aqueous space and high capture by reverse-phase evaporation, *Proc. Natl. Acad. Sci. U.S.A.,* 75, 4194, 1978.
16. **McClare, C. W. F.,** An accurate and convenient organic phosphorus assay, *Anal. Biochem.,* 39, 527, 1971.
17. **Chen, P. S., Jr., Toribara, T. Y., and Warner, H.,** Microdetermination of phosphorus, *Anal. Chem.,* 28, 1756, 1956.
18. **Poorthuis, B. J. H. M., Yazaki, P. J., and Hostetler, K. Y.,** An improved two dimensional thin-layer chromatography system for the separation of phosphatidyl-glycerol and its derivatives, *J. Lipid Res.,* 17, 433, 1976.
19. **Spanner, S.,** Separation and analysis of phospholipids, in *Form and Function of Phospholipids,* Ansell, G. B., Hawthorne, J. N., and Dawson, R. M. C., Eds., Elsevier, Amsterdam, 1973, chap. 3.
20. **Hauser, H. O.,** The effect of ultrasonic irradiation on the chemical structure of egg lecithin, *Biochem. Biophys. Res. Commun.,* 45, 1049, 1971.

21. **Van Echteld, C. J. A., De Kruijff, B., Mandersloot, J. G., and De Gier, J.,** Effects of lysophosphatidylcholines on phosphatidylcholine and phospatidylcholine/cholesterol liposome systems as revealed by ^{31}P-NMR electron microscopy and permeability studies, *Biochim. Biophys. Acta,* 649, 211, 1981.
22. **Röschlau, P., Bernt, E., and Gruber, W.,** Enzymatische Bestimmung des Gesamtcholesterius im Serum, *Z. Klin. Chem. Klin. Biochem.,* 12, 403, 1974.
23. **Johnson, S. M.,** A new specific cholesterol assay gives reduced cholesterol/phospholipid molar ratios in cell membranes, *Anal. Biochem.,* 95, 344, 1979.
24. **Juliano, R. L. and Stamp, D.,** Effect of particle size and charge on the clearance rate of liposomes and liposome encapsulated drugs, *Biochem. Biophys. Res. Commun.,* 63, 651, 1975.
25. **Kao, Y. J. and Juliano, R. L.,** Interaction of liposomes with the reticuloendothelial system, *Biochim. Biophys. Acta,* 677, 453, 1981.
26. **Abra, R. M. and Hunt, C. A.,** Liposome disposition *in vivo.* III. Dose and vesicle-size effects, *Biochim. Biophys. Acta,* 666, 493, 1981.
27. **Ellens, H., Mayhew, E., and Rustum, Y. M.,** Reversible depression of the reticuloendothelial system by liposomes, *Biochim. Biophys. Acta,* 714, 479, 1982.
28. **Guiot, P., Baudhuin, P., and Gotfredsen, C.,** Morphological characterization of liposome suspensions by stereological analysis of freeze-fracture replicas from spray-frozen samples, *J. Microsc.,* 120, 159, 1980.
29. **Koppel, D. E.,** Analysis of macromolecular polydispersity in intensity correlation spectroscopy: the method of cumulants, *J. Chem. Phys.,* 57, 4814, 1972.
30. **Brown, J. C., Pursey, P. N., and Dietz, R.,** Photon correlation study of polydisperse samples of polystyrene in cyclohexan, *J. Chem. Phys.,* 62, 1136, 1975.
31. **Selser, J. C., Yeh, Y., and Baskin, R. J.,** A light-scattering characterization of membrane vesicles, *Biophys. J.,* 16, 337, 1976.
32. **Bezot, P., Ostrowsky, N., and Hesse-Bezot, C.,** Light scattering data analysis for samples with large polydispersities, *Opt. Commun.,* 25, 14, 1978.
33. **Chu, B.,** *Laser Light Scattering,* Academic Press, New York, 1974.
34. **Cummins, H. Z. and Pike, E. R.,** *Photon Correlation and Light Beating Spectroscopy,* Plenum Press, New York, 1974.
35. **Kirkland, J. J., Yau, W. W., and Szoka, F. C.,** Sedimentation field flow fractionation of liposomes, *Science,* 215, 296, 1982.
36. **Hunt, C. A. and Tsang, S.,** α-Tocopherol retards autoxidation and prolongs the shelf-life of liposomes, *Int. J. Pharmaceut.,* 8, 101, 1981.

Chapter 18

PREPARATION OF LIPOSOMES ON THE INDUSTRIAL SCALE: PROBLEMS AND PERSPECTIVES

L. S. Rao

TABLE OF CONTENTS

I. INTRODUCTION

It is evident that a wide variety of techniques has been developed and adapted to suit the individual requirements in terms of size, width of size distribution, electrical charge, number of lamellae, and lipid composition of the liposomes. Practical applications of liposomes have been explored in three main areas of research, namely in model membrane studies, in controlled and targeted drug delivery in vivo, and in transfer of genetic and other material into cells in culture. Of these, the drug delivery concept, including the immunological adjuvant potential of liposomes, has attracted a predominant share of the interest. Very promising results have been obtained so far in specific applications in experimental therapy related to the reticuloendothelial system as in the accumulation of toxic metals in the liver[1] and in visceral leishmaniasis.[2-4] Successful use of macrophage activating factors in liposomes for the activation of alveolar macrophages to a tumoricidal state for the eradication of lung metastases in laboratory animals has also been reported.[5] Therapeutic benefits have also been reported in cancer chemotherapy by utilizing the liposome carrier with cytotoxic drugs in laboratory animals by several workers.

Whereas the use of liposomes as a model for biological membranes is confined to the research laboratory, the successful application of liposomes for drug delivery and/or genetic engineering will depend not only on a demonstration of the superiority of the liposome carrier for the intended purpose, but also upon the technical and economic feasibility of the system in practice. One of the fundamental questions will be whether it is possible to prepare, isolate, and characterize the particular liposomal system on an industrial scale with clearly defined and reproducible properties. Furthermore, the physical and chemical stability of the liposomes on storage is of vital importance.[6] An attempt is made in this chapter to examine some widely used procedures in the light of these requirements, especially of a liposomal drug delivery system incorporating hydrophilic drugs.

II. FEASIBILITY AS A DRUG CARRIER

The translation of the successes at the laboratory level in the use of liposomes for drug delivery into viable products for clinical use is a task facing many problems. It is possible at the outset to identify some of the critical factors in this process:

1. The availability of the lipid raw materials of adequate and consistent degree of purity
2. The need to identify and develop processes for the preparation and isolation of liposomes on a large scale with high and consistent degrees of drug encapsulation
3. The need for a sterile product for parenteral use
4. Storage stability of the product, both in terms of the chemical stability of the lipids and the drug and the physical stability of the liposomes in regard to their size and leakage of the encapsulated drug
5. Rapid and meaningful methods for the estimation of the size and size distribution of the liposomes
6. The need for the cost of the product to be reasonable in relation to the therapeutic benefits

III. PURITY OF LIPIDS

A recent review[7] has dealt with the various lipid components of liposomes, their purity, and handling techniques. Although purification of naturally occurring phospholipids and other lipids to the required degree of purity is possible, there are distinct advantages in the use of synthetic components in an industrial operation for assurance on purity and constitution on the one hand and supply on the other. Synthetic phospholipids are expensive at present, but the possibility of large-scale usage might make them more economical.

IV. PREPARATION OF LIPOSOMES

When considering the preparative methodology of liposomes, it has to be recognized that different applications require different types of liposomes. Variation of the lipid composition and electrical charge of the liposomes was shown to alter the in vivo distribution and clearance rates of otherwise similar liposomes.[8,9] It is possible to distinguish two main types of liposomes, namely, large uni- or multilamellar liposomes, which are cleared rapidly from the circulation after i.v. administration and small unilamellar liposomes, which remain in circulation for long periods of time. These two main types also represent the potentials for targeting drugs to the reticuloendothelial cells and for sustained drug release in the blood stream or possible targeting to other tissues and organs, respectively.

Szoka and Papahadjopoulos[10] have recently reviewed the currently available methods for the preparation of various kinds of liposomes. In addition, the contributions to the present volume should elaborate on the basic methods and their variations. The following discussion will, therefore, omit detailed description of the methods, and, instead, concentrate on the technogical implications of the general procedures.

A. Large Liposomes

1. Multilamellar Vesicles (MLV)

Multilamellar liposomes were first described by Bangham et al.[11] The procedure used by them still remains the simplest and the most widely used. This involves the deposition of a thin lipid film from an organic solvent medium on the walls of a container, followed by agitation with an aqueous solution of the material to be encapsulated. Provided the agitation is carried out at temperatures above the characteristic gel-liquid crystal phase transition temperature of the phospholipid and the lipid film is indeed thin, MLV should form spontaneously with little effort. However, the degree of agitation required to form a good dispersion of MLV (without poorly hydrated lipid globules) will depend upon the thickness of the lipid film. With relatively thicker films produced by more concentrated lipid solutions and/or smaller evaporation vessels, more vigorous agitation is required. Thus various publications describe types of agitation ranging from gentle hand-shaking through vortex mixing with or without glass beads, to gentle sonication.

This need for the generation of a thin lipid film by evaporation procedures makes this technique difficult to operate on a large scale. An ICI patent[12] presents an alternative means of producing an easily hydratable lipid matrix by freeze-drying solutions of lipids in organic solvents such as *t*-butanol. The physical structure of the lipid in the lyophilized mass appears to be conducive to easy and rapid hydration by aqueous media.[13] An additional drawback with the MLV is the low degree of drug encapsulation reported by many authors. Furthermore, this low encapsulation degree seems to be variable from batch to batch under otherwise similar conditions.[13,14] However, these disadvantages of low and inconsistent drug encapsulation may not be critical for lipophilic materials, which can be incorporated into the bilayer phase of the liposomes.[15] The MLV appear to be attractive because of the ease of their preparation and relative stability on storage and in presence of blood components.[16] However, the disadvantages include low encapsulation efficiency, inconsistency of properties, and technical problems in scale-up.

2. Large Uni- or Oligolamellar Vesicles

Relatively large liposomes with either one or a few lamellae, as for example the reverse-phase-evaporation vesicles (REV), occupy an intermediate position in regard to their clearance rate from the circulation between the MLV and small unilamellar vesicles (SUV). Their clearance rate, however, appears to be more suited for drug delivery to the reticuloendothelial cells than for prolonged drug release in the circulation. Therefore, these types of liposomes are included in this section.

a. Ether Infusion

Ether infusion techniques[17,18] have been described for the preparation of large unilamellar liposomes. The basic principle of the injection of a solution of lipids in ether into an aqueous solution of the material to be encapsulated at a high enough temperature for the rapid evaporation of the solvent appears suitable for scale-up. Large molecules such as proteins and DNA have been encapsulated into liposomes using this technique. However, the encapsulation efficiency is very low (however, see Chapter 2, Volume I), although relatively large aqueous space volumes have been reported.[19] Additionally, the size distribution of the liposomes is heterogeneous.

b. Calcium-Induced Fusion

This method, developed by Papahadjopoulos et al.[20] relies on the effect of calcium ions on acidic phospholipids to form initially cochleate cylindrical structures, which are then transformed into large unilamellar vesicles (LUV) by the addition of the chelating agent EDTA. Although the preparative procedure and efficiency of encapsulation are attractive,[10] the presence of calcium ions and the sequestrant during the formation of the liposomes and consequent contamination of the liposomes with these may be unacceptable. Furthermore, this rather specialized technique for acidic lipids has not been explored sufficiently in drug delivery applications.

c. Reverse-Phase-Evaporation (REV)

This method was introduced by Szoka and Papahadjopoulos[21] for the preparation of large uni- or oligolamellar liposomes incorporating a variety of drugs and biologically active materials. The efficiency of drug encapsulation, generally around 50%, is very good. The method appears suitable for use with a variety of lipids and lipid mixtures and for the encapsulation of even large macromolecular materials. The aqueous space in the liposomes at 7 to 11 ℓ/mol is also very high in comparison with that in MLV or SUV.[21] Liposomal dispersions prepared in this manner are said to be mixtures of large unilamellar and oligolamellar liposomes with the size and number of lamellae being dependent upon the nature of the lipid mixture, and possibly the exact procedure.[10,22]

The advantages of this procedure are obvious. The relatively high encapsulation efficiency for a variety of compounds and the basic principle of going from an emulsion to a liposomal dispersion are very attractive from scale-up and processing considerations. Greater reproducibility of liposome size range and drug encapsulation can be expected due to the absence of the "dry lipid" stage. One of the disadvantages of this method is the need for sonication of the mixture of the lipid solution and the aqueous drug solution to form an "adequate" emulsion. This step will be difficult to operate on a large scale in a reproducible manner. Although Szoka and Papahadjopoulos[10] stress the need for sonication in order to ensure a homogeneous emulsion, and thus a primarily unilamellar liposomal dispersion, other emulsification techniques may be equally effective. Other drawbacks of this procedure include the presence of the organic solvent in contact with the material to be encapsulated and the temperature needed for the evaporation of the solvent under reduced pressure. These conditions may be too severe for some labile materials.

B. Small Unilamellar Liposomes

Small unilamellar liposomes are of great interest in drug delivery because of their potential for controlled drug release in the blood stream and for targeting to nonreticuloendothelial cells and tissues. The slow clearance of SUV from circulation after i.v. administration[8] offers these possibilities. SUV also have a potential for fusion with target cell membranes. Methods for the preparation of SUV are examined in the following sections.

1. Sonication

The most widely used procedure for the preparation of SUV consists of sonicating MLV using either a probe or a bath sonicator.[23,24] Electron microscopic techniques are used to demonstrate the unilamellar structure and to estimate the size, typically about 50 nm as the mean diameter. In general, adequate ultrasonic treatment leads to a homogeneous preparation in terms of size. One of the consequences of the small size of the SUV is the limited aqueous space inside the liposomes and the low encapsulation efficiency.[8] Even more importantly, application of ultrasonic energy as a means of converting MLV to SUV is unlikely to be a reliable and consistent operation in large-scale manufacture.

2. Ethanol Injection

This procedure, introduced by Batzri and Korn,[25] is reported to produce relatively small unilamellar liposomes. Although this technique is, in principle, similar to the ether injection method,[18] the size and degree of heterogeneity of the liposomes are dependent upon the concentration of the lipids and the relative volumes of the alcoholic and the aqueous phases. The basic procedure is amenable to scale-up and needs to be examined further. Two major drawbacks of this method are the low drug encapsulation levels and the yield of liposomes as a very dilute dispersion at the end of the process. An additional concentration step is required to produce an adequate liposomal dispersion.

3. Detergent Dialysis

The detergent dialysis procedure was developed by Milsmann et al.[26] In this method, mixed micelles are first formed by dispersing a phospholipid and a surfactant in an aqueous solution of the material to be encapsulated. This dispersion of mixed micelles is then dialyzed extensively, when the surfactant is gradually removed and the phospholipid assumes a unilamellar liposome structure. Homogeneous dispersions of SUV are formed by this technique. An additional advantage is the absence of an evaporation step and elevated temperatures. This allows encapsulation of labile materials under mild conditions.

The degree of encapsulation is, however, very low because of the loss of material during dialysis (while the liposomes are being formed), except in the case of encapsulation of macromolecular compounds. Another disadvantage of this method is the presence of considerable amounts of residual surfactant in the final liposomal dispersion at the end of dialysis.

4. Reverse Micelle Method

The reverse micelle method was described in a patent by the Battelle Memorial Institute.[27] In this method, a small quantity of an aqueous solution of the material to be encapsulated is initially dispersed in a volatile organic solvent containing the lipid mixture. The inverted micellar dispersion is then redispersed in an aqueous buffer medium to form a water in oil in water type emulsion. On evaporation of the solvent under reduced pressure or by means of a stream of nitrogen, unilamellar liposomes are obtained.

Very high degrees of drug encapsulation, in terms of the encapsulated drug as a percentage of the total drug used, are claimed. However, it should be pointed out that the drug solution only forms the interior aqueous phase of the emulsion. Consequently, the final liposomal dispersion is dilute with respect to the encapsulated drug, although the apparent degree of encapsulation is high. This technique has potential for large-scale application and should be explored further.

V. REMOVAL OF UNBOUND DRUG

When lipophilic drugs of appropriate structure are associated with liposomes by inclusion in the bilayer phase, the degree of "encapsulation" is dependent upon the saturation of the lipid phase. Under these circumstances, it is possible to achieve degrees of encapsulation of over 90%, making it unnecessary to remove the unbound drug. However, the encapsulated drug is only a fraction of the total drug used in the case of water-soluble drugs. An additional process step is required to remove the unbound drug from the drug-loaded liposomes in dispersion. Dialysis, centrifugation, and gel filtration have been used for this purpose.

A. Dialysis

Dialysis is probably the simplest and most widely used procedure for the removal of the unbound drug, except when macromolecular compounds are involved. It is a technique requiring no complicated or expensive equipment and is capable of being scaled up. Dialysis is effective in removing nearly all of the free drug with a sufficient number of changes of the dialyzing medium. Dialysis is, however, a slow process. Typically, removal of over 95% of the free drug in a liposomal dispersion might require a minimum of 3 changes of the external medium over 10 to 24 hr at room temperature. Additionally, the volume of the liposomal dispersion will alter during dialysis, unless care is taken to balance the osmotic strengths of the liposomal dispersion and the dialyzing medium. It is also possible that the presence of the external dialyzing medium in equilibrium with the liposomal dispersion might induce leakage of the encapsulated drug.

B. Centrifugation

Centrifugation at various g values is an effective means of isolating various kinds of liposomes from the free drug in the suspending medium. Two or more resuspension and centrifugation steps are usually included to effect a complete removal of the free drug. The centrifugal force required to pull liposomes down into a pellet is dependent upon the size of the liposomes, and to a certain extent, on the flocculation state of the dispersion. High g values and conditions of refrigeration are required for liposomes in the small to medium size ranges. On the other hand, low speeds of around 2000 to 4000 r/min may be adequate to spin down medium to large liposomes.

Clearly, the use of refrigerated centrifuges operating at high speeds with large volumes of liposomal dispersions is energy intensive and expensive, even if it were feasible. Therefore, this method may not be suitable for the isolation of small liposomes. For relatively large liposomes, low speed centrifugation offers the advantages of a short time scale of operation and an opportunity to concentrate or dilute the original dispersion to the required extent. It is essential to ensure that the osmotic strength of the resuspending medium is matched with that of original liposomal dispersion in order to avoid osmotic shock and rupture of liposomes.

C. Gel Filtration

Gel permeation chromatographic technique is used extensively both to separate liposomes from unbound drug and also to fractionate heterogeneous liposomal dispersions. The technique is very effective and rapid at the laboratory level. Although gel filtration is used in the purification of biological materials such as insulin on a large scale, the technique is difficult and expensive. Additionally, dilution of the liposomal dispersion with the eluting medium may necessitate another concentration step. Much remains to be established in regard to recovery levels and lipid losses on the column materials.

VI. STABILITY

Clearly any liposomal formulation must have adequate stability over the time period between its preparation and ultimate use. Stability in this context includes the chemical stability of the component materials (lipids and drugs) and the physical stability of the liposomes in terms of the integrity of the encapsulated material and the size parameters. Thus a liposomal drug product, for instance, must show an adequate shelf life under normal conditions of storage. In addition, the liposomes should maintain their integrity in vivo until incorporation into the target tissue or until their sustained drug release function is completed.

The stability of various types of liposomes in presence of blood components has been reviewed.[8,28] The chemical nature of the phospholipid, the proportion of cholesterol, and the size (and number of lamellae) of the liposomes all appear to influence their in vivo stability. It is, therefore, necessary to design and test liposomal systems for adequate stability in vivo. Small unilamellar liposomes for prolonged residence in the circulation will clearly require greater stability than will a multilamellar liposomal dispersion for targeting to the reticuloendothelial cells.

The in vitro storage stability of drug-loaded liposomes is of great importance, if eventual commercial development of a liposomal product for clinical use is to occur. The following discussion will, therefore, concentrate on the storage and presentation aspects of liposomal drug formulations.

A. Chemical Stability

The chemical stability of the drug component of liposomes is outside the scope of this discussion. The lipid components of a liposomal system are subject to various degradative processes on storage. Hunt and Tsang[29] showed that the oxidative decomposition of egg PC in liposomes can be retarded drastically by addition of α-tocopherol (α-T) to the lipid mixture. Particular care is required in the purification and use of unsaturated phospholipids subject to this oxidative degradation.

Both unsaturated and saturated phospholipids are subject to hydrolysis in aqueous media, resulting initially in the formation of the corresponding lysophospholipid and fatty acid. Frøkjaer et al.[6] reported that the hydrolytic degradation of distearoyl phosphatidylcholine (DSPC) was rapid at both acidic and alkaline pH values. The rate of hydrolysis was minimum at a pH of 6.5. This degradative reaction has implications both for the integrity of the liposomes and consequently for the leakage of encapsulated drugs and for the physical quality and toxicity of the liposomes. The extent to which this hydrolytic degradation and the consequent formation of the lysophospholipid and fatty acid can be accommodated by the liposomal bilayer phase without obvious effects remains to be established.

B. Physical Stability

Liposomes can become physically unstable on storage either due to leakage of the encapsulated drug into the suspending medium and/or due to aggregation or fusion of the liposomes to form larger entities. It is particularly important that a liposomal drug product remains stable over a reasonable storage period, preferably under ambient conditions of storage. Frøkjaer et al.[6] reported that the increase in size and drug leakage on storage of small unilamellar liposomes were dependent upon the nature both of the lipid components and of the encapsulated marker compounds. Furthermore, the continuous and rapid leakage of a variety of encapsulated materials from liposomes on storage is well documented in the literature. Drug retention behavior of a range of liposomal compositions under various conditions of storage was cited by Szoka and Papahadjopoulos.[7] It is commonly accepted that the size stability of multilamellar liposomes is greater than that of small unilamellar liposomes.

It is possible to conceive of a liposomal drug carrier system for storage either as a normal dispersion or in a form to be reconstituted before use. If the problem of drug leakage can be resolved, presentation as a ready to use dispersion has obvious advantages from convenience and cost considerations. A reconstitutable form will, on the other hand, avoid both chemical and physical stability problems in storage. Work on reconstitutable forms has been limited to multilamellar liposomes. It has not been possible, as far as can be ascertained, to convert a unilamellar liposomal dispersion into a reconstitutable form, which on eventual reconstitution yields a unilamellar system again.

C. Reconstitutable Forms

A liposomal drug product for reconstitution before use can be in one of three possible forms: dry lipid film with or without the drug, lyophilized lipid mixture with or without the drug, or lyophilized drug-loaded liposomes.

With a dry lipid film or lyophilized lipid mixture, the degree of drug encapsulation achievable on reconstitution is a critical parameter. As discussed earlier, the degree of encapsulation of hydrophilic drugs is usually not very high. Therefore, the reconstituted dispersion may contain a variable and high free drug content. This situation may be unacceptable in many cases. On the other hand, when lipophilic drugs of appropriate structure are involved, the reconstituted product may be satisfactory in this respect.[30] However, the type and degree of agitation required during the reconstitution process, e.g., ultrasonic treatment at above the ambient temperatures,[30] may be inconvenient, if not impractical, in clinical use.

For hydrophilic drugs, ''lyophilized drug-loaded liposomes'' are an attractive option of a reconstitutable form. Schneider and Lamy[31] reported that the addition of hydrophilic polymeric materials such as dextran to a liposomal dispersion prior to lyophilization produced a free-flowing powder. This product reconstituted readily in aqueous media. When insulin-containing liposomes were treated in this manner, the reconstituted liposomes were found to contain about 70% of the original insulin content in the encapsulated state. Thus, leakage of the encapsulated insulin during the lyophilization and reconstitution processes was about 30%. It is possible that, when drug compounds of lower molecular weight are involved, this leakage may be considerably higher. However, this technique offers a potentially useful means of achieving storage stability for a liposomal formulation and should be examined in greater detail.

VII. METHODS OF CHARACTERIZATION

Methods of characterization of liposomes immediately after preparation and following storage are required for adequate quality control of the product. These methods have to be meaningful and rapid in the context of their use in an industrial environment. Of particular importance will be the size and size distribution of the liposomes.

Electron microscopic methods are widely used in research work for establishing the lamellar composition and estimating the size of liposomes. In addition to the technical questions regarding the effect of sample preparation procedures on the liposome integrity and the uncertainties in the sizing techniques, it is rarely recognized that a statistically meaningful analysis would require examination and counting of several thousands of liposomes from each batch. Such a task is clearly difficult and too labor intensive to form a component of a quality control exercise.

In practice, a single parameter such as a mean diameter may be sufficient, if this were validated by at least two independent methods of measurement. Methods such as laser light scattering, light obscuration (Hiac), and resistive pulse (Coulter) which measure the liposome size in the aqueous dispersion state are more appropriate than electron microscopy. Dynamic laser light scattering can size even small liposomes down to 30 nm, although it can only yield an average diameter value with an indication of the polydispersity.[32] When relatively large liposomes are involved, light obscuration and resistive pulse techniques can also provide quantitative size distribution data.

VIII. SUMMARY AND PERSPECTIVES

Various concepts and procedures in current use for the preparation of drug-encapsulating liposomes have been examined for suitability for development to industrial scale. Clearly, the needs and restraints of large-scale preparation and processing are quite different from those at the research laboratory level.

Medium to large liposomes (MLV and LUV) are cleared rapidly from circulation after i.v. administration and targeted (passively) to the reticuloendothelial cells. For the preparation of such liposomes on a large scale in a reproducible manner, the widely used thin lipid film method[11] does not appear to be suitable. Ether infusion[18] procedure has the drawback of low drug encapsulation. The REV procedure[21] offers the advantages of high efficiency of drug encapsulation and amenability to scale-up. This method may not be suitable for use with organic solvent- or heat-sensitive materials.

Small unilamellar liposomes offer the potential for sustained drug release in the blood stream and possible targeting to tissues other than the reticuloendothelial cells. The preparation of SUV is much more complicated than that of large liposomes. The extended ultrasonic treatment to convert MLV to SUV, a widely used technique,[24] will be difficult to operate on a large scale in a reproducible manner. Ethanol injection method[25] suffers from the disadvantages of a low drug-encapsulation efficiency and a dilute product requiring a further concentration step. The reverse micelle procedure[27] appears to offer scale-up potential and a good chance of reproducibility, although the absolute amounts of encapsulated drug per unit mass of lipid appear to be low. The detergent dialysis procedure[26] leads to low drug-encapsulation levels and contamination of the product by residual surfactant.

For the removal of the unbound drug from the drug-loaded liposomes, dialysis may be the most suitable procedure with SUV. When larger liposomes are involved, dialysis or low speed centrifugation may be applicable. High speed centrifugation and gel permeation chromatographic procedures, widely used at the laboratory level, may prove difficult to scale-up and too expensive to operate.

Storage stability is an important prerequisite for eventual development and utilization of the liposome drug carrier concept. In this context, SUV will probably need to be stable in their original dispersion state, since a reconstitutable form of SUV seems unlikely. The stability of SUV was found to be dependent not only on the lipid composition, but also upon the nature of the encapsulated material.[6] Therefore, each liposomal formulation will need to be designed and tested for adequate storage stability in regard to size and encapsulated drug content.

For multilamellar liposomes, storage stability in the dispersion state will be a highly desirable attribute, but may not be critical. It is possible to provide a solid form of these liposomal compositions for reconstitution before use. With either a lipid film or a lyophilized lipid matrix, reconstitution can yield a satisfactory product, where encapsulation of lipophilic drugs of appropriate structure is involved.[12,30] On the other hand, lyophilization of the drug-loaded liposomes is a viable technique for use with hydrophilic drugs. Drug leakage during the lyophilization and reconstitution steps will need to be considered.[31]

An additional important factor in the successful development of liposomal drug formulations is the probable cost of the product in relation to the demonstrable therapeutic benefits. Selection of suitable procedures and optimization of drug-encapsulation level and the final drug-to-lipid ratio are vital for minimizing the cost of the product.

REFERENCES

1. **Rahman, Y. E.,** Liposomes and chelating agents, in *Liposomes in Biological Systems,* Gregoriadis, G. and Allison, A. C., Eds., John Wiley & Sons, Chichester, 1980, chap. 10.
2. **Black, C. D. V., Watson, G. J., and Ward, R. J.,** The use of Pentostam liposomes in the chemotherapy of experimental leishmaniasis, *Trans. R. Soc. Trop. Med. Hyg.*,71, 550, 1977.
3. **New, R. R. C., Chance, M. L., Thomas, S. C., and Peters, W.,** Antileishmanial activity of antimonials entrapped in liposomes, *Nature (London),* 272, 55, 1978.
4. **Alving, C. A., Steck, E. A., Chapman, W. L., Waits, V. B., Hendricks, L. D., Swarz, G. M., and Hanson, W.L.,** Therapy of leishmaniasis: superior efficacies of liposome-encapsulated drugs, *Proc. Natl. Acad. Sci. U.S.A.*,75, 2959, 1978.
5. **Poste, G. and Fidler, I. J.,** Active non-specific immunotherapy of lung metastases by macrophage activating agents encapsulated in liposomes, in *Optimization of Drug Delivery,* Bundgaard, H., Hansen, A. B., and Kofod, H., Eds., Munksgaard, Copenhagen, 1982, 418.
6. **Frøkjaer, S., Hjorth, E. L., and Wørts, O.,** Stability and storage of liposomes, in *Optimization of Drug Delivery,* Bungaard, H., Hansen, A. B., and Kofod, H., Eds., Munksgaard, Copenhagen, 1982, 384.
7. **Szoka, F. and Papahadjopoulos, D.,** Comparative properties and methods of preparation of lipid vesicles (liposomes), *Ann. Rev. Biophys. Bioeng.*, 9, 467, 1980.
8. **Gregoriadis, G., Kirby, C., Meehan, A., and Senior, J.,** Targeting of liposomes: some prerequisites for success, in *Liposomes, Drugs and Immunocompetent Cell Functions,* Nicolau, C. and Paraf, A., Eds., Academic Press, London, 1981, 29.
9. **Fidler, I. J., Raz, A., Fogler, W. E., Kirsch, R., Bugelski, P., and Poste, G.,** Design of liposomes to improve delivery of macrophage-activating agents to alveolar macrophages, *Cancer Res.*, 40, 4460, 1980.
10. **Szoka, F. and Papahadjopoulos, D.,** Liposomes: preparation and characterization, in *Liposomes: From Physical Structure to Therapeutic Applications,* Knight, C. G., Ed., Elsevier/North-Holland, Amsterdam, 1981, chap. 3.
11. **Bangham, A. D., Standish, M. M., and Watkins, J. C.,** Diffusion of univalent ions across the lamellae of swollen phospholipids, *J. Mol. Biol.*, 13, 238, 1965.
12. Imperial Chemical Industries Ltd., Belgian Patent 866697, 1978.
13. **Lashmar, U. and Rao, L. S.,** unpublished observations, 1979.
14. **Alving, C. A., Steck, E. A., Chapman, W, L., Waits, V. B., Hendricks, L. D., Swartz, G. M., and Hanson, W. L.,** Liposomes in leishmaniasis: therapeutic effects of antimonial drugs, 8-amino-quinolines, and tetracycline, *Life. Sci.*, 26, 2231, 1980.
15. **Knight, C. J.,** Hydrophobic pro-drugs in liposomes, in *Liposomes: From Physical Structure to Therapeutic Applications,* Knight, C. J., Ed., Elsevier/North-Holland, Amsterdam, 1981, chap. 13.
16. **Scherphof, G., Roerdink, F., Hoekstra, D., Zboworski, J., and Wisse, E.,** Stability of liposomes in presence of blood constituents: consequences for uptake of liposomal lipid and entrapped compounds by rat liver cells, in *Liposomes in Biological Systems,* Gregoriadis, G. and Allison, A. C., Eds., John Wiley & Sons, Chichester, 1980, chap. 6.
17. **Deamer, D. W. and Bangham, A. D.,** Large volume liposomes by an ether vaporization method, *Biochim. Biophys. Acta,* 443, 629, 1976.
18. **Schieren, H., Rudolph, S., Finkelstein, M., Coleman, P., and Weissman, G.,** Comparison of large unilamellar vesicles prepared by a petroleum ether vaporization method with multilamellar vesicles: ESR, diffusion and entrapment analyses, *Biochim. Biophys. Acta,* 542, 137, 1978.
19. **Deamer, D. W.,** Preparation and properties of ether injection liposomes, *Ann. N.Y. Acad. Sci.*, 308, 250, 1978.
20. **Papahadjopoulos, D., Vail, W. J., Jacobson, K., and Poste, G.,** Cochleate lipid cylinder formation by fusion of unilamellar lipid vesicles, *Biochim. Biophys. Acta,* 394, 483, 1975.
21. **Szoka, F. and Papahadjopoulos, D.,** Procedure for preparation of liposomes with large internal aqueous space and high capture by reverse-phase evaporation, *Proc. Natl. Acad. Sci. U.S.A.*, 75, 4194, 1978.
22. **Szoka, F., Olson, F., Heath, T., Vail, J., Mayhew, E., and Papahadjopoulos, D.,** Preparation of unilamellar liposomes of intermediate size (0.1—0.2μm) by a combination of reverse phase evaporation and extrusion through polycarbonate membranes, *Biochim. Biophys. Acta,* 601, 559, 1980.
23. **Papahadjopoulos, D. and Miller, N.,** Phospholipid model membranes. I. Structure characteristics of hydrated liquid crystals, *Biochim. Biophys. Acta,*135, 624, 1967.
24. **Huang, C. H.,** Studies of phosphatidylcholine vesicles: formation and physical characteristics, *Biochemistry,* 8, 344, 1969.
25. **Batzri, S. and Korn, E. D.,** Single bilayer liposomes without sonication, *Biochim. Biophys. Acta,* 298, 1015, 1973.

26. **Milsmann, M. H. W., Schwendener, R. A., and Weder, H. G.,** The preparation of large single bilayer liposomes by a fast and controlled dialysis, *Biochim. Biophys. Acta*, 512, 147, 1978.
27. Battelle Memorial Institute, British Patent Application 2001929A, 1979.
28. **Scherphof, G., Damen, J., and Hoekstra, D.,** Interactions of liposomes with plasma proteins and components of the immune system, in *Liposomes: From Physical Structure to Therapeutic Applications*, Knight, C. G., Ed., Elsevier/ North-Holland, Amsterdam, 1981, chap 10.
29. **Hunt, C. A., and Tsang, S.,** α-Tocopherol retards auto-oxidation and prolongs the shelf-life of liposomes, *Int. J. Pharm.*, 8, 101, 1981.
30. Imperial Chemical Industries Ltd. and National Research Development Corporation, British Patent 1523965, 1977.
31. **Schneider, M. and Lamy, B.,** British Patent Application 2002319A, 1979.
32. **Gaehwiller, Ch.,** A new method for the rapid determination of the average size and an index of polydispersity of submicron particles in liquids using laser light scattering spectroscopy, *Powder Technol.*, 25, 11, 1980.

INDEX

D

E

F

G

H

I

L

M

N

O

P

Q

R

S

T

U

V

W

X

Z

For Product Safety Concerns and Information please contact our EU representative GPSR@taylorandfrancis.com Taylor & Francis Verlag GmbH, Kaufingerstraße 24, 80331 München, Germany

Batch number: 10731074

Printed by Printforce, the Netherlands